Nonequilibrium Phase Transitions in Lattice Models

This book provides an introduction to nonequilibrium statistical physics via lattice models.

Beginning with an introduction to the basic driven lattice gas, the early chapters discuss the relevance of this lattice model to certain natural phenomena and examine simulation results in detail. Several possible theoretical approaches to the driven lattice gas are presented. In the next two chapters, absorbing-state transitions are discussed in detail. The later chapters examine a variety of systems subject to dynamic disorder before returning to look at the more surprising effects of multi-particle rules, nonunique absorbing-states and conservation laws. Examples are given throughout the book, the emphasis being on using simple representations of nature to describe ordering in real systems. The use of methods such as mean-field theory, Monte Carlo simulation, and the concept of universality to study and interpret these models is described. Detailed references are included.

The book will be of interest to graduate students and researchers in statistical physics and also to researchers in areas including mathematics, chemistry, mathematical biology, and geology interested in collective phenomena in complex systems.

After receiving a PhD in theoretical physics from the University of Barcelona in 1972, Joaquín Marro moved to the Yeshiva University of New York where he received a second PhD for work on computational statistical physics in 1975. In 1976 he returned to the University of Barcelona to do research in statistical mechanics. In 1983 he was appointed Professor of Physics and in 1986 Director of the Department of Fundamental Physics. In 1987 he joined the faculty of Granada University where he is presently professor of condensed matter physics (and subdirector) at the Institute Carlos I for Theoretical and Computational Physics. His main research interests are statistical and computational physics.

Ronald Dickman grew up in New York City, has an undergraduate degree in art, and a PhD in physics from the University of Texas at Austin (1984). He has written over seventy research papers on nonequilibrium statistical physics, polymer physics, the theory of liquids, and disordered systems. Since 1986, Prof. Dickman has taught at Lehman College, City University of New York, and at the CUNY Graduate Center, and has given talks on his work at many universities and institutes in the US, Europe, Japan and Brazil.

Collection Aléa-Saclay:
Monographs and Texts in Statistical Physics

General editor: Claude Godrèche

C. Godrèche (ed.): *Solids Far from Equilibrium*
P. Peretto: *An Introduction to the Modeling of Neural Networks*
C. Godrèche and P. Manneville (eds.): *Hydrodynamics and Nonlinear Instabilities*
D. H. Rothman and S. Zaleski: *Lattice-Gas Cellular Automata*
A. Pimpinelli and J. Villain: *The Physics of Crystal Growth*
B. Chopard and M. Droz: *Cellular Automata Modeling of Physical Systems*
J. Marro and R. Dickman: *Nonequilibrium Phase Transitions in Lattice Models*

Nonequilibrium Phase Transitions in Lattice Models

Joaquín Marro
Universidad de Granada

Ronald Dickman
City University of New York

CAMBRIDGE
UNIVERSITY PRESS

CAMBRIDGE UNIVERSITY PRESS
Cambridge, New York, Melbourne, Madrid, Cape Town, Singapore, São Paulo

Cambridge University Press
The Edinburgh Building, Cambridge CB2 2RU, UK

Published in the United States of America by Cambridge University Press, New York

www.cambridge.org
Information on this title: www.cambridge.org/9780521480628

First published 1999
This digitally printed first paperback version 2005

A catalogue record for this publication is available from the British Library

Library of Congress Cataloguing in Publication data

Marro, Joaquín, 1945–
Nonequilibrium phase transitions in lattice models / Joaquín Marro, Ronald Dickman.
p. cm. – (Collection Aléa-Saclay. Monographs and texts in statistical physics)
Includes bibliographical references and index.
ISBN 0 521 48062 0 (hardbound)
1. Phase transformations (Statistical physics) 2. Lattice gas. 3. Lattice dynamics.
I. Dickman, Ronald. II. Title. III. Series.
QC175.16.P5M37 1999
530.13–dc21 98-29461 CIP

ISBN-13 978-0-521-48062-8 hardback
ISBN-10 0-521-48062-0 hardback

ISBN-13 978-0-521-01946-0 paperback
ISBN-10 0-521-01946-X paperback

Art is the lie that helps us see the truth
Pablo Picasso

To Julia and Adriana

Contents

Preface

Nature provides countless examples of many-particle systems maintained out of thermodynamic equilibrium. Perhaps the simplest condition we can expect to find such systems in is that of a nonequilibrium steady state; these already present a much more varied and complex picture than equilibrium states. Their instabilities, variously described as nonequilibrium phase transitions, bifurcations, and synergetics, are associated with pattern formation, morphogenesis, and self-organization, which connect the microscopic level of simple interacting units with the coherent structures observed, for example, in organisms and communities.

Nonequilibrium phenomena have naturally attracted considerable interest, but until recently were largely studied at a macroscopic level. Detailed investigation of phase transitions in lattice models out of equilibrium has blossomed over the last decade, to the point where it seems worthwhile collecting some of the better understood examples in a book accessible to graduate students and researchers outside the field. The models we study are oversimplified representations or caricatures of nature, but may capture some of the essential features responsible for nonequilibrium ordering in real systems.

Lattice models have played a central role in equilibrium statistical mechanics, particularly in understanding phase transitions and critical phenomena. We expect them to be equally important in nonequilibrium phase transitions, and for similar reasons: they are the most amenable to precise analysis, and allow one to isolate specific features of a system and to connect them with macroscopic properties. Equilibrium lattice models are typically specified by their energy function or 'Hamiltonian' on configuration space; here, the models (we restrict our attention to lattice Markov processes or particle systems) are defined by a set of transition probabilities. Unlike in equilibrium, the stationary probability distribution is not known *a priori*.

In fact, lattice models of nonequilibrium processes have lately begun to multiply at a dizzying pace. Sandpiles, driven lattice gases, traffic models, contact processes, surface catalytic reactions, branching annihilating random walks, and sequential adsorption are just a few classes of nonequilibrium lattice models that have become the staple of the statistical physics literature. Despite the absence of general unifying principles for this varied set of models, it turns out that many of them fall naturally into one of a small number of classes. That one can now recognize 'family resemblances' amongst models encouraged us to attempt the present work. Some sort of schema, however incomplete and provisional, to the bewildering array of models under active investigation, should help to establish connections between seemingly disparate fields, and avert unnecessary duplication of effort. Since a general formalism, analogous to equilibrium statistical mechanics, is lacking for nonequilibrium steady states, the field presents a particular challenge to theoretical physics.

This book is based largely upon research by the authors, and their students and colleagues, during the past few years. We have by no means attempted a comprehensive survey of nonequilibrium phase transitions, not even of lattice models of such. We have little to say, for example, about self-organized criticality or surface growth problems. Clearly these subjects demand books in themselves. On the other hand, we have tried to do more than provide a compendium of recent results. We hope to present a set of examples with sufficient vividness and clarity that the reader will be convinced of their intrinsic interest, and be drawn to think about them, or to devise his or her own.

At certain points in the book we express our attitude regarding various issues, some of them controversial. But we have tried to point out the weakness or controversial nature of the arguments, within the limitations posed by our own lack of familiarity with certain methods and results. In other words, we don't that claim ours is the definitive account of all problems considered here, and therefore encourage others to address the gaps or misconceptions they find in the present work.

It is a pleasure for us to thank many colleagues whose comments, suggestions, and corrections were valuable to us in producing this presentation. One or both of us have enjoyed discussions with Abdelfattah Achahbar, Juanjo Alonso, Dani ben-Avraham, Martin Burschka, John Cardy, Michel Droz, José Duarte, Richard Durrett, Jim Evans, Julio Fernández, Hans Fogedby, Pedro Garrido, Jesús González-Miranda, Peter Grassberger, Geoffrey Grinstein, Malte Henkel, Iwan Jensen, Makoto Katori, Peter Kleban, Norio Konno, Eduardo Lage, Joel Lebowitz, Roberto Livi, Antonio López-Lacomba, Maria do Céu Marques, José Fernando Mendes, Adriana Gomes Moreira, Miguel Angel Muñoz, Mário de Oliveira, Maya Paczusky, Vladimir Privman, Sid Redner, Maria Augusta dos Santos,

Beate Schmittmann, Tânia Tomé, Raúl Toral, Alex Tretyakov, Lorenzo Vallés, Royce Zia, and Robert Ziff. Our work during the last years has been partially supported by grants from the DGICYT (PB91-0709) and the CICYT (TXT96-1809), from the *Junta de Andalucía*, and from the European Commission.

Granada and New York

Joaquín Marro
Ronald Dickman

1
Introduction

The subject of this book lies at the confluence of two major currents in contemporary science: phase transitions and far-from-equilibrium phenomena. It is a subject that continues to attract scientists, not only for its novelty and technical challenge, but because it promises to illuminate some fundamental questions about open many-body systems, be they in the physical, the biological, or the social realm. For example, how do systems composed of many simple, interacting units develop qualitatively new and complex kinds of organization? What constraints can statistical physics place on their evolution?

Nature, both living and inert, presents countless examples of *nonequilibrium* many-particle systems. Their simplest condition — a nonequilibrium steady state — involves a constant flux of matter, energy, or some other quantity (de Groot & Mazur 1984). In general, the state of a nonequilibrium system is not determined solely by external constraints, but depends upon its *history* as well. As the control parameters (temperature or potential gradients, or reactant feed rates, for instance) are varied, a steady state may become unstable and be replaced by another (or, perhaps, by a periodic or chaotic state). Nonequilibrium instabilities are attended by ordering phenomena analogous to those of equilibrium statistical mechanics; one may therefore speak of *nonequilibrium phase transitions* (Nicolis & Prigogine 1977, Haken 1978, 1983, Graham 1981, Cross & Hohenberg 1993). Some examples are the sudden onset of convection in a fluid heated from below (Normand, Pomeau, & Velarde 1977), switching between high- and low-reactivity regimes in an open chemical reactor (Field & Burger 1985, Gray & Scott 1990), and the transition to high field intensity as a laser is pumped beyond a threshold value (Graham & Haken 1971).

Phase transitions and critical phenomena have captivated statistical physicists for many decades (see, for example, Stanley (1971), and the Domb & Green (1972–6) and Domb & Lebowitz (1977–95) series). Much theoretical progress has resulted from the parallel application of varied approaches, including exact solutions, mean-field theories, computer simulations, series expansions, and renormalization group methods, to simple lattice models and their continuum analogs. It is now possible to understand the phase diagram of many systems using these methods, and to fathom the remarkable universality of critical behavior (Ma 1976, Amit 1984, Zinn-Justin 1990).

The *dynamics* of phase transitions and critical points presents new challenges, since macroscopic properties are no longer given by averages over a known, stationary probability distribution. The same applies to systems that are kept out of equilibrium. In this book we illustrate, by means of simple examples, some of the diversity of nonequilibrium phase transitions. The examples are sufficiently detailed to show that 'nonequilibrium phase transitions' represents not a superficial resemblance to equilibrium phenomena, but a precise use of the terms 'phase' and 'phase transition' in their statistical mechanics sense. Equilibrium phase diagrams can be determined on the basis of the free energy; out of equilibrium, the free energy is not defined, in general. However, even in its absence, we can recognize a *phase* of a many-particle system from well-defined, reproducible relations between its macroscopic properties (or, more formally, the many-particle distribution functions) and the parameters governing its dynamics. A phase transition is characterized by a singular dependence of these attributes upon the control parameters. From this vantage, equilibrium is a special case, in which the dynamics happens to be derivable from an energy function, thereby permitting an analysis with no reference to time. A key point of difference between equilibrium and nonequilibrium statistical mechanics is that whereas in the former case the stationary probability distribution is known, out of equilibrium one must actually find the time-independent solution(s) of the master equation for the process. This is a formidable task and can only be carried out approximately for most models.

We restrict our attention to lattice models and, in particular, to lattice Markov processes or *interacting particle systems* (Griffeath 1979, Liggett 1985, Durrett 1988, Konno 1994). A well-known disadvantage of lattice models is that they are usually too crude to be directly comparable with experiment. In fact, if one is interested in predicting a nonequilibrium phase diagram, it is better, for cases in which fluctuations are of minor significance, to employ a macroscopic description, i.e., a set of (deterministic) partial differential equations. This approach finds success in applications to hydrodynamics (convection, couette flow), oscillations and waves in surface catalysis, and chemical reactions (Field & Burger 1985, Bär *et al.* 1994, Cross & Hohenberg 1993). If we ignore these avenues in favor of lattice models, it is largely because macroscopic descriptions hold little surprise in the way of criticality; mean-field behavior is implicit at this level. The range of critical phenomena exhibited by the models considered here, by contrast, is at least as interesting as in equilibrium. Another reason for focusing on lattice models is that as statistical physicists, we prefer to see macroscopic behavior emerge from interactions among a multitude of simple units, rather than to adopt a macroscopic picture as our starting point. Clearly, a microscopic approach becomes essential for describing systems in which fluctuations are significant. Continuum models of nonequilibrium phenomena that do incorporate fluctuations (Langevin equations or stochastic partial differential equations) are also being developed. These approaches generally pose even greater conceptual and computational challenges than the lattice models we focus on in this book.

One can distinguish three broad categories of nonequilibrium problems. One class comprises Hamiltonian systems out of equilibrium by virtue of their preparation; one is then interested in how the system approaches equilibrium. This subject, which subsumes such active areas of investigation as critical dynamics, nucleation theory, and spinodal decomposition, lies outside the scope of this book. Our focus is rather on models that are intrinsically out of equilibrium — models that violate the principle of detailed balance. It is useful to draw a further distinction between perturbations of an equilibrium model by a nonequilibrium 'driving force,' and models that have no equilibrium counterpart. An example of the former is the driven diffusive system considered in chapters 2 and 3. When the 'electric field' that biases hopping vanishes, we recover the familiar lattice gas. In the 'perturbed equilibrium' models one may still speak of energy and temperature. The transition rates may obey a *local* detailed-balance condition, but cannot be derived from a potential energy function. Or the rates may represent competing processes, proceeding at different temperatures, as illustrated in chapters 4 and 8. The catalytic and population models considered in chapters 5, 6, and 9 have no equilibrium analog; their dynamics involves events that are strictly irreversible. The concepts of energy and temperature are not pertinent to such models.

Many of the methods brought to bear on equilibrium lattice models — mean-field theories, Monte Carlo simulations (coupled with finite-size scaling ideas), series expansions, and renormalization group methods — have been applied successfully to nonequilibrium systems. All of these methods are described in some detail in this work, with the exception of renormalization group methods, which would seem to require a book in itself.

1.1 Two simple examples

Out of equilibrium, the well-known prohibition against phase transitions in one-dimensional systems with short-range interactions is lifted. There are many examples of nonequilibrium phase transitions in one space dimension (Privman 1997), and one might have hoped that some would permit an exact solution. In fact, very few exactly soluble models with nonequilibrium phase transitions are known. To provide an easily-worked example, we study instead the *branching process:* a 'zero-dimensional' model, or more precisely, one lacking spatial structure.

Consider a population $n(t) \geq 0$ of individuals that give birth at rate λ, and die at unit rate. The individuals might represent organisms in a colony, with a population well below the carrying capacity, neutrons in a chain reaction, or photons in a lasing medium. $n(t)$ is a Markov process with transition rates

$$c(n \to n+1) = \lambda n \tag{1.1}$$

and

$$c(n \to n-1) = n. \tag{1.2}$$

Note that $n=0$ is absorbing; if all individuals die the population remains zero at future times. The probability, $P_n(t)$, of having exactly n individuals at time t is governed by the master equation,

$$\frac{dP_n}{dt} = \lambda(n-1)P_{n-1} + (n+1)P_{n+1} - (1+\lambda)nP_n. \tag{1.3}$$

We solve this using the generating function

$$g(x,t) = \sum_{n=0}^{\infty} x^n P_n(t), \tag{1.4}$$

which satisfies

$$\frac{\partial g}{\partial t} = (1-x)(1-\lambda x)\frac{\partial g}{\partial x}, \tag{1.5}$$

subject to the boundary condition $g(1,t) = 1$, which reflects normalization. If we assume a single individual at time zero, then $g(x,0) = x$. For $\lambda \neq 1$ the solution is readily found to be

$$g(x,t) = \frac{(1-\lambda x)e^{(1-\lambda)t} - (1-x)}{(1-\lambda x)e^{(1-\lambda)t} - \lambda(1-x)}. \tag{1.6}$$

Setting $x = 0$ we find the extinction probability, $P_0(t)$, and hence the *survival probability*

$$P(t) = 1 - P_0(t) = \frac{\lambda - 1}{\lambda - e^{(1-\lambda)t}}. \tag{1.7}$$

For $\lambda < 1$, $P(t)$ decays exponentially, while for $\lambda > 1$ it approaches a nonzero value. The ultimate survival probability is

$$P_\infty \equiv \lim_{t\to\infty} P(t) = \begin{cases} 0 & \lambda < 1 \\ 1 - \lambda^{-1} & \lambda > 1 \end{cases} \tag{1.8}$$

which shows that $\lambda = 1$ is a critical value marking the boundary between possible survival and certain extinction. Thus P_∞ (the *order parameter* for this model) has a singular dependence upon λ, justifying our labeling this a phase transition. For $\lambda = 1$ the generating function is

$$g(x,t)|_{\lambda=1} = \frac{1+(1-x)(t-1)}{1+(1-x)t}, \tag{1.9}$$

from which we find

$$P(t) = \frac{1}{1+t}, \quad \lambda = 1. \tag{1.10}$$

Thus the relaxation time for the survival probability diverges $\propto |1-\lambda|^{-1}$ as $\lambda \to 1$, and when $\lambda = 1$ relaxation follows a power law rather than an exponential decay. These features — a sharp boundary between extinction and survival, a diverging relaxation time, and power-law relaxation — are typical of critical points, and will be found in more complex and interesting models.

The model considered here does not allow for a nontrivial steady state. It is easy to show that the mean population size is $\langle n \rangle = n(t=0)e^{(\lambda-1)t}$, hence for $\lambda > 1$ the population grows out of all bounds.[1] To prevent this we require a saturation term (the birth rate should decline for large n), which renders the analysis much more difficult. We return to this issue in chapter 6, which is devoted to the *contact process*, a birth-and-death model of particles on a lattice, with offspring appearing at nearest-neighbor *vacant* sites. This prevents exponential population growth: since sites may not be doubly occupied, the density can never exceed unity, and the model possesses a nontrivial or *active* stationary state for large enough λ.

There is one variant of the contact process for which we can obtain exact results.[2] Consider a one-dimensional lattice, with the rule that a vacant site adjacent to an occupied one becomes occupied at rate λ, while an occupied site adjacent to a vacant one becomes vacant at unit rate. Suppose we start with only the origin occupied. Then at later times, the system is either in the absorbing state (no particles) or it consists of a string of n occupied sites, with no gaps. The number of particles, $n(t)$, is a continuous-time lattice random walk, starting from $n = 1$, with an absorbing boundary at $n = 0$, and with a bias to the right $\propto \lambda - 1$. That is, for $\lambda \leqslant 1$ n must eventually hit zero, while for $\lambda > 1$ there is a nonzero probability of survival as $t \to \infty$. Well-known results for random

---

[1] The situation is reminiscent of the *Gaussian approximation* to Ising/ϕ^4 field theory. Neglect of the term $\propto \phi^4$ yields a soluble model with pathological low-temperature behavior; see Binney *et al.* (1992).

[2] This is the continuous-time version of so-called *compact directed percolation*; see Essam (1989).

walks (Feller 1957, Barber & Ninham 1970) imply that $P_\infty \propto \lambda - 1$, and, for $\lambda = \lambda_c = 1$, $P(t) \sim t^{-1/2}$, while the mean-square population over *surviving* trials $\langle n^2 \rangle_{surv} \sim t$. Since the particles are arrayed in a compact cluster, $R^2(t)$, the mean-square distance of particles from the origin, also grows $\sim t$. The asymptotic time-dependence of $P(t)$, $\langle n \rangle$, and $R^2(t)$ in critical systems is discussed extensively in chapters 5, 6, and 9. The present examples provide some of the rare instances in which the power laws governing this evolution are known exactly.

Next we describe a simple model that illustrates several themes associated with nonequilibrium steady states: dynamic competition, spatial structure (pattern formation), and anisotropy. It is closely related to the physically motivated driven lattice gas analyzed in detail in chapters 2 and 3. Consider a simple cubic (sc) lattice in two or more dimensions, with toroidal boundary conditions, and a fraction n of its sites occupied by particles, the rest vacant. The evolution proceeds via nearest-neighbor (NN) particle–hole exchanges. With probability q the exchange is in the *longitudinal* direction (defined by the unit vector $\hat{\mathbf{x}}$) and involves a driving field; with probability $1 - q$ it takes one of the transverse directions, via a *thermal mechanism*. The field introduces a bias: particle displacements of $\hat{\mathbf{x}}$ are accepted with probability p, while displacements of $-\hat{\mathbf{x}}$ are accepted with probability $1 - p$. (In other words, along this direction we have an *asymmetric exclusion process*.) In contrast with longitudinal exchanges, which do not involve NN interactions, exchanges in a transverse direction are accepted with probability b if the second neighbor along the jump direction is occupied, and with probability $1 - b$ if it is vacant. (Table 1.1 gives the rates for a two-dimensional system with equal *a priori* probabilities for longitudinal and transverse jumps.) These processes mimic the effects of field-driven motion (for $p \neq \frac{1}{2}$), and, for $b > \frac{1}{2}$, of a tendency toward cluster formation, but in a manner that cannot be reconciled with a potential energy function. As confirmed in simulations, $b - \frac{1}{2}$ is a temperature-like variable, analogous to the inverse temperature β, while $p - \frac{1}{2}$ represents a longitudinal driving field. The latter has no equilibrium analog, but given the periodic boundaries, leads rather to a nonequilibrium steady state with a longitudinal current.

This simple system exhibits the great variety and some of the difficulties characterizing the phenomena studied in this book. Its behavior is best illustrated by 'snapshots' of typical configurations, as in figure 1.1, which is for b sufficiently large that the system segregates into a particle-rich and a particle-poor phase; it depicts the process of phase separation from a random initial configuration. (This simulation employs $p = 1$, but we find similar results for other values, even $p = \frac{1}{2}$.) Simulations suggest that phase separation occurs discontinuously, by a series of 'avalanches.' Another interesting observation is that the pair correlation function exhibits self-

Table 1.1. The rate for a two-dimensional version of the lattice gas with $q = \frac{1}{2}$ (i.e., no *a priori* bias), assuming that the preferred direction, $+\hat{x}$, is vertical upwards. The symbols • and ○ stand for occupied and vacant sites, respectively (Marro & Achahbar 1998).

Process	Rate
● ○ ● → ○ ● ●	b
● ○ ● → ● ● ○	b
● ○ ○ → ○ ● ○	$1 - b$
○ ○ ● → ○ ● ○	$1 - b$
○ → ● ● → ○	p
● → ○ ○ → ●	$1 - p$

similarity or time-scale invariance if one scales time by the mean width of the strips. Figure 1.2 illustrates the kind of order found in this system, namely, anisotropic segregation at large b, and a linear interface for any value of p. (For $b < \frac{1}{2}$ the tendency appears to be towards chess-board configurations, as in an antiferromagnet, independent of p.) We refer the reader to Marro & Achahbar (1998) for further details.

1.2 Perspective

The examples of the previous section give some of the flavor of the models we consider in this book. The population model exhibits a phase transition between an active state (survival) and a kind of trap — an empty state with no further evolution. This transition has no equilibrium analog. The second example displays phase separation at a particular value of a temperature-like variable, like the equilibrium lattice gas, but in a non-Hamiltonian model with highly anisotropic dynamics (also — unlike for equilibrium — the resulting two phases, *liquid* and *gas*, are not symmetric here, in general).

Our point of view in this book is, quite naturally, strongly influenced by our awareness of the theory of (equilibrium) phase transitions and critical phenomena. The latter appears sufficiently powerful and broadly applicable to guide at least our initial questions about nonequilibrium models. Since a number of key ideas from equilibrium theory provide touchstones for our discussion, it is well to mention them briefly.

The central result in the modern theory of critical phenomena is *uni-*

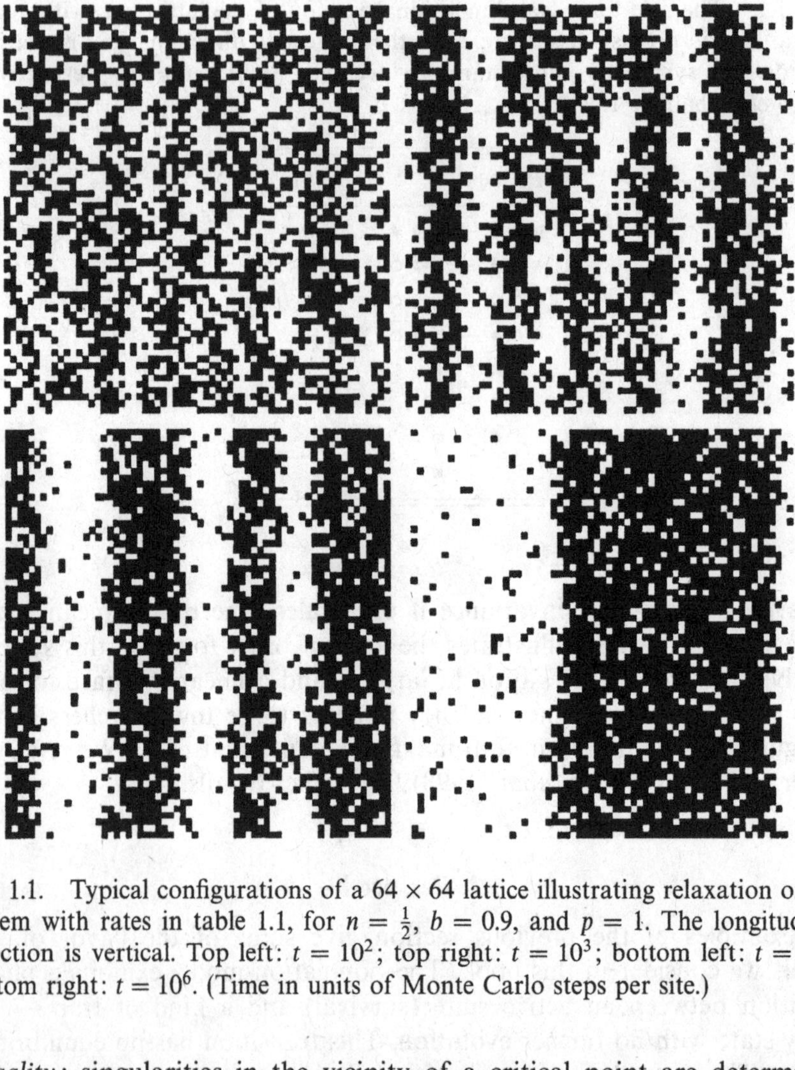

Fig. 1.1. Typical configurations of a 64 × 64 lattice illustrating relaxation of the system with rates in table 1.1, for $n = \frac{1}{2}$, $b = 0.9$, and $p = 1$. The longitudinal direction is vertical. Top left: $t = 10^2$; top right: $t = 10^3$; bottom left: $t = 10^4$; bottom right: $t = 10^6$. (Time in units of Monte Carlo steps per site.)

versality: singularities in the vicinity of a critical point are determined by a small set of basic features — spatial dimensionality, dimensionality and symmetry of the order parameter, and whether the interactions are long- or short-range (Stanley 1971, Ma 1976, Fisher 1984, Amit 1984, Zinn-Justin 1990). The insensitivity to details of molecular structure or interactions is evident when we note that the Ising model, simple materials such as argon or carbon dioxide, and binary liquid mixtures all exhibit the same critical behavior. When we turn to critical *dynamics*, conservation laws enter as another possible determinant (Hohenberg & Halperin 1977).

Universality reflects the existence of a diverging correlation length ξ in a system at its critical point. For example, if $\epsilon = (T - T_c)/T_c$ (the *reduced temperature*) then in the absence of an external magnetic field we expect

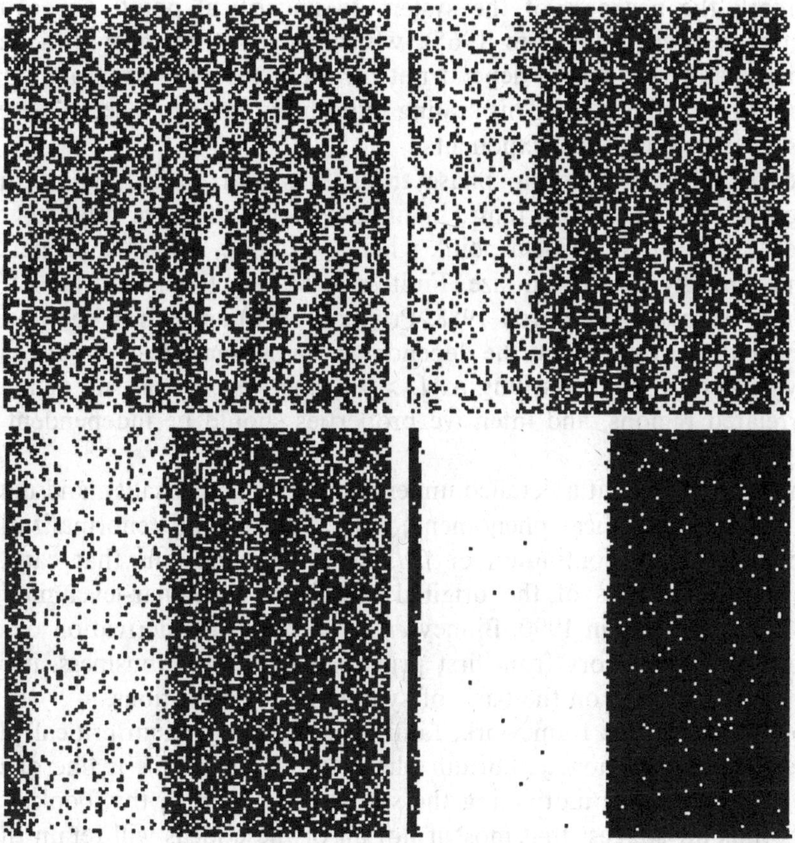

Fig. 1.2. Typical configurations in the stationary regime of the system with rates in table 1.1, for a 128×128 lattice with $n = \frac{1}{2}$ and (from left to right) $b = 0.84$ and 0.865 (top), and 0.877 and 0.98 (bottom). The configurations shown here, for $p = \frac{1}{2}$, are indistinguishable from those with different p but the same value of $\epsilon \equiv 1 - b/b_c(p)$, where $b_c(p)$ denotes the critical line.

the correlation length of a ferromagnet to diverge as $\xi \sim \epsilon^{-\nu}$. Assigning a model to a particular *universality class* fixes certain quantities, for example critical exponents such as ν. Similarly, the relaxation time for fluctuations diverges at the critical point. Thus we expect static correlation functions to depend (asymptotically) on distance r through the ratio r/ξ, and dynamic correlations to depend on t through the ratio t/τ, where τ is a relaxation time.

Another powerful result in critical phenomena is a general scheme for classifying thermodynamic variables. Thus in the simplest cases we expect to find just two relevant parameters, one 'temperature-like' (conjugate to energy), the other 'field-like' (conjugate to magnetization or density) that mark the nearness of the system to its critical point. Associated with this is *scaling* (Widom 1965), which asserts that, near criticality, thermodynamic properties depend not on these variables separately, but only in a particular ratio. This scaling hypothesis leads to a set of *scaling relations* amongst critical exponents.

The singularities marking phase transitions and critical points only emerge in the infinite-size limit; in finite systems they are rounded off. In the vicinity of the critical point, where ξ is large, intensive properties depend strongly on system size. Finite-size scaling theory (Fisher 1971, Fisher & Barber 1972, Cardy 1988, Privman 1990) appeals to the notion that near the critical point, the dependence on system size L should only involve the ratio L/ξ. As long as $L >> \xi$ the system consists of many uncorrelated regions, and intensive properties should be independent of system size.

Finally we note that a detailed understanding of universality, and many other aspects of critical phenomena, rests on applying renormalization group methods to continuum or *field theory* descriptions that capture the essential features of the original lattice or molecular-level models (Ma 1976, Zinn-Justin 1990, Binney *et al.* 1992). While derivation of the appropriate field theory from first principles is a subtle business, it can often be constructed on the basis of symmetry considerations.

How much of this framework, familiar from equilibrium critical phenomena, applies to nonequilibrium phase transitions? This is one of the main open questions motivating the studies described in this book. The results thus far suggest that most if not all of these ideas will retain their validity in the context of nonequilibrium lattice models. (What is missing, in a sense, is *thermodynamics*![3]) We will also see that many of the analytical and numerical methods used to study equilibrium phase transitions remain useful here.

What, then, is so different about phase transitions out of equilibrium? In connection with lifting the detailed balance condition there appear a number of new possibilities: phase transitions to an absorbing state, transitions in one dimension (even in models with short-range interactions), novel spatial structures, highly dependent upon the *history* of the system, and unexpected interfacial properties. One encounters an enormous richness of steady states as one varies the dynamical rules. Switching between different forms of the spin flip rate, or varying the relative frequency of exchanges in different directions — changes that have no effect on the equilibrium state — can produce completely different phase diagrams in

[3] Indeed, it has been appreciated for some time that in percolation, which possesses neither a Hamiltonian nor dynamics, a phase transition arises purely from the interplay of statistics and geometry; see Stauffer & Aharony (1992).

the driven lattice gas, for example. The range of patterns generated under anisotropic dynamics is particularly surprising.

1.3 Overview

Clearly, the opportunities for devising unusual systems are enormous, and one can envision many more possibilities than can be investigated systematically. Thus we have tried to discipline our imaginations by starting with simple models, and focusing on their phase diagrams and issues of universality. These studies raise questions that lead us to introduce additional complications in a number of cases. We begin in chapter 2 by introducing the basic driven lattice gas, discussing its relevance to materials such as fast ionic conductors, and then turn to a detailed examination of simulation results. Several theoretical approaches to the driven lattice gas are discussed in chapter 3, which closes with an account of the curious properties of a layered system.

Having dealt, in chapters 2 and 3, with an anisotropic nonequilibrium perturbation of the usual lattice gas, we turn in chapter 4 to an isotropic one involving competition between ordering and diffusion, and consider mean-field theories and hydrodynamic-like descriptions as well as simulation results.

Chapters 5 and 6 bring us to a detailed consideration of absorbing-state transitions, first in the context of models for surface catalysis, then in a simpler population model, the contact process, which is closely related to directed percolation. Chapter 6 closes with a brief examination of the effects of quenched disorder on the contact process and directed percolation, and a sketch of operator and series formalisms. This theme is then developed in chapters 7 and 8, which treat a variety of systems subject to *dynamic* disorder. In chapter 9 we return to absorbing-state transitions, and explore the surprising effects of multi-particle rules, nonunique absorbing-states, and conservation laws.

2

Driven lattice gases: simulations

The *driven lattice gas* (DLG) devised by Katz, Lebowitz, & Spohn (1984) is a physically motivated, precisely defined model which serves as an introduction to the subject of nonequilibrium phase transitions. The DLG is a kinetic lattice gas whose particles are driven by an external electric field $E\hat{\mathbf{x}}$, where $\hat{\mathbf{x}}$ defines a principal lattice direction. In addition to the field, which is constant both in space and time, a parameter denoted Γ allows the speeding up of the jumps of particles along $\hat{\mathbf{x}}$ as compared to jumps transverse to the field. When $\Gamma \to \infty$ two distinct microscopic time scales emerge. There is also a heat bath at temperature T. For a wide range of parameter values, the result is a net current of particles that generates heat which is absorbed by the bath. This particle-conserving irreversible dynamics drives the system toward a nonequilibrium steady state that resembles natural driven systems in certain respects. Eventually phase separation occurs, as in equilibrium ($E = 0$), but exhibiting more complex behavior. In particular, the field modifies the low-temperature interface which becomes quite anisotropic for large E, inducing novel critical properties. This chapter highlights these facts by describing the behavior of different realizations of the DLG observed in Monte Carlo (MC) simulations.[1]

There are several motivations for our detailed examination of simulation results. First, systematic experimental studies of nonequilibrium phase transitions in nature are difficult, and numerical results allow meaningful comparison with some experimental observations. Computer studies may provide hints that finally lead to a realistic coherent theory for the DLG and its variations that is still lacking. This is illustrated below (and in

[1] We mainly describe the work in Marro *et al.* (1985), Vallés & Marro (1986, 1987), Marro & Vallés (1987), and Achahbar & Marro (1995). We refer the reader also to work by others which is quoted below.

chapter 3), where a number of fundamental questions are raised. On the other hand, the study of nonequilibrium phenomena is less developed than the equilibrium case, where statistical mechanics is well established. This makes the interpretation of nonequilibrium MC data, and even its production, challenging.

The chapter is organized as follows. The basic version of the DLG is formulated in §2.1, and we discuss in §2.2 its use as a model of various natural phenomena. The description of MC results is initiated in §2.3. §2.4 examines a layered DLG. We discuss in §2.5 the behavior of correlations, and §2.6 is devoted to critical behavior, including some finite-size scaling ideas. The rather phenomenological description in this chapter is complemented in chapter 3 by mean-field theory and other approaches.

2.1 The basic system

Consider a lattice gas (equivalently an Ising spin system with conserved magnetization) on the simple *cubic* lattice in *d* dimensions with configurations $\vec{\sigma} = \{\sigma_{\mathbf{r}}; \mathbf{r} \in \mathbf{Z}^d\}$. The occupation variable has two states, $\sigma_{\mathbf{r}} = 1, 0$, corresponding to the presence or not of a particle at site $\mathbf{r}$; the particle density is n. The prohibition against multiple occupancy simulates a repulsive core; in addition, the particles interact via a pair NN potential. Therefore, each $\vec{\sigma}$ has a potential or configurational energy given (except for a constant) by

$$H(\vec{\sigma}) = -4J \sum_{NN} \sigma_{\mathbf{r}} \sigma_{\mathbf{s}}\,, \tag{2.1}$$

where the sum is over all NN pairs of sites; $|\mathbf{r} - \mathbf{s}| = 1$. This is the lattice gas studied by Yang & Lee (1952). In addition, the DLG involves an external uniform electric field, $\mathbf{E} = E\hat{\mathbf{x}}$, and the particles behave as positive ions (only) in relation to it. We assume periodic boundary conditions. (While simulations of necessity employ finite lattices, the infinite-size, thermodynamic-limit properties are of prime interest.) The field together with the heat bath at temperature T induces the time evolution of $\vec{\sigma}$ according to the *master equation*:[2]

$$\frac{\partial P_E(\vec{\sigma};t)}{\partial t} = \sum_{\vec{\sigma}^{\mathbf{rs}}} \left[c_E(\vec{\sigma}^{\mathbf{rs}}; \mathbf{r}, \mathbf{s}) P_E(\vec{\sigma}^{\mathbf{rs}};t) - c_E(\vec{\sigma}; \mathbf{r}, \mathbf{s}) P_E(\vec{\sigma};t) \right] \,. \tag{2.2}$$

Here, $\vec{\sigma}^{\mathbf{rs}}$ represents the configuration $\vec{\sigma}$ with the occupation variables at NN sites $\mathbf{r}$ and $\mathbf{s}$ exchanged, $P_E(\vec{\sigma};t)$ is the probability of configuration

[2] This is the familiar Markovian description that has been considered for different situations by Pauli (1928), Glauber (1963), Kawasaki (1972), van Kampen (1981), and Liggett (1985), for example.

$\vec{\sigma}$ at time t, and $c_E(\vec{\sigma}, \mathbf{r}, \mathbf{s})$ stands for the transition probability per unit time (*rate*) for the exchange $\sigma_\mathbf{r} \rightleftarrows \sigma_\mathbf{s}$ when the configuration is $\vec{\sigma}$; the density n remains constant during time evolution. The exchange $\sigma_\mathbf{r} \rightleftarrows \sigma_\mathbf{s}$ corresponds to a *particle jumping to a NN hole* if $\sigma_\mathbf{r} - \sigma_\mathbf{s} \neq 0, |\mathbf{r} - \mathbf{s}| = 1$; we assume $c_E(\vec{\sigma}; \mathbf{r}, \mathbf{s}) = 0$ otherwise, i.e., only NN particle–hole exchanges are allowed. One further assumes that jumps along $\hat{\mathbf{x}}$ are performed with an *a priori* frequency Γ times larger than transverse ones. (Except for a few cases mentioned below, most MC simulations involve $\Gamma = 1$. The large-Γ limit is analyzed in chapter 3.)

The rate in (2.2) is

$$c_E(\vec{\sigma}; \mathbf{r}, \mathbf{s}) = \phi\,[\beta\delta H - \mathbf{E}\cdot(\mathbf{r} - \mathbf{s})(\sigma_\mathbf{r} - \sigma_\mathbf{s})]\ , \tag{2.3}$$

where ϕ is an arbitrary function, $\delta H \equiv H(\vec{\sigma}^{\mathbf{rs}}) - H(\vec{\sigma})$, and $\beta = (k_B T)^{-1}$. This means that $c_E(\vec{\sigma}; \mathbf{r}, \mathbf{s}) = \phi(\beta\delta H)$ for jumps of particles perpendicular to the field, while the field influences, even dominates, otherwise, for example, $c_E(\vec{\sigma}; \mathbf{r}, \mathbf{s}) = 1$ and 0 for jumps along directions $+\hat{\mathbf{x}}$ and $-\hat{\mathbf{x}}$, respectively, for large enough values of E. A motivation for (2.3) is that the rates that are familiar in the literature have this structure when $E = 0$. Letting $X \equiv \beta\delta H$, the *Metropolis rate* (Metropolis *et al.* 1953) corresponds to $\phi(X) = \min\{1, \mathrm{e}^{-X}\}$, the *Kawasaki rate* (Kawasaki 1972) to $\phi(X) = 2(1 + \mathrm{e}^X)^{-1}$, and a rate considered by van Beijeren & Schulman (1984) to $\phi(X) = \mathrm{e}^{-\frac{1}{2}X}$. These cases have the symmetry $\phi(X) = \mathrm{e}^{-X}\phi(-X)$ which corresponds to the *detailed-balance* condition.[3] It guarantees that the model for $E = 0$ correctly describes the Gibbs (equilibrium) state for given values of $H(\vec{\sigma})$, n, and T. In other words, the asymptotic steady solution of equation (2.2) for $E = 0$ and given n is the canonical one,

$$P_{eq}(\vec{\sigma}) = \exp[-\beta H(\vec{\sigma})] \left\{ \sum_{\vec{\sigma}} \exp[-\beta H(\vec{\sigma})] \right\}^{-1}\ , \tag{2.4}$$

if detailed balance (with respect to the microscopic dynamics) holds, and the DLG reduces to the ordinary lattice gas. As is well known, (2.4) exhibits a second order phase transition for $n = \frac{1}{2}$ at $T_C^0 = 0$, 2.27, and 4.5 (units of J/k_B) for $d = 1$, 2, and 3, respectively; therefore, a unique state exists for $T \geq T_C^0$, while for $d > 1$, phase separation occurs below the phase coexistence curve $T_C^0(n)$.

When $E > 0$, the situation is qualitatively different, as there is a preferential direction for hopping. This results in a net dissipative current, if permitted by the boundary conditions. Consequently, the steady state

[3] See §2.3 for an explicit example of a rate, and §§4.4 and 7.4 for a systematic discussion of rates.

$P_E(\vec{\sigma})$ defined as the limit as $t \to \infty$ of $P_E(\vec{\sigma};t)$ in (2.2) is not an equilibrium state for $E \neq 0$. For periodic boundaries, the electric force is nonconservative and so cannot be represented in the Hamiltonian; the work done by the field during a jump has been added in (2.3) to δH. The rate (2.3) with property $\phi(X) = e^{-X}\phi(-X)$ satisfies a detailed-balance condition *locally* but not globally so that microscopic reversibility of the process (Kolmogorov 1934) as in equilibrium is not guaranteed. An important consequence is that even the general form of $P_E(\vec{\sigma})$ is unknown: the stationary solution of the master equation (2.2) does not assume the form (2.4), involving a short-range energy function. (If the boundaries stop the particles, imposing maximum and minimum coordinates along the field direction, then of course E may be derived from a potential, and we have an equilibrium problem. However, this is not the situation studied in the DLG.)

Unlike the equilibrium ($E = 0$) case, $P_E(\vec{\sigma})$ is defined simply as the time-independent solution of the master equation, and depends on the specific rate function ϕ. Thus the nature of $P_E(\vec{\sigma})$ as one varies ϕ and the other parameters is of primary interest. As one may imagine, its study requires a nontrivial generalization of the methods of equilibrium statistical mechanics. The latter, however, serves as a reference in some cases.

2.2 Models of natural phenomena

The DLG is relevant to the understanding of several phenomena. Consider first the process of surface growth that corresponds to the hexagonal packing of discs in figure 2.1 (Gates 1988, Krug & Spohn 1992). One may represent the rough surface profile by a line connecting the center of the discs along the top layer, as indicated. Each profile corresponds to a certain $\vec{\sigma} = \{\sigma_x\}$, such that $\sigma_x = 1$ or 0 for the xth line segment if the slope is negative or positive, respectively. The *surface shape* $\vec{\sigma}$ changes if a disc is attached to the surface or removed from it according to the rules of hexagonal packing. It induces changes from $\cdots 10 \cdots$ to $\cdots 01 \cdots$ for deposits, and from $\cdots 01 \cdots$ to $\cdots 10 \cdots$ for evaporations. Assuming these occur with rates α_i and β_i, respectively, where i is the number of NNs of the disc involved, one may describe this process of jumps to the right and to the left by means of equation (2.2) for the one-dimensional DLG. (An interesting if somewhat arbitrary analogy may also be devised in two dimensions.) The density n is related to the inclination of the surface, which is expected to attain a steady value.

The ordinary lattice gas may be endowed with a particle current by imposing suitable boundary conditions (Spohn 1991) instead of modifying the rate as in (2.3). This corresponds to a system that exchanges matter as

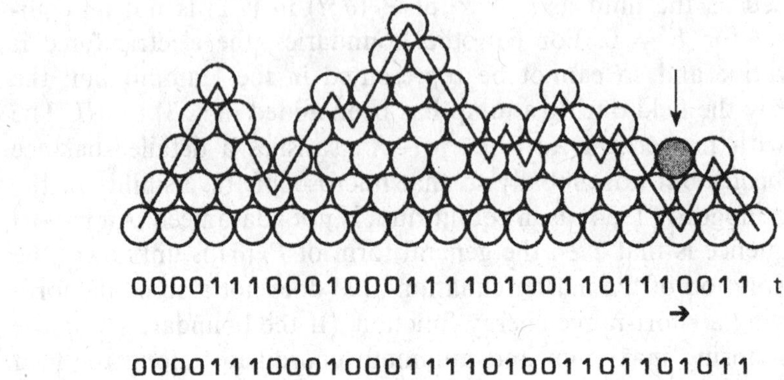

Fig. 2.1. Packing of discs in a section of the free surface in a hexagonal crystal. The line is a representation of the (rough) surface profile that may in turn be represented (bottom of diagram) as a set of values for $\vec{\sigma} = \{\sigma_x; x = 1, 2, \ldots\}$. Arrival of a new disc corresponds to a *particle* hopping to the right, as indicated.

well as energy with its surroundings. Imposing unequal chemical potentials at the boundaries prevents the approach to equilibrium. The DLG and its variations may also be used to model pedestrian dynamics, a message passing through a network, traffic flowing in a city (phenomena similar to a phase transition between a situation of fluid traffic and a traffic jam have been described), and *single-file diffusion* (in which particles propagating through channel geometries cannot pass each other).[4] On the other hand, a simple relation exists between the DLG and the *asymmetric exclusion process* in one dimension (Evans & Derrida 1994). The latter corresponds to $H(\vec{\sigma}) = 0$, i.e., there is no particle interaction, except for the hard core exclusion (at most one particle per site), and particles jump with probability p to the right and with probability $1 - p$ to the left. As indicated above, this may be thought of as due to the action of an electric field on charged particles, or may represent cars in a queue or steps of an interface. Also interesting is the relation to the continuum model equation by Kardar, Parisi, & Zhang (1986), and with self-organized criticality.[5]

For $H(\vec{\sigma}) \neq 0$, the DLG in §2.1 serves as an idealized model of *fast-* or *super-ionic conductors* (FICs). These are solid electrolytes in which one species of ions moves through a solid matrix set up by other ions. Thus, in contrast to more familiar electrolytes, FICs do not need to

[4] See, for example, Campos *et al.* (1995), Nagel & Paczuski (1995), Kukla *et al.* (1996), and Krug & Ferrari (1996).

[5] See Hwa & Kardar (1989), and Grinstein, Lee & Sachdev (1990), for example. The concept of self-organized criticality is introduced in Bak, Tang & Weisenfeld (1988); Peliti (1994) gives a brief review including some references.

be liquids, for example, molten salts or aqueous solutions, to present a high ionic conductivity. Examples include inorganic crystals such as β- or β''-alumina, glasses such as AgI–Ag_2MoO_4, and polymers such as (polyethylene oxide)–$NaBF_4$. They have been very actively studied. In fact, these materials are suitable for many technological applications, for example, high-temperature batteries. Moreover, a full understanding of their properties involves some challenging fundamental problems in solid state and statistical physics, for example, some related to diffusion and mechanisms of migration in a disordered state.[6]

Experiments reveal two well-defined classes of ionic conductivity in any of these materials, namely, poor ionic conduction at low temperature and good ionic conduction at higher temperature. This and similar observations have suggested that FICs undergo a phase transition. This is a *nonequilibrium* phase transition, however, given that it is strongly influenced by the external electric field which produces the (dissipative) current of ions. One may compare the properties of FIC materials to those of the DLG originally developed to deal with more general problems in statistical mechanics. It is shown below that such comparisons reveal some of the basic ingredients which determine experimentally observed behavior, and suggest further experiments. It is argued that the latter might confirm the existence of peculiar, anisotropic critical behavior.

Diffusion in FIC materials, in which ionic motion is greatly influenced by the immediate surroundings, cannot be modeled as a simple random walk. The DLG, by contrast, offers a convenient framework for studying collective diffusion. In fact, the DLG and its variants simplify many of the microscopic details in real materials (which, in light of the theory of cooperative phenomena, one should expect to be rather irrelevant for macroscopic behavior) while incorporating the main feature: the possibility of nonequilibrium steady states in a collection of driven interacting particles. The way in which interactions are represented deserves comment. One may assume that, due to the familiar screening effect (Debye & Hückel 1923, Lieb 1976), Coulomb interactions are reduced in practice to a hard core repulsion plus an effective short-range interaction which is neglected beyond NN positions in the model. At first glance, the interactions would appear to be repulsive, i.e., $J < 0$ in (2.1), since the carriers all bear like charges. Nevertheless, models corresponding to the cases $J < 0$ *and* $J > 0$, have been analyzed, and it now appears that the latter, the case of attractive interactions, allows one to understand some of the behavior

[6] For the pioneering study of FICs, and for some relevant experimental information see Sato & Kikuchi (1971), Boyce & Huberman (1979), Salomon (1979), Dieterich, Fulde, & Peschel (1980), Bates, Wang, & Dudney (1982), Dixon & Gillan (1982), Olson & Adelman (1985), and Marro, Garrido, & Vallés (1991).

reported for FIC materials. One may argue that this is related to lattice deformations producing elastic forces that mediate the repulsions, finally leading to an effective attraction in some cases (Dieterich *et al.* 1980). Other possibilities are that only the sign of the transverse interactions matters, or even that the basic phenomena of interest do not depend essentially on most details of interactions (§§1.1 and 2.3; see also §3.3 for the effect of the sign of particle interactions on the steady state). This is an aspect of the relation between the DLG and FICs whose clarification is worth further experimental analysis.

Study of the model indicates that as the field strength E increases, the interesting new features of the steady state become more pronounced, while the current tends toward saturation. The limit of saturating fields ($E \to \infty$ or, equivalently, motion contrary to E is forbidden) is therefore of particular interest. In addition to its physical relevance, this limit is convenient for theoretical analysis, since E is no longer a parameter. In experiments the electric field is typically weak, so that varying E is also interesting; this is done in §2.3, and in chapter 3. In any event, the parameter E simply serves to control the drift; its relation to the actual, macroscopic field, $\mathscr{E}$, is unknown in general. That is, to obtain the function $\mathscr{E}(E)$ involves a coarse-graining procedure (§4.3) which depends on specific details of the transport process in the system.

The description below mainly refers to $d = 1, 2$. The case $d = 3$ has been studied less systematically because of some added numerical difficulties. Furthermore, FIC materials often exhibit low-dimensional effects. That is, a number of real compounds have layer or channel structures that suggest that the conductivity is confined to one or two dimensions; see Heeger, Garito, & Iterrande (1975), Kennedy (1977), Beyeler (1981), and Kukla *et al.* (1996). For example, the sodium ions in β''-alumina and the silver ions in $AgCrS_2$ seem to remain basically within lattice planes, β-eucryptite and potassium hollandite are *quasi one-dimensional* conductors, i.e., ions are compelled to move in channels, and organic conductors and semiconductors sometimes display a sort of spatially restricted or highly anisotropic conductivity. Therefore, one- and two-dimensional models are of more than academic interest within the present context. Some attention is also devoted below to the three-dimensional case, however.

Some versions of (continuum) *driven-diffusive systems* (Wannier 1951, Eyink, Lebowitz, & Spohn 1996) admit the DLG or a variation of it as a microscopic model, in principle. In fact, most observable behavior which has a cooperative origin, for example, phase transitions, critical phenomena and hydrodynamic laws, is only determined by some very general features, such as the dimensions of the system and its order parameter, range and symmetry of the interactions, etc., and by the existence of very different spatial and temporal scales at the microscopic and macroscopic levels of description (§4.3). Consequently, microscopic models that involve a drastic oversimplification of nature may still accurately reproduce emergent, macroscopic properties of natural systems. *All* that is necessary is that they contain the basic ingredients that are responsible for the phenomena of interest, a condition which seems to hold in the present case. The claim that a simple model may capture many essential features of nature is consistent with the observation that quite different (FIC) materials, including inorganic crystals, glasses, polymers, etc., behave in a similar manner. Of course, this does not necessarily mean that DLG behavior will be experimentally accessible, as discussed later.

2.3 Monte Carlo simulations

Having introduced the DLG and discussed its relevance to observable phenomena, we examine its properties as revealed in MC simulations performed by a number of different investigators. The studies described here involve lattices with periodic, i.e., toroidal boundary conditions. In one dimension, rings of $N = 200$ sites were used, in two dimensions, square and rectangular systems with $10^2 < N < 10^6$ were employed, and in three dimensions the system consisted of sc lattices of $10^2 < N \leqslant 1.25 \times 10^5$ sites. A random initial configuration, $\vec{\sigma}^0 = \{\sigma_x = 1,0 \mid x = 1,2,\ldots,N\}$, with particle density n evolves according to (2.1)–(2.3) with temperature T and field E, and with an exchange rate ϕ of the Metropolis form, unless otherwise indicated. In most cases a saturating field is employed, so that jumps against E are prohibited. This procedure is found to yield steady-state properties which are practically indistinguishable from the ones obtained by gradually turning on the field E after reaching equilibration of $\vec{\sigma}^0$ with $E = 0$. Simulations were also used to investigate the effect of changing ϕ, and of varying Γ, on the steady state.

The main results may be summarized as follows. For $d = 1$, there seems to be no phase transition, but rather a unique translation invariant stationary state.[7] The spin–spin correlation function is found to decay exponentially with distance, as for equilibrium systems. The simulations (Katz *et al.* 1984) employ a modification of the Metropolis rate with a nontrivial E dependence; see table 2.1. This is an example of (2.3) except that ϕ depends on E.

A quantity of interest is the steady-state current, $j(T,E) \equiv \langle j_+ \rangle - \langle j_- \rangle$, where j_+ (j_-) is the number of jumps along (against) the field, per site

[7] It has been suggested, however, that a parallel updating of configurations may induce one-dimensional ordered states in a sense (Bagnoli, Droz, & Frachebourg 1991); Evans *et al.* (1995) have studied the asymmetric exclusion process (with short-range interactions and unbounded noise) of two types of charges moving in opposite directions which exhibits symmetry breaking in one dimension; see also Nagel & Paczuski (1995).

Table 2.1. The rate $c_E(\vec{\sigma}; x, x+1)$ used in MC simulations of the one-dimensional DLG by Katz *et al.* (1984). The symbols • and ∘ stand for occupied and vacant sites, respectively.

Process	Rate
∘ • ∘∘ → ∘ ∘ •∘	1
• • ∘• → • ∘ ••	1
∘ ∘ •∘ → ∘ • ∘∘	e^{-E}
• ∘ •• → • • ∘•	e^{-E}
• • ∘∘ → • ∘ •∘	$\frac{1}{2}\min\left(1, e^{-4\beta J}\right)$
∘ ∘ •• → ∘ • ∘•	$\frac{1}{2}e^{-E}\min\left(1, e^{-4\beta J}\right)$
∘ • ∘• → ∘ ∘ ••	$\frac{1}{2}\min\left(1, e^{4\beta J}\right)$
• ∘ •∘ → • • ∘∘	$\frac{1}{2}e^{-E}\min\left(1, e^{4\beta J}\right)$

and per unit time, and $\langle\cdots\rangle$ denotes a stationary average, realized in practice by the time average over a long simulation. The current vanishes in equilibrium, and increases linearly with E for small E. With a further increase in the driving field, j saturates at $\frac{1}{2}\Gamma(\Gamma + d - 1)^{-1}$ which is simply the probability of attempting a move along E. For both attractive and repulsive interactions the current is suppressed at low T, but this occurs more strongly in the former case because of clustering. (This is illustrated below and in chapter 3.) Interestingly enough, a comparison between numerical and experimental data reveals a close qualitative similarity between the temperature dependence of the conductivity in FIC materials and $j(T)$ in the model for saturating field and $d \geq 1$ (Marro 1996). The one-dimensional data conform to the empirical formula

$$j(T, E) = j(T, \infty)\left\{1 - \exp\left[-E\frac{\gamma(T, 0)}{j(T, \infty)}\right]\right\}, \tag{2.5}$$

where

$$\gamma(T, 0) = \lim_{E \to 0}\left[\frac{\partial j(T, E)}{\partial E}\right] \tag{2.6}$$

represents the zero-field conductivity. The relation (2.5), however, is not satisfied for $d > 1$ except at very high temperature. Further interesting behavior for $d = 1$ is revealed by the methods described in chapter 3.

For $d = 2$, most simulations employ attractive interactions, $\Gamma = 1$, and saturating fields. The limit $N \to \infty$ is sometimes obtained by analysis of size dependence and extrapolations from several lattice sizes. Two main qualitative conclusions have emerged (Marro *et al.* 1985, Marro & Vallés 1987, Vallés & Marro 1986, 1987). On one hand, for high temperature

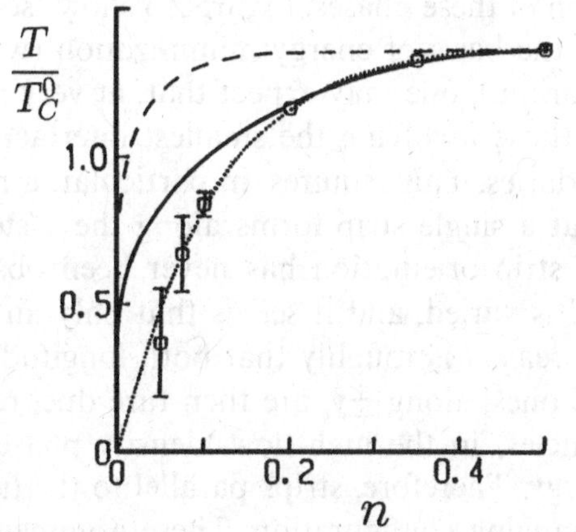

Fig. 2.2. Phase diagram for the DLG, as obtained by the MC method for $d = 2$, $\Gamma = 1$, Metropolis rate, attractive interactions, and saturating electric field. The graph shows the dependence of the transition temperature T_C^∞ on particle density n, and the associated error bars that locate approximately the limit of metastability or the spinodal line. The dashed line is the equilibrium Onsager result, and the solid line is the corresponding (equilibrium) mean-field result, both normalized to their respective critical temperatures. The dotted line fitting the data is a guide to the eye. (We note that in this and all subsequent figures, the uncertainty associated with a given data point is of the same magnitude as the symbol size, unless the point bears error bars.) (From Marro *et al.* (1991).)

for a given density n there is a unique translation invariant stationary state which is anisotropic and has interesting spatial correlations; this is discussed in §2.5. On the other hand, for any $E > 0$, there is anisotropic phase separation for $T < T^*(E, n)$. The transition temperature $T^*(E, n)$ increases with E, saturating as $E \to \infty$, for $n = \frac{1}{2}$ at $T_C^\infty \approx 1.4T_C^0$, where T_C^0 denotes the equilibrium critical temperature. The nature of this phase transition changes as one varies n. It appears to be continuous for $n = n_C = \frac{1}{2}$, and is clearly discontinuous for small n, $|n_C - n| \geq 0.3$. It is difficult to determine the order of the transition in the intermediate range on the basis of simulation data. In addition to the shift of the transition temperature, the phase diagram for $E > 0$ deviates significantly from the one for the (equilibrium) lattice gas, and also from a mean-field approximation to the latter, see figure 2.2.

Anisotropic phase segregation

Simulations of the DLG reveal that in phase separated systems, regions of a given phase (i.e., *liquid*, or particle-rich phase, and *vapor*, or particle-poor phase) are highly anisotropic, being elongated parallel to the field; §2.4

contains a discussion of these phases. Figure 2.3 shows some representative configurations. On the basis of energy minimization (which properly applies only in equilibrium), one may expect that, at very low T, the typical configurations are those involving the smallest interfacial area consistent with toroidal boundaries. This requires, in particular, a minimum number of interfaces so that a single strip forms along the system. On the other hand, a change of strip orientation has never been observed during the steady regime as E is varied, and it seems that only an interface parallel to E is stable. The reason is roughly that both longitudinal jumps, along $\pm\hat{\mathbf{x}}$, and transverse ones, along $\pm\hat{\mathbf{y}}$, are then rare due, respectively, to the lack of holes (particles) in the high-(low-)density phase and to the high cost of surface energy. Therefore, strips parallel to the field correspond to the most slowly changing configuration. These arguments are elaborated further in §3.5.

The essential anisotropy of the problem is also evident from the internal energy per site, corresponding to (2.1), i.e., $u \equiv -\langle H\rangle (JN)^{-1} = u_{\hat{\mathbf{x}}} + u_{\hat{\mathbf{y}}}$ where $u_{\hat{\mathbf{x}}(\hat{\mathbf{y}})}$ is the longitudinal (transverse) component. This is simply related to the average number of particle–hole bonds, say $e = e_{\hat{\mathbf{x}}} + e_{\hat{\mathbf{y}}}$, as $u = 1 - 2e$, $u_{\hat{\mathbf{x}}(\hat{\mathbf{y}})} = 1 - 2e_{\hat{\mathbf{x}}(\hat{\mathbf{y}})}$. For a random configuration in a half filled lattice, one has $e_{\hat{\mathbf{x}}} \approx e_{\hat{\mathbf{y}}} \approx \frac{1}{2}$, which implies that $u_{\hat{\mathbf{x}}}$ and $u_{\hat{\mathbf{y}}}$ are zero. The system with $n = \frac{1}{2}$ tends rapidly to this limit as T increases. This is depicted in figure 2.4, which also suggests that $T_C^\infty \approx 1.4T_C^0$, if one assumes the sudden change of slope marks the critical point. The data for $u_{\hat{\mathbf{x}}(\hat{\mathbf{y}})}$ are expected to reflect the behavior of the currents. The relation is only simple for saturating fields, however. That is, the fraction of attempted moves along the $+\hat{\mathbf{x}}$ direction is $f(\Gamma) = \frac{1}{2}\Gamma(\Gamma + d - 1)^{-1}$ for a d-dimensional lattice if longitudinal jumps are attempted with a frequency Γ times larger than transverse ones, and one has for saturating field that $j(T) = e_{\hat{\mathbf{x}}}(T)f(\Gamma)$, or $j(T) = \frac{1}{4}e_{\hat{\mathbf{x}}}(T)$ and $j(\infty) = \frac{1}{8}$ for $\Gamma = 1$ ($d = 2$).

A *nonequilibrium specific heat* may be defined as $C_E \equiv \partial u(T)/\partial T$. One may consider C_E for each of the two directions separately, but no systematic differences between the two cases are revealed by the data. The behavior of C_E averaged over the two directions is depicted in figure 2.5. For $n = \frac{1}{2}$, C_E exhibits a maximum near $1.4T_C^0$. On extrapolating the finite-size data for $n = \frac{1}{2}$ to estimate the infinite lattice specific heat, one finds a very pronounced peak, similar to that observed in equilibrium. The data for $n \ll \frac{1}{2}$ exhibit clear discontinuities at the transition temperature. Qualitatively similar peaks and discontinuities have been observed in FIC materials (Marro & Vallés 1987).

For an equilibrium system, the specific heat is related to energy fluctuations by $k_B T^2 C_{E=0} = \langle (H - \langle H\rangle)^2\rangle$, but there is no reason to expect this

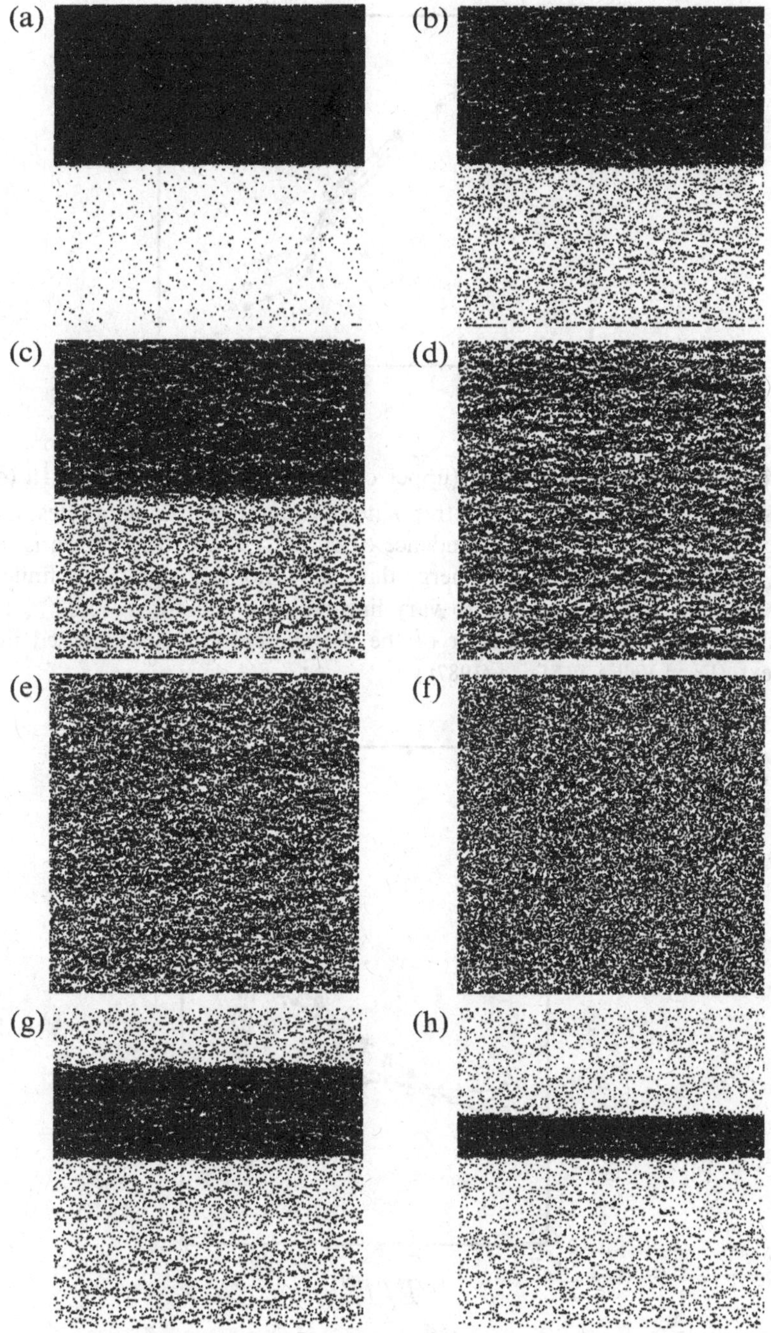

Fig. 2.3. Typical MC steady-state configurations of the two-dimensional system under a saturating (horizontal) electric field, for $L = 300$. The different graphs correspond to $n = \frac{1}{2}$ and $T/T_C^0 = 0.8$ (a), 1.2 (b), 1.3 (c), 1.45 (d), 2.0 (e), and 4.0 (f), and $n = 0.35$ and $T = 1.2T_C^0$ (g), and $n = 0.2$ and $T = 1.13T_C^0$ (h).

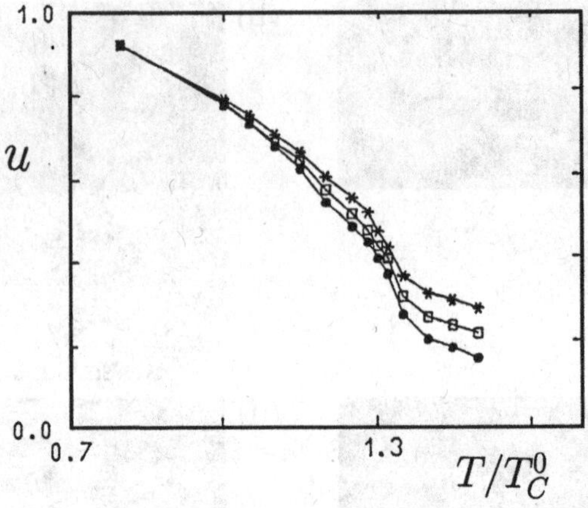

Fig. 2.4. The energy densities $u_{\hat{x}}$ (upper curve), $u_{\hat{y}}$ (lower curve) and $\frac{1}{2}u$ (open squares) for the *infinite lattice* as extrapolated from data for finite lattices; $n = \frac{1}{2}$. (Figure 2.5 illustrates the size dependence of $\partial u/\partial T$.) The extrapolation is based on the observation that the raw energy data show no very significant finite-size effects for $T > T_C^\infty$, and appear to vary linearly with L^{-1} for $T < T_C^\infty$. These graphs also represent the behavior of the particle current, as discussed in the main text. (From Vallés & Marro (1987).)

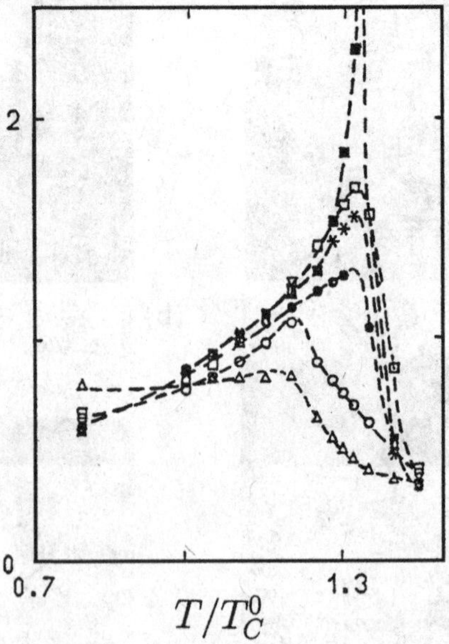

Fig. 2.5. The *specific heat,* defined as the temperature derivative of the energy u, for $n = \frac{1}{2}$ and different sizes (represented by different symbols). The upper curve with full squares corresponds to the *infinite lattice.*

to hold far from equilibrium. That is, lacking a detailed-balance condition and a Hamiltonian there is no proof of the microscopic reversibility that leads to the familiar fluctuation–dissipation relations. As for nonequilibrium systems in general, the situation concerning the validity of such relations is not definite for the DLG (this question is discussed further in the last paragraph of §2.4). A preliminary study of energy fluctuations in the DLG by Vallés & Marro (1987) gave no indication of the phase transition. However, theoretical arguments (§3.2) and higher-quality simulation data (for the layered model version presented below) suggest the aforementioned result is affected by substantial finite-size effects. We raise the issue here as an illustration of the difficulties encountered when attempting to draw firm conclusions from MC data for certain quantities, for example, energy fluctuations in the present case.

Order parameter

Identifying an order parameter is an essential part of understanding a continuous phase transition. While the choice is rather obvious in more familiar models, the fact that the DLG is endowed with particle-conserving dynamics, essential anisotropy, and a nonequilibrium condition (precluding the use of different ensembles) makes definition of an order parameter a more subtle problem here. One quantity that has been used as a measure of order in the two-dimensional DLG and some related systems is

$$m = \frac{1}{2\sqrt{n(1-n)}}\sqrt{|\langle M_{\hat{\mathbf{x}}}\rangle - \langle M_{\hat{\mathbf{y}}}\rangle|}\,, \tag{2.7}$$

where $M_{\hat{\mathbf{x}}}$ and $M_{\hat{\mathbf{y}}}$ are squared longitudinal and transverse *magnetizations* given by

$$M_{\hat{\mathbf{x}}(\hat{\mathbf{y}})} = \frac{1}{L_{\hat{\mathbf{x}}} \times L_{\hat{\mathbf{y}}} \times L_{\hat{\mathbf{x}}(\hat{\mathbf{y}})}} \sum_{y(x)} \left[\sum_{x(y)} (1 - 2\sigma_{x,y}) \right]^2 , \tag{2.8}$$

respectively, with $\sigma_{x,y} \equiv \sigma_{\mathbf{r}}$, where x and y denote the principal lattice directions. m is a measure of the difference between the densities of the liquid (strip-like) and gas phases: one has $\langle M_{\hat{\mathbf{x}}}\rangle = \langle M_{\hat{\mathbf{y}}}\rangle$ at infinite T as $N \to \infty$, which implies $m = 0$, while $\langle M_{\hat{\mathbf{x}}}\rangle \to 1$ and $\langle M_{\hat{\mathbf{y}}}\rangle \to 0$ in the limit of zero T at critical n, leading to $m \to 1$, and m is observed to change monotonically between these two limits.

Certain components of the structure factor, $S(\mathbf{k})$, are related to m, and may afford a more complete characterization of order. This observation led to Wang, Binder, & Lebowitz (1989) to study

$$m' = \sqrt{\left\langle M_{\hat{\mathbf{x}}}^{sin}\right\rangle^2 + \left\langle M_{\hat{\mathbf{x}}}^{cos}\right\rangle^2 + \left\langle M_{\hat{\mathbf{y}}}^{sin}\right\rangle^2 + \left\langle M_{\hat{\mathbf{y}}}^{cos}\right\rangle^2}\ , \tag{2.9}$$

where

$$M_{\hat{\mathbf{y}}}^{sin} \equiv \frac{\pi}{2L_{\hat{\mathbf{x}}}L_{\hat{\mathbf{y}}}} \sum_{x=1}^{L_{\hat{\mathbf{x}}}} \sum_{y=1}^{L_{\hat{\mathbf{y}}}} (1 - 2\sigma_{x,y}) \sin\left(\frac{2\pi}{L_{\hat{\mathbf{y}}}} y\right)\ , \quad \text{etc.} \tag{2.10}$$

The computation of m is more convenient in practice. As an alternative, one may define for $n = \frac{1}{2}$ the *density*

$$\varphi = \left\langle \frac{1}{2L_{\hat{\mathbf{x}}}} \sin\left(\frac{\pi}{L_{\hat{\mathbf{y}}}}\right) \left| \sum_{x=0}^{L_{\hat{\mathbf{x}}}-1} \sum_{y=0}^{L_{\hat{\mathbf{y}}}-1} (1 - 2\sigma_{x,y})\, \mathrm{e}^{i\mathbf{k}\cdot\mathbf{r}} \right| \right\rangle \tag{2.11}$$

with $\mathbf{r} = (x, y)$ and $\mathbf{k} = \left(0, 2\pi/L_{\hat{\mathbf{y}}}\right)$.

As one might expect, m, m', and φ reflect essentially the same steady-state behavior in the condensed phase. Aside from m being more appropriate when monitoring the development of order with time, they only seem to differ from each other in their respective dependence on lattice size at high temperature (for example, m, φ, and u have been reported to vary with size for the $L \times L$ lattice above the critical point as $L^{-\omega}$ with different values for ω). Figure 2.6 depicts the behavior of $m(n, T)$; it confirms the statements above about details of the phase transitions in this system. A qualitative difference exists between the curve $m(T)$ for $n = \frac{1}{2}$ in figure 2.6 and the phase diagram in figure 2.2; this is a peculiarity of the system, i.e., $m(T)$ is not a true magnetization, and no Gibbs ensemble exists. See §2.4 for further discussion of order parameters.

To obtain quantitative results one must analyze the size dependence of the data (see Fisher (1971), Cardy (1988), Privman (1990), for instance). It has been reported (figure 2.6, for example) that m exhibits large finite-size effects both above (unlike the energy) and below T_C^∞. This is not the case for the ordinary (equilibrium) Ising model with periodic boundary conditions, but it is consistent with the slow decay of correlations that seems to characterize the present problem (§2.5). One observes that m_L, i.e., the value of m estimated for the $L \times L$ lattice, follows to a good approximation (for the sizes investigated):

$$m_L \sim L^{-1} \quad \text{for} \quad T < T_C^\infty\ . \tag{2.12}$$

This kind of size dependence has been reported for the Ising system *with free edges* (Landau 1976). On this assumption, one may obtain values of m for $L \to \infty$ below T_C^∞ ($m = 0$ for the infinite lattice above T_C^∞). It seems

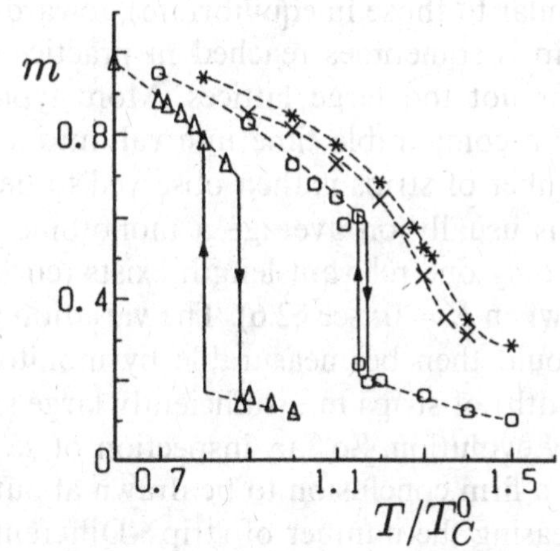

Fig. 2.6. The order parameter (2.7) for the system in figure 2.2 as a function of temperature, for $n = 0.5$ (*), 0.35 ($\times$), 0.2 (o), and 0.1 (Δ). The arrows represent transitions between the two branches, as observed in MC simulations for $n \leqslant 0.2$ as the system is heated up from zero temperature or cooled down from infinite temperature, respectively.

natural to use an approach which works well in equilibrium, and try the power law behavior

$$m \sim \mathscr{B}|T - T_C^\infty|^\beta \quad \text{as} \quad T \to T_C^\infty \text{ (from below)}. \qquad (2.13)$$

With this aim, one may plot $\ln m$ versus $\ln|T - T_C^\infty|$, and try to identify a value of T_C^∞ which yields a linear region. If this is possible, the slope near T_C^∞ corresponds to β. Alternatively, one may plot m raised to the power of $1/\beta$ versus T for different trial values of β, looking also for straight lines. The latter procedure has the advantage that no guess for T_C^∞ that might introduce further errors is involved. Both methods indicate that the data are consistent (over a rather large T interval near T_C^∞, as shown below) with (2.13) for $\beta \approx 0.3$, apparently excluding both the familiar $2-d$ Ising value of $\beta = 0.125$, and the classical or mean-field value of 0.5. The estimates obtained from this analysis, namely, $T_C^\infty \simeq 1.38 T_C^0$, $0.25 \leqslant \beta \leqslant 0.33$, and $\mathscr{B} \simeq 1.2$, are also supported by the rest of the data. For example, the same follows from the analysis of m' and φ, and no inconsistency has been detected so far on the assumption (2.13). Scaling behavior is discussed in §2.6.

Relaxation dynamics

The manner in which the DLG relaxes toward the steady state deserves comment. Below T_C^∞, time evolutions are monotonic, except for fluctuations (apparently similar to those in equilibrium), towards one-strip states. That is, a single strip is sometimes reached in practice after a relatively short (MC) time, for not too large lattices. More typically, the system decays instead, after a comparable time interval, into states with two or more strips. The number of strips is then observed to decrease with time. The time evolution is usually on average a monotonic process in which one may argue that only one relevant length exists (corresponding to the mean cluster radius when $E = 0$; see §2.6). The variation with time of such a relevant length would then be measurable by monitoring the number (alternatively, the width) of strips in a sufficiently large system. Figure 2.7 shows a typical early evolution. So far, inspection of several runs of this sort has not allowed a firm conclusion to be drawn about the mechanisms responsible for decreasing the number of strips. Different runs follow distinct progressions: monomer evaporation and deposition resulting in some strips increasing their width at the expense of others; two or more strips coalescing; diffusion of small anisotropic clusters from a rough surface to the neighboring strip; and, more generally, a combination of these (and other) mechanisms. In some cases after a rather extended period, there is a final relaxation into a one-strip state. A one-strip state has been never observed to split and produce several strips. (We discuss the nature of multi-strip states in more detail in §3.4.) For $n \leqslant 0.2$, there are not enough particles in the system to build up the transient, multi-strip states, and the evolution proceeds directly through one-strip states. In all cases, the single strip becomes more compact with time. Multi-strip states show a dependence on system size. In fact, it is likely that the corresponding escape time diverges as $L \to \infty$. States with 2, 7, 15, and 30 strips lasting for large times have been observed for $L = 50$, 100, 300, and 600, respectively, in systems operating at comparable parameter values. In spite of some similarity, multi-strip states are not metastable in the sense stated in Penrose & Lebowitz (1971), i.e., they are segregated states with no counterpart in classical theory.

The two-dimensional DLG exhibits, at least for $n < 0.35$ at a saturating field, well-defined finite discontinuities of the order parameter, energy, current, and specific heat, and conventional metastable states, as in an equilibrium phase transition of first order; see figures 2.2 and 2.6. The temperature at which these transitions occur is denoted $T^*(E,n)$, with $T^*(E \to \infty, n = \frac{1}{2}) \equiv T_C^\infty$ and $T^*(E = 0, n) \equiv T_C^0(n)$. Table 2.2 gives some MC estimates for $T^*(\infty, n)$; the error bars for $n \ll \frac{1}{2}$ are a (rather rough) estimate for the location of the closest metastable states observed during the evolution.

The fact that $T^*(\infty, n) \gg T_C^0(n)$ for $n \approx \frac{1}{2}$ while $T^*(\infty, n) < T_C^0(n)$ for sufficiently small n, which reflects the strong (interface) anisotropy, has been interpreted in Garrido, Marro, & Dickman (1990) as the result of

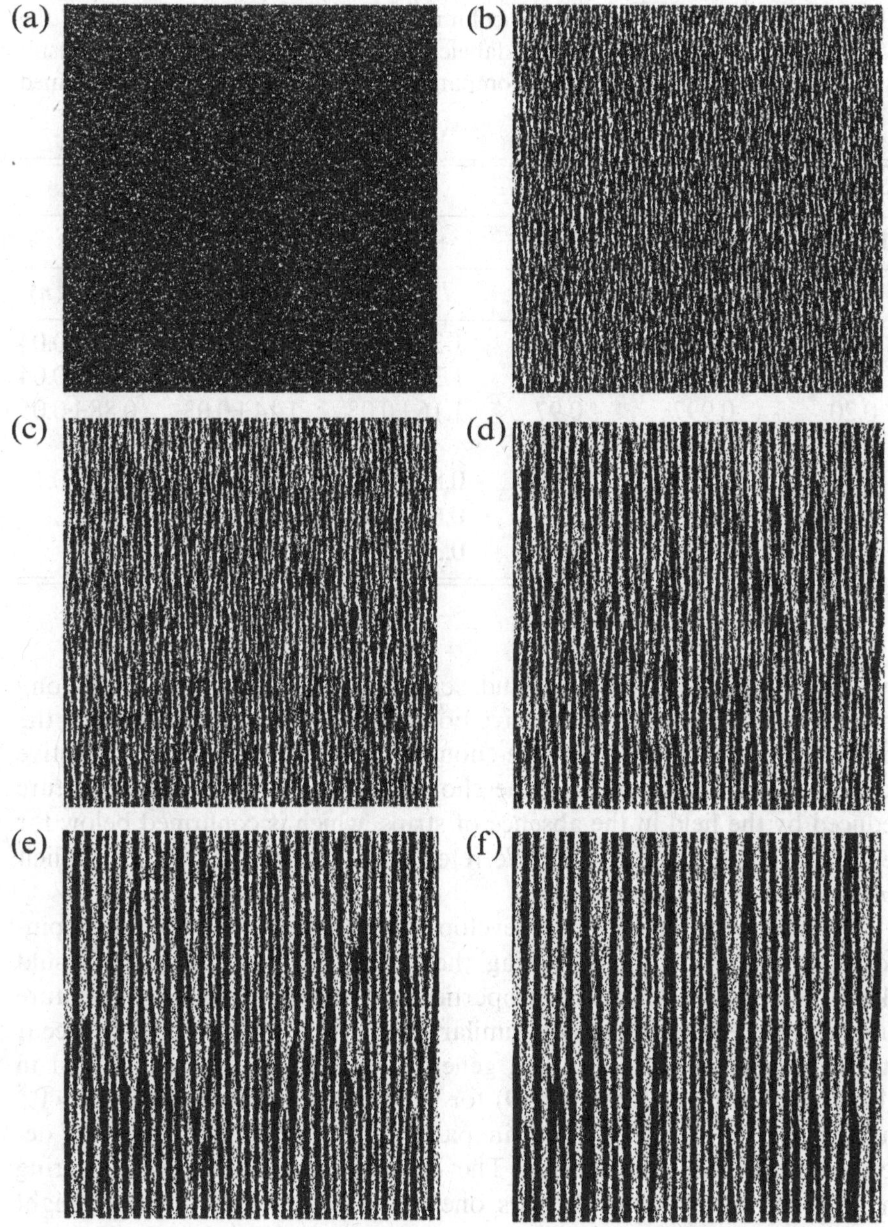

Fig. 2.7. Typical MC configurations during the very early stage of evolution of a 600×600 system under a saturating field with $n = \frac{1}{2}$ and $T = T_C^0$ following a quench from infinite temperature. This illustrates the formation of strips whose number decreases with time. The graphs are for time $t = 0$ (a) , 200 (b), 500 (c), 1400 (d), 3000 (e), and 6000 (f) in units of MC steps (per site). (Courtesy of Jorge Jiménez.)

Table 2.2. MC estimates of transition temperatures, $T^*(E,n)$ and $T'(E,n)$, for different systems, as indicated. The column labeled λ_∞ shows MC data for the two-dimensional DLG. The column labeled λ shows the exact equilibrium result by Onsager, which is included for comparison. The cases Λ and Λ_∞ are defined in §2.4.

System:	λ	Λ	λ_∞	Λ_∞	Λ_∞
Method:	exact	MC	MC	MC	MC
n	$T^*(0,n)$	$T^*_\Lambda(0,n)$	$T^*(\infty,n)$	$T^*_\Lambda(\infty,n)$	$T'(\infty,n)$
0.50	1	1	1.38±0.02	1.30±0.02	0.95±0.04
0.35	0.999986	~ 1	1.32±0.03	1.24±0.02	0.89±0.04
0.20	0.997	0.97	1.16±0.03	1.14±0.03	0.88±0.05
0.15	0.988	0.95	—	—	—
0.10	0.964	0.93	0.84±0.04	0.93±0.06	0.8±0.1
0.075	0.939	—	0.67±0.13	—	—
0.050	0.895	—	0.37±0.18	—	—

two competing effects. On one hand, some correlations are destroyed along the field, which reduces the effective bond strength. On the other hand, the moving strip that forms for large enough n acts in practice as an effective cohesion force (consequently, one should expect the critical temperature reduced by the field in the absence of strips, which is confirmed below for repulsive interactions, $J < 0$). We refer the reader to §3.5 for an explicit formulation of this competition.

It would be interesting to develop theories of nucleation and spinodal decomposition incorporating these facts. In particular, one would like to analyze in detail the properties of the time-dependent structure function, for example, the self-similarity that one may suspect to occur at low enough temperature as a generalization of the effect observed in Marro, Lebowitz, & Kalos (1979) for isotropic segregation when $E = 0$.[8] Such a theory should include, in particular, the peculiar behavior described below for correlations. The existence of anisotropic clustering in the one-phase region suggests one that the structure function might also exhibit interesting behavior at high temperature for small wavevectors.

[8] We refer the reader to Yeung *et al.* (1992), Puri *et al.* (1992, 1994), and Alexander *et al.* (1996) for some efforts along this line.

Experiments

Substances such as α-AgI exhibit conductivities of order 1 Ω^{-1} cm^{-1} below a certain temperature. This appears to derive from the high mobility of Ag^+ ions, which move rather freely through the lattice composed of I^- ions. The onset of such high ionic conductivity is usually regarded as marking a phase transition. Many of the observations remain puzzling, however. In an attempt to simplify a somewhat confusing set of experimental results, Boyce & Huberman (1979) identified three kinds of FIC materials. We turn now to a review of their classification, and its possible relation to the observations above for the DLG with $J > 0$.

FIC materials of *type I* exhibit abrupt discontinuities, for example, a jump of four decades in the ionic conductivity of AgI at a certain temperature. Although experiments are not quite clear cut (due to competing polymorphic transitions and other effects), this looks similar to the discontinuous behavior of the DLG with $n \ll \frac{1}{2}$. In fact, the unusually high mobility of the conducting ions would be impossible without a high vacancy density, as in the DLG with small n. In some substances vacancies seem to originate in the 'melting' of one of the sublattices at the transition temperature; in fact, a change of entropy which is comparable to the one for melting has been measured (O'Keeffe & Hyde 1976, Hayes 1978). The materials of *type II* exhibit a conductivity–temperature curve which is continuous with a change of slope at the transition temperature. This slope has been reported to change rather abruptly for $AgCrS_2$ (a more gradual change reported for PbF_2 might be related to a complex experimental situation producing a *diffuse transition*). One is tempted to identify this with a second-order transition, as in the model for $n = \frac{1}{2}$. However, a *weak* first-order transition cannot be ruled out in these materials (Gurevich & Kharkats 1986), for example, as described above for the DLG when $n = 0.35$. To resolve this ambiguity, systematic studies of the specific heat, of the structure function, and of the possible occurrence of metastable states are needed. Finally, *type III* FIC materials are characterized by an exponential increase in conductivity over a broad temperature range. Assuming ideal experimental conditions, this would reflect the absence of (anisotropic) segregation. A mean-field treatment of the one-dimensional DLG, presented in chapter 3, yields a similar result. This suggests that a useful classification of FIC materials should incorporate both the sign of the effective interactions between mobile ions, and the number of dimensions in which they are free to move.

The relevance of DLG simulations to FIC materials was explored in some detail in Marro *et al.* (1991; see also Marro 1996). The similarity between the specific heat and the current in the DLG and the corresponding quantities as observed in certain FICs is amazing, as mentioned above. It should be remarked, however, that the relation between the DLG and FIC materials is far from obvious, and that the models here are not intended to be a realistic portrayal of nature; see also, for example, Eyink *et al.* (1996), Markowich, Ringhofer, & Schmeisser (1990). As a further example, we mention the case of silver sulfide, Ag_2S, whose ionic conductivity is essentially modified by structural changes; see Ray & Vashishta (1989). This material is known as argentite when it behaves as a typical FIC, within the temperature range 450–866 K at atmospheric pressure. In this state the sulfur ions form a bcc lattice while silver ions are delocalized, diffusing among interstitial positions. When the temperature is reduced below 450 K there is a transformation from a bcc to a monoclinic lattice, and the diffusion constant for the silver ions drops by four orders of magnitude. (This transition from argentite to acantite is of great interest in geology.)

Influence of the rate

MC studies have also investigated the effect of kinetics on steady properties. No new qualitative facts have been reported so far for kinetics that satisfy detailed balance locally in the sense specified in §2.1. The details are interesting, however.

Consider first a saturating field with $d = 2$, $n = \frac{1}{2}$, and varying Γ. For $\Gamma \to \infty$, only longitudinal jumps occur in a MC simulation (and, conceptually, two time scales can be distinguished for the evolutions associated with longitudinal and transverse jumps; see §3.4 for a detailed discussion of this). Within the shortest time scale, the system can be regarded as a set of longitudinal chains whose particles cannot hop between chains but which interact when located at NN positions, as may occur in *quasi* one-dimensional conduction. This seems to be the case in hollandite, in which K^+ ions move in channels; see also the *single-file* diffusion reported for zeolites such as $AlPO_4 - 5$ (Meier & Olson 1992). A related solvable model devised by van Beijeren & Schulman (1984) has a mean-field critical point.[9] Therefore, one might expect a crossover of the DLG towards classical behavior as Γ is increased. There is only a qualitative study for $1 \leqslant \Gamma < 400$ (see Vallés & Marro (1986) for details of the implementation of transition probabilities; see also §§3.4 and 3.5). The data indicate a second-order phase transition, with the transition temperature, $T_C^\infty(\Gamma)$, decreasing monotonically with increasing Γ. The

[9] The normalization of the rate in this model is independent of H (equation (2.1)), and the particles within each longitudinal line *rapidly* return to the stationary condition between consecutive transverse jumps; this corresponds to the two time scales of the limit $\Gamma \to \infty$ for the DLG; see chapter 3 for further details.

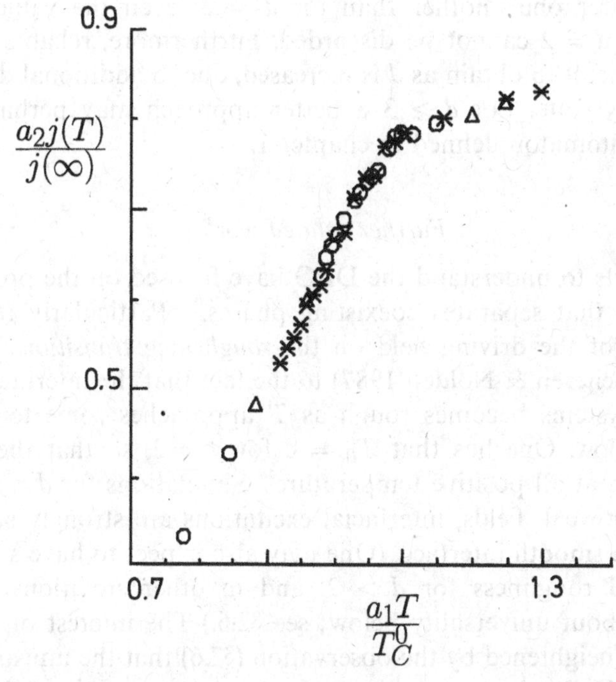

Fig. 2.8. The net particle current (normalized to the infinite temperature value) for $d = 3$, $n = \frac{1}{2}$, $J > 0$, a saturating field, and different transition rates (represented here by different symbols). The parameters a_1 and a_2 (Marro *et al.* 1991) depend only on the rate.

order parameter for not too large Γ scales in the familiar manner, i.e. $m(\Gamma) = c(\Gamma)\left[1 - T/T_C^{\infty}(\Gamma)\right]^{\beta}$ with β roughly equal to the value for $\Gamma = 1$. It is true that the data for $\Gamma \geq 80$ cannot be scaled with the rest, but the case $\Gamma \approx 400$ still seems too far away in practice from the theoretical prediction for the limit $\Gamma \to \infty$ (and more data would be needed) to draw any definite conclusion.

The case of $d = 3$, with a saturating field, attractive interactions, $L \leqslant 50$, $n = \frac{1}{2}$, and $\Gamma = 1$ has been studied for several rates. The Metropolis and Kawasaki cases (see equations (3.10) and (3.11)), and the case in which $c_E = 1$ for jumps in the direction of the field, $c_E = 0$ for jumps in the opposite direction, and $c_E = \exp\left(-\delta H/2k_B T\right)$ for any transverse jump have been considered. The results confirm that the order parameter (and the current) for different rates may be scaled together, as for varying Γ above, by renormalizing units appropriately; see figure 2.8. There is also some evidence that all the data fit equation (2.13) with $\beta = 0.4 \pm 0.1$. Thus, one cannot discriminate so definitely in three dimensions between the equilibrium ($\beta \simeq 0.3$) and classical ($\beta = 0.5$) values, in part because they are nearer one another than for $d = 2$; even the value $\beta \approx 0.3$ reported for $d = 2$ cannot be discarded. Furthermore, reliable MC data are more difficult to obtain as d is increased, due to additional degeneracy of the steady state. For $d \geq 3$ a better approach may perhaps be the lattice-gas automaton defined in chapter 1.

Further related work

Further efforts to understand the DLG have focused on the properties of the interface that separates coexisting phases.[10] Particularly remarkable is the effect of the driving field on the *roughening transition.* The latter refers (van Beijeren & Nolden 1987) to the fact that the interface in many segregated systems becomes rough as T approaches some temperature, T_R, from below. One has that $T_R = 0$ for $d = 2$, so that the interface is then rough at all positive temperatures. Simulations for $d = 2$ indicate that even for weak fields, interfacial excitations are strongly suppressed, resulting in a smooth interface. (One may also expect to have suppression of interfacial roughness for $d > 2$, and in other situations given the conclusion about universality below; see §2.6.) The interest of interfacial properties is heightened by the observation (§2.6) that the unusual β value in the $2-d$ DLG as reported above may be a consequence of the peculiar anisotropic interface in this system.

The case of repulsive interactions (i.e., $J < 0$ in equation (2.1)) has also been studied. The field breaks gauge symmetry, and so the behavior may be fundamentally different from the $J > 0$ case discussed above. Indeed, the pioneering work by Katz *et al.* (1984) suggested that the field tends to lower the transition temperature for $J < 0$, and that the DLG becomes disordered at all temperatures for strong enough fields. Simulations and field-theoretic arguments suggest that E is irrelevant to critical behavior in the DLG with repulsive interactions. Simulations of the two-dimensional system (Leung, Schmittmann, & Zia 1989) indicated that continuous phase transitions are characterized by $\beta = 0.125$. This study provides an outline of the phase diagram in a corner of the (E, T) plane. It suggests there is a line of Ising-like critical points extending from $(0, T_C^0)$ to (roughly) $\left(E = J,\ T = \frac{1}{2}T_C^0\right)$, beyond which there is a line of first-order transitions. At finite temperatures there is a unique (disordered) phase for all $E > 2J$; for $T = 0$ the possibility of an ordered state persists up to $E = 3J$. (One expects this to hold for $n \neq \frac{1}{2}$ as well.) This picture is consistent with the mean-field theory described in chapter 3. Since the repulsive system is arranged into sublattices rather than exhibiting

[10] Leung *et al.* (1988), Vallés, Leung, & Zia (1989), Hernández-Machado *et al.* (1989), Schmittmann & Zia (1995); see also §3.4.

coexisting phases of differing density, there is no well-defined interface and no dramatic anisotropy.

The above is puzzling because FIC materials apparently containing repulsive couplings are reported to undergo phase transitions that are similar to the ones above for $J > 0$. This has motivated our previous comment that materials might have effective attractive interactions in some cases. Perhaps, however, the clue is in Szabó (1994; see Szabó & Szolnoki 1990, Szabó, Szolnoki, & Antal 1994; see also chapter 1) claiming that, contrary to the conclusion drawn from previous MC work, a sort of order (which cannot be detected by the approximations in chapter 3) seems to occur for $J < 0$ at low temperature; also relevant to this problem are the observations in Helbing & Molnár (1995) and in Campos *et al.* (1995) concerning the closely related phenomena of traffic flow.

The case in which one allows for both positive and negative ions in the lattice, so that a particle current occurs in both directions, has been studied by Aertsens & Naudts (1990) in an effort to understand the unusual electrical conductivity of water-in-oil microemulsions; see also Evans *et al.* (1995). Korniss, Schmittmann, & Zia (1997) have studied a three-state lattice gas that simulates two species with charge exchange; see also Vilfan *et al.* (1994). One may think of simple variations of this model, for example, to describe dynamics of polymers by reptation (Duke 1989) and electrophoresis. On the other hand, it would be interesting to pursue the effort in Lauritsen & Fogedby (1993) and Schmittmann & Bassler (1996) for constant and random **E**, respectively, to determine whether quenched impurities modify critical properties.

2.4 Quasi two-dimensional conduction

Let us denote by λ, λ_E, and λ_∞ the ordinary two-dimensional lattice gas, the corresponding DLG, and the DLG for saturating fields, respectively. Define Λ (and similarly Λ_E), as the union of a pair of copies of λ, $\lambda^{(1)}$, and $\lambda^{(2)}$, with corresponding sites, $\mathbf{r}^{(1)}$ and $\mathbf{r}^{(2)}$, in the two copies connected insofar as hopping is concerned, but not energetically.[11] In other words, a particle at $\mathbf{r}^{(i)}$, $i = 1, 2$, may jump to one of five sites, one of which lies in the other copy, but the energy is

$$H_\Lambda(\vec{\sigma}) = H(\vec{\sigma}^{(1)}) + H(\vec{\sigma}^{(2)}), \tag{2.14}$$

where $\vec{\sigma}^{(i)}$ represents a configuration of $\lambda^{(i)}$, and $H(\vec{\sigma})$ is given by (2.1); we shall only consider the case $J > 0$.

[11] We summarize in this section some of the results in Achahbar & Marro (1995), and Achahbar, Garrido, & Marro (1996).

The systems Λ and Λ_E exhibit an extra degree of freedom, i.e., particles can leave a given plane, which is interesting from several points of view. In particular, the extremely slow evolution typical of λ may be accelerated in Λ, because density fluctuations have an additional mode for relaxation, which is expected to be important at least for sufficiently large k-modes. (The difficulty in attaining a steady state when evolution proceeds exclusively via diffusion is well known.) Consideration of Λ may therefore help in the study of phase coexistence, for instance. On the other hand, the layered geometry may be appropriate to model low-dimensional conduction (as in some FIC materials, see §2.2), the phenomenon of staging in intercalated compounds (Carlow & Frindt 1994), surface magnetism (Hu & Kawazoe 1994), etc. In any case, it seems interesting to study the novel features, for example, phase transitions, that may be exhibited by Λ and Λ_E, and by some natural generalizations of them that have been described in Achahbar *et al.* (1996), and the possible relation between Λ_∞ and λ_∞. The closest precursor of this idea is probably the *Gibbs ensemble.* This refers to a computational procedure in which particles have access to several boxes for which the pressure, volume, number of particles, etc. may be variables.[12] The case of Λ is somewhat simpler both conceptually and in practice. The behavior of two *coupled* planes studied by Hansen *et al.* (1993; see also Hill, Zia, & Schmittmann 1996) differs essentially from that of Λ.

Equilibrium phases

The equilibrium properties of Λ, i.e., the layered case when *the field is turned off*, are well understood. There is segregation below a temperature $T^*(E=0,n)$ into a liquid phase of density $n_\ell(T) = n_0(T)$ in one of the planes, and gas of density $n_g(T) = 1 - n_0(T)$ that fills the other plane. Here $n_0(T)$ is the Onsager solution for the pure phase, namely

$$2n_0(T) - 1 = \left\{1 - \left[\sinh\left(\frac{2J}{k_B T}\right)\right]^{-4}\right\}^{1/8}. \tag{2.15}$$

As illustrated in figure 2.9, the gas always occupies one of the planes, and liquid the other for $n = \frac{1}{2}$, whilst it forms a *liquid drop* (cluster) of density $n_0(T)$ coexisting in the same plane with gas of density $1 - n_0(T)$ for $n < \frac{1}{2}$. Then $n = \frac{1}{2}n_g + \frac{1}{2}\left[xn_\ell + (1-x)\,n_g\right]$ and, consequently, the fraction of

---

[12] See, for instance, Panagiotopoulos (1987, 1992), Amar (1989), Smit, Smedt, & Frenkel (1989), Vega *et al.* (1992), Mon (1993), and Rovere, Nielaba, & Binder (1993).

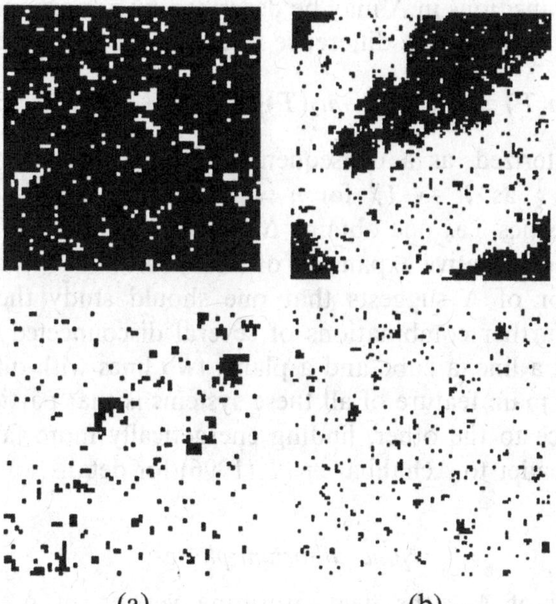

Fig. 2.9. Typical equilibrium MC configurations in the two planes of Λ for (a) $n = \frac{1}{2}$ and $T < T_C^0$, and (b) $n = 0.2$ and $T = 0.9T_C^0$ below the temperature for phase segregation at that density. (From Achahbar *et al.* (1996).)

liquid phase is

$$x(T) = 2\frac{n - 1 + n_0(T)}{2n_0(T) - 1} \quad \text{for} \quad \Lambda, \tag{2.16}$$

which is just twice the liquid-phase fraction for λ. The particles are uniformly distributed within each plane and between the two planes of Λ for $T > T^*(0, n)$. Therefore, $T^*(0, n)$ is the temperature for which $x(T^*) = 0$ holds or, using (2.16), $n_0(T^*) = 1 - n$. The same is true for λ, so that both systems have the same transition temperature for any density and, in particular, $T^*(0, n = \frac{1}{2}) = T_C^0$, the Onsager critical temperature (Hansen *et al.* 1993). Simulations confirm this; it seems worth mentioning, however, that the MC estimate for $T^*(0, n)$ for small n systematically differs (beyond typical errors) from the corresponding Onsager value, as shown in table 2.2. This is related to the fact that one has $x_\Lambda(n) < x_\lambda(n)$ for finite systems at given density, because some particles from the liquid droplet in Λ must go to the other plane to fill out the gas phase. The consequences of this become more dramatic as n is decreased; of course, one should expect the more familiar finite-size effects (see below) to add to this peculiar effect in Λ. It is likely that *Gibbs ensemble* simulations are similarly affected.

The phase transitions in Λ may be described by a (nonconserved) order parameter, for example, the difference in density between the planes,

$$\Delta n(T) \equiv \frac{1}{n} |n_1(T) - n_2(T)| = \frac{1}{n} |n - 1 + n_0(T)| \ . \qquad (2.17)$$

This is characterized, as a consequence of (2.15), by the Ising critical exponent $\beta = \frac{1}{8}$ as $T \to T_C^0$ for $n = \frac{1}{2}$. There is no discontinuity for off-critical densities, i.e., one obtains $\Delta n \sim |T - T^*(0, n)|$ sufficiently near $T^*(0, n)$ by a perturbative expansion of (2.17) for $n \neq \frac{1}{2}$.

The behavior of Λ suggests that one should study the equilibrium properties of further combinations of several disconnected lattices, such as a plane and a line, a cube and a plane, two lines with different lattice spacing, etc. A main feature of all these systems is that particles may hop from one lattice to the other, finding energetically more favorable sites. We refer the reader to Achahbar *et al.* (1996) for details.

Nonequilibrium phases

Detailed study of Λ_∞, the field-saturating version of Λ, has revealed novel phase transitions and helped to clarify certain aspects of the DLG. The model is based on equations (2.2) and (2.3) where, for attempted jumps perpendicular to the field, i.e., along $\pm\hat{\mathbf{y}}$ and $\pm\hat{\mathbf{z}}$ (the latter correspond to jumps to the other lattice), one uses the Metropolis probability, $\min\{1, \exp(-\beta\delta H)\}$, where δH does not involve any interaction between planes; see equation (2.14). Exchanges which transfer a particle along (against) the field are always (never) accepted.

The kind of anisotropic clustering observed at high T in λ_∞ occurs in Λ_∞ also; see figures 2.10(a) and 2.10(d), for instance. As T is reduced, two phase transitions are observed. Immediately below $T_\Lambda^*(\infty, n)$, the system segregates into a liquid phase consisting of two approximately equal strips, occupying adjacent regions in the two planes — see figures 2.10(b) and 2.10(e) — and the gas fills the rest of the system.[13] This phase transition is similar to the one in λ_∞ except for the existence of the second plane. This situation persists as T is further reduced, until at $T'(\infty, n) < T_\Lambda^*(\infty, n)$ there is a transition to a state in which liquid exists in only one of the planes; see figures 2.10(c) and 2.10(f). More subtle is the relation between Λ_∞ and Λ : the kind of segregation in figure 2.10(c), which happens to exhibit a tricritical point for $E = E_C \simeq 2$, so that the transition is discontinuous for large fields and becomes of second

[13] While MC experiments indicate that segregation in the two planes of Λ_∞ corresponds to the only stable situation within the indicated temperature range, which is confirmed by mean-field theory in §3.6, it corresponds to a (rare) sort of (segregated) *metastable* condition in Λ, i.e., for $E = 0$ (Achahbar *et al.* 1996).

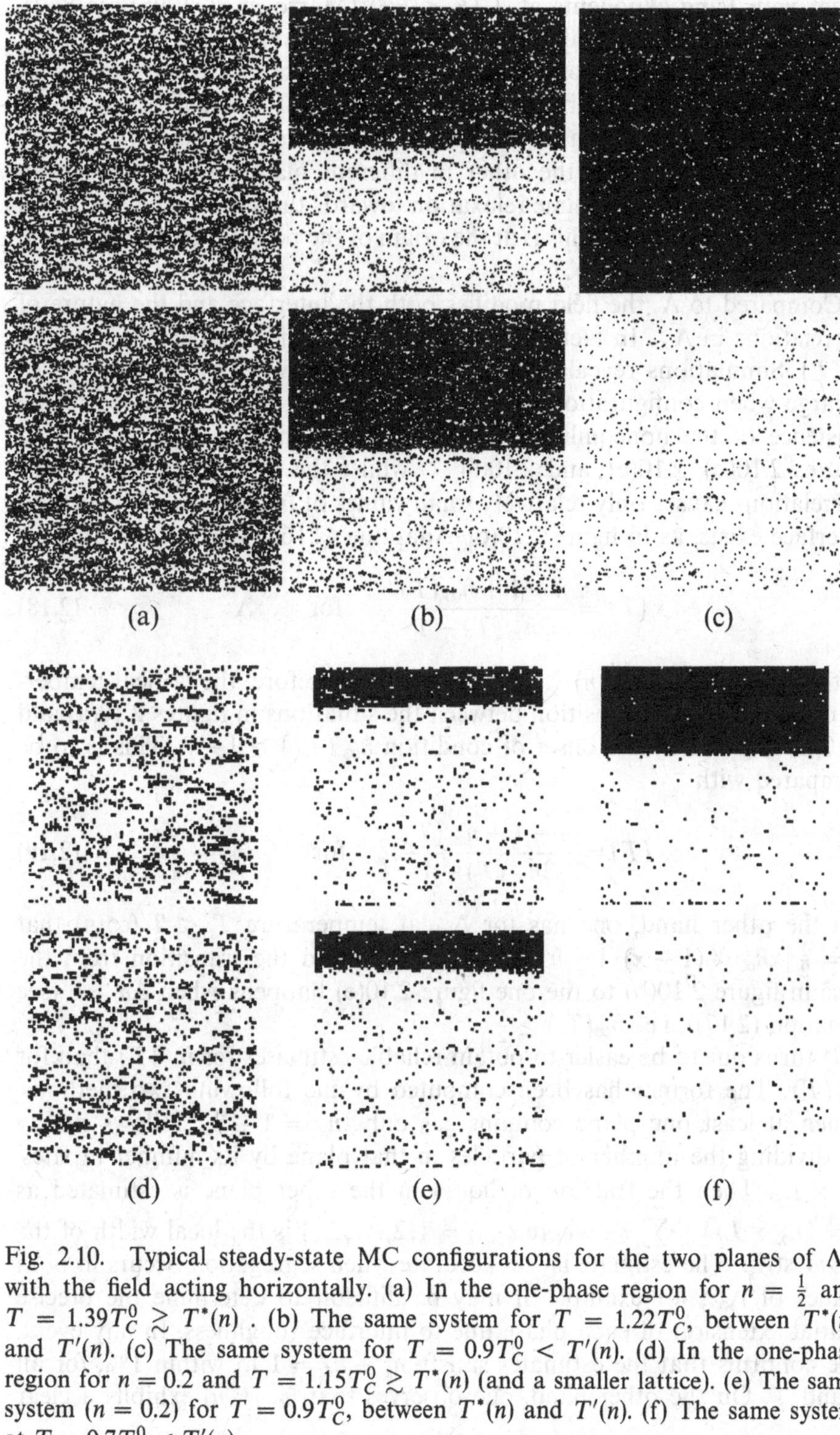

Fig. 2.10. Typical steady-state MC configurations for the two planes of Λ_∞ with the field acting horizontally. (a) In the one-phase region for $n = \frac{1}{2}$ and $T = 1.39T_C^0 \gtrsim T^*(n)$. (b) The same system for $T = 1.22T_C^0$, between $T^*(n)$ and $T'(n)$. (c) The same system for $T = 0.9T_C^0 < T'(n)$. (d) In the one-phase region for $n = 0.2$ and $T = 1.15T_C^0 \gtrsim T^*(n)$ (and a smaller lattice). (e) The same system ($n = 0.2$) for $T = 0.9T_C^0$, between $T^*(n)$ and $T'(n)$. (f) The same system at $T = 0.7T_C^0 < T'(n)$.

order with Ising exponents at $T'(E < E_C)$ (Marro *et al.* 1996; see §3.7), is comparable to the equilibrium one in figure 2.9(a), and the situation in figure 2.10(f) corresponds to the one in figure 2.9(b). However, the situations in figures 2.10(b) and 2.10(e) seem a full nonequilibrium effect, i.e., the only equilibrium counterparts are metastable states with two droplets, one on top of the other in different planes, which have been observed during the MC time relaxation towards the states in figures 2.9(a) and 2.9(b), respectively. This is at the origin of the peculiar critical behavior reported below at $T^*(\infty)$.

Compared to Λ, the field modifies both the interface and the nature of correlations in Λ_∞. In fact, the density of the liquid changes to $\hat{n}_\infty(T) \neq n_0(T)$. Simulations reveal that the difference between $\hat{n}_\infty(T)$ and $n_0(T)$ is large when configurations are striped, which is to be associated to the existence of the nonequilibrium interface all along the system — as in figures 2.10(b), 2.10(e), and 2.10(f) — whilst the different nature of the correlations causes only relatively small values of $\hat{n}_\infty(T) - n_0(T)$ when no interface exists, as in figure 2.10(c). Adapting (2.16), we may write that

$$x(T) = \frac{n - 1 + \hat{n}_\infty(T)}{2\hat{n}_\infty(T) - 1} \qquad \text{for} \qquad \Lambda_\infty \tag{2.18}$$

within the range $T'(\infty, n) < T < T^*_\Lambda(\infty, n)$. Therefore, the critical temperature for the phase transition between the situations in figures 2.10(a) and 2.10(b) is given by the onset of condition $\hat{n}_\infty(T^*_\Lambda) \geq 1 - n$. This is to be compared with

$$x(T) = \frac{n - 1 + n_\infty(T)}{2n_\infty(T) - 1} \qquad \text{for} \qquad \lambda_\infty\,. \tag{2.19}$$

On the other hand, one has for Λ_∞ at temperature $T < T'(\infty, n)$ that $n = \frac{1}{2}[x\hat{n}_\infty + (1 - x)(1 - \hat{n}_\infty)] + \frac{1}{2}(1 - \hat{n}_\infty)$, and the transition from the case in figure 2.10(b) to the one figure 2.10(e) happens when $\Delta n \geq 0$ (see equation (2.17)), i.e., $\hat{n}_\infty(T') \geq \frac{1}{2}$.

It turns out to be easier to obtain reliable estimates for $\hat{n}_\infty(T)$ than for $n_\infty(T)$. The former has been computed by the following two methods. When at least one plane contains gas only, $n_g = 1 - \hat{n}_\infty$ follows simply by dividing the number of particles in that plane by its number of sites, $L_{\hat{\mathbf{x}}} \times L_{\hat{\mathbf{y}}}$. Then the fraction of liquid in the other plane is estimated as $x = (L_{\hat{\mathbf{x}}} \times L_{\hat{\mathbf{y}}})^{-1} \sum_i \ell_i$, where ℓ_i, $(i = 1, 2, \ldots, L_{\hat{\mathbf{x}}})$ is the local width of the *liquid* strip. The estimate is less accurate when segregation occurs in both planes of Λ_∞; for example, it may be difficult to determine the precise spatial extension of each phase due to interface roughness. In any event, one confirms that the estimates satisfy $n_g + n_\ell = 1$ to within 1%, for all n and T. On the other hand, one observes that $\hat{n}_\infty(T, n)$ exhibits a clear

Table 2.3. Estimates for the difference of density between the liquid and gas phases, namely, $2n_\ell - 1$, for the systems indicated. Therefore, $\hat{n}_\infty(T, n = \frac{1}{2})$ is larger than $n_0(T, n = \frac{1}{2})$ for $T > 0.6T_C^0$, and the difference increases with T, while $\hat{n}_\infty - n_0 \approx 0$ in practice for $T \leqslant 0.6T_C^0$.

T/T_C^0	λ $n = 0.50$	Λ_∞ $n = 0.50$	Λ_∞ $n = 0.35$
0.6	0.993	0.993	0.984
0.7	0.981	0.982	0.967
0.8	0.955	0.959	0.940
0.9	0.895	0.906	0.893
0.93	0.863	0.870	—
0.98	0.748	0.842	—

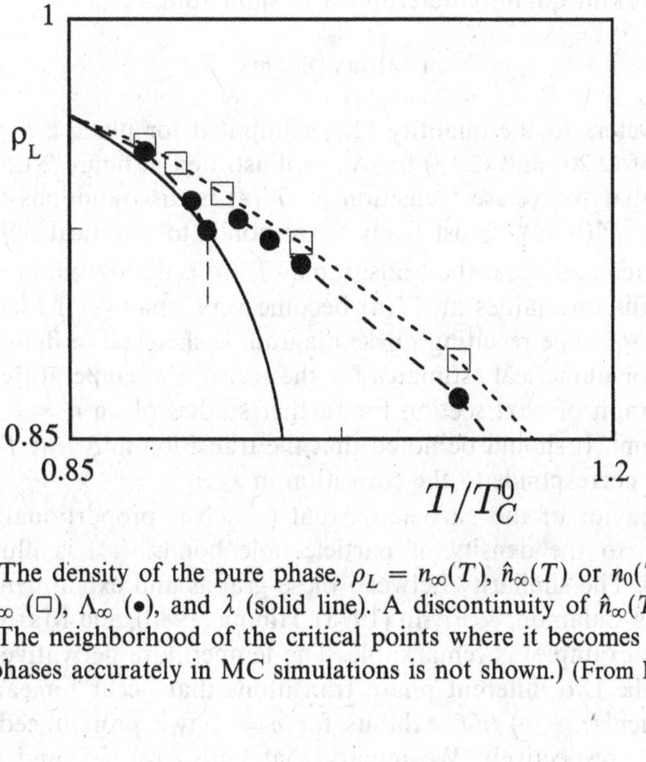

Fig. 2.11. The density of the pure phase, $\rho_L = n_\infty(T)$, $\hat{n}_\infty(T)$ or $n_0(T)$, respectively for λ_∞ (□), Λ_∞ (•), and λ (solid line). A discontinuity of $\hat{n}_\infty(T)$ at T' is indicated. (The neighborhood of the critical points where it becomes difficult to define the phases accurately in MC simulations is not shown.) (From Marro *et al.* (1996).)

discontinuity at $T^*(n)$ for $n \ll \frac{1}{2}$, and also at $T'(n)$, where $\hat{n}_\infty$ increases with T for any n. Moreover, one confirms that $\hat{n}_\infty \neq n_0$ and $\hat{n}_\infty \neq n_\infty$ in general; these two facts are illustrated in table 2.3 and in figure 2.11, respectively.

The latter observation indicates that contrary to the situation for $E = 0$ (where the behavior of Λ is simply related to that of λ, as described above), the properties of Λ_∞ differ from those of λ_∞. This is confirmed by the analysis of correlations in §2.5, and also by the mean-field study in chapter 3. (As a consequence of equations (2.18) and (2.19), the same is implied by the fact that $T^*_\Lambda(\infty, n) \neq T^*(\infty, n)$ in table 2.2.) The peculiar behavior of Λ_∞ may reflect interplane currents along the interface for $T'(n) < T < T^*(n)$; in fact, these currents may induce the phase transition at $T'(n)$ which does not occur in λ_∞. (The existence of two transitions in Λ_∞ is further explained in Alonso *et al.* (1995); see §3.7.) In spite of these differences, it is natural to expect Λ_∞ to have the same critical behavior at T^* as λ_∞; this expectation (which is supported by data below) is based upon the high degree of universality associated with critical phenomena (Λ and λ have the same critical behavior for $E = 0$), and on the fact that no new microscopic symmetries are introduced by dynamics in Λ_∞ as compared to those already present in λ_∞.

An important quantity determined in simulations is

$$m = \max\{m_1, m_2\} , \tag{2.20}$$

where m_i refers to the quantity (2.7) computed for plane i, $i = 1, 2$. The behavior of (2.20) and (2.17) for Λ_∞ is illustrated in figure 2.12. One may conclude that the phase transition at $T'(n)$ is discontinuous for any n, that $T_C^\infty \equiv T^*(n = \frac{1}{2})$ most likely corresponds to a critical point for the infinite system, and that the transition at $T^*(n)$ is discontinuous for $n \ll \frac{1}{2}$ while the discontinuities at $T^*(n)$ become very small — if there are any — for $n > 0.3$. The resulting phase diagram is sketched in figure 2.13; see table 2.2 for numerical estimates for the transition temperatures, and the last paragraph of next section for further studies of an $n = \frac{1}{2}$ section of this diagram. (It should be noted that the transition in Λ_∞ at $T^*(n)$ is the one which corresponds to the transition in λ_∞.)

The behavior of the particle current (which is proportional, for large enough E, to the density of particle–hole bonds, $\langle e \rangle$) is illustrated in figure 2.14. The similarity between these graphs and experimental results (see Vargas, Salamon, & Flynn (1978), Hibma (1980), and Marro & Vallés (1987), for example) is remarkable. The temperature derivative $\partial \langle e \rangle / \partial T$ indicates the two different phase transitions that occur for each density n. In particular, $\partial \langle e \rangle / \partial T$ exhibits for $n = \frac{1}{2}$ two pronounced peaks at T' and T^*, respectively. We remark that both $\partial \langle e \rangle / \partial T$ and the mean-square fluctuations $\langle (e - \langle e \rangle)^2 \rangle$ exhibit qualitatively similar structures for sufficiently large lattices, namely, $L \geq 128$. The situation is puzzling, however: whilst the two curves coincide near T' (Marro *et al.* 1996), they are distinct near T^*, so that it is not clear whether the DLG obeys a

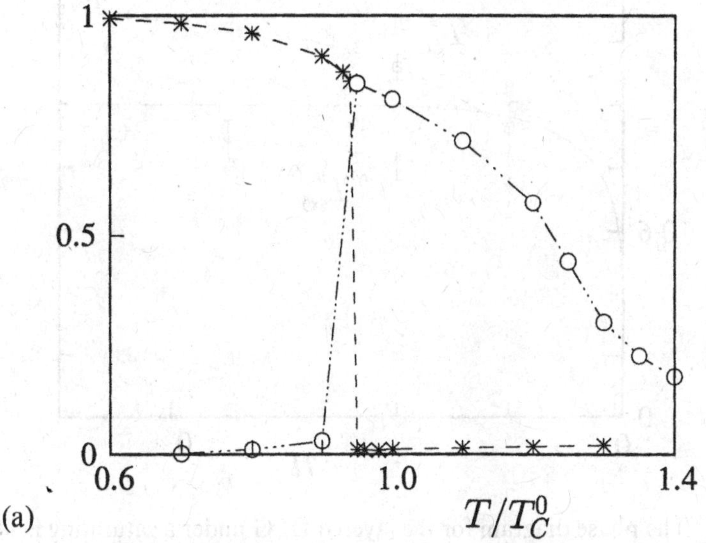

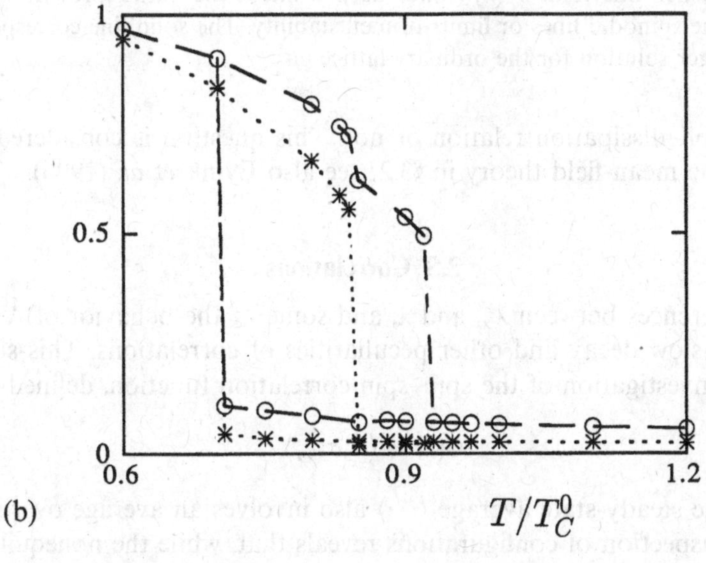

Fig. 2.12. Temperature dependence of the order parameters m (o) and Δn (*) for $n = \frac{1}{2}$ (a), and 0.1 (b). The indicated transitions in these graphs as one increases T are between states with one strip and with two strips, one at each plane, and between the latter and states with no strips, respectively; see figure 2.3. (From Achahbar & Marro (1995).)

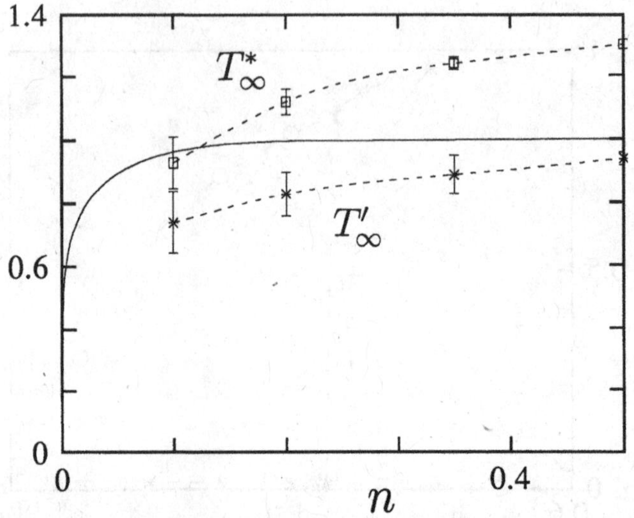

Fig. 2.13. The phase diagram for the layered DLG under a saturating field: $T^*(n)$ (upper dashed line) and $T'(n)$ (lower dashed line). The bars represent, approximately, the spinodal lines or limits of metastability. The solid line corresponds to the Onsager solution for the ordinary lattice gas.

fluctuation–dissipation relation or not. This question is considered again within the mean-field theory in §3.2; see also Eyink *et al.* (1996).

2.5 Correlations

The differences between λ_∞ and λ, and some of the behavior of Λ_∞, may reflect a slow decay and other peculiarities of correlations. This suggests further investigation of the spin–spin correlation function, defined as

$$G(\mathbf{r}) = \langle \sigma_{\mathbf{s}} \sigma_{\mathbf{s+r}} \rangle \ , \tag{2.21}$$

where the steady-state average $\langle \cdots \rangle$ also involves an average over $\mathbf{s} \in \mathbf{Z}^d$. Direct inspection of configurations reveals that, while the nonequilibrium systems are essentially homogeneous (on a large scale) above T^*, they are anisotropic, with clusters predominantly directed along the field. This is clear even for a temperature as high as $T = 2T_C^0$ (figure 2.3(e)) where equilibrium ($E = 0$) configurations would in practice look similar to the ones for infinite temperature (figure 2.7(a)). This effect was reported in a study of λ_∞ by Vallés & Marro (1987), who concluded that above T^* the longitudinal correlation function decays more slowly than exponentially with $\mathbf{r}$. The fact that $T^*(E,n) > T_C^0(n)$ for $n \approx \frac{1}{2}$ is a direct consequence of this effect of the field. It should also be reflected in the structure function,

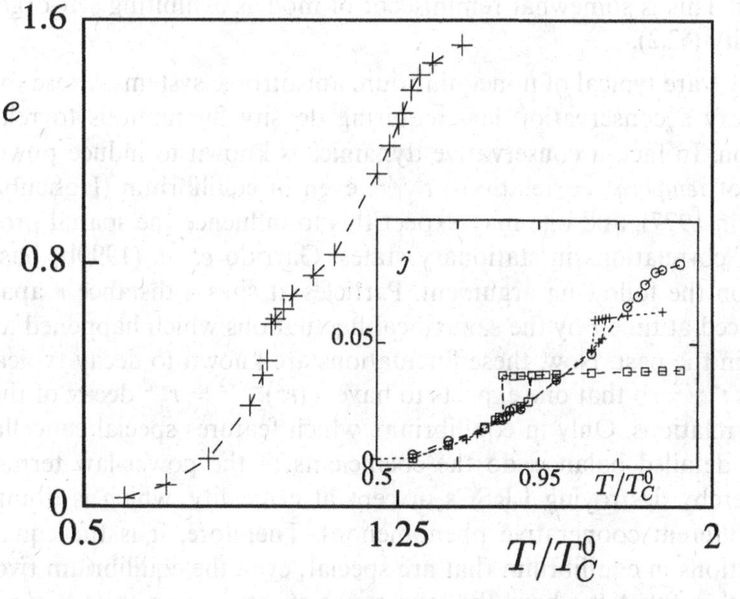

Fig. 2.14. The energy (averaged over the two planes) as a function of temperature for a finite lattice with $n = \frac{1}{2}$. The two changes in slope correspond to phase transitions. The inset shows the particle current in the field direction for $n = 0.35$ (o), 0.20 (×), and 0.10 (□).

defined as

$$S(\mathbf{k}) \equiv \sum_{\mathbf{r}} G(\mathbf{r}) \exp(i\mathbf{k} \cdot \mathbf{r}) \; ; \qquad (2.22)$$

for example, such behavior of correlations may produce experimentally-observable peculiar nonanalyticities near the origin $\mathbf{k} = 0$. Correlations have been investigated in detail by several authors, using MC simulations, series expansions of the master equation, and further theory for the DLG and other related systems; a brief account of this work is reported in this section.[14]

In equilibrium, spatial correlations are long-range under special occasions, for example, at critical points; otherwise, one observes exponential decay. Nonequilibrium systems, by contrast, seem to exhibit long-range spatial correlations (LRSCs) of the kind $G(\mathbf{r}) \sim \mathbf{r}^{-d}$ generically. In particular, LRSCs occur in λ_∞ at high T, where correlations assume (in two dimensions) a quadrupolar form, $G(\mathbf{r}) \sim r^{-4}\left(ax^2 - by^2\right)$, for $r \gg 1$. (Here a and b are positive constants.) This causes strong anisotropy of $S(\mathbf{k})$ at

[14] We describe, in particular, some work by Zhang *et al.* (1988), Garrido *et al.* (1990a), Cheng *et al.* (1991), and Alonso *et al.* (1995); we also refer the reader to the papers Hwa & Kardar (1989), and Grinstein *et al.* (1990).

small **k**. This is somewhat reminiscent of models exhibiting self-organized criticality (§2.2).

LRSCs are typical of nonequilibrium anisotropic systems whose dynamics satisfy a conservation law requiring density fluctuations to relax via diffusion. In fact, a conservative dynamics is known to induce power-law decay of *temporal* correlations, $t^{-d/2}$, even in equilibrium (Hohenberg & Halperin 1977), and one may expect this to influence the spatial propagation of correlations in stationary states. Garrido *et al.* (1990b) base this claim on the following argument. Particles at sites a distance r apart are influenced at time t by the same local fluctuations which happened at time $t \approx r^2$ in the past. Now, these fluctuations are known to decay typically in time as $t^{-d/2}$, so that one expects to have a $(r^2)^{-d/2} = r^{-d}$ decay of the spatial correlations. Only in equilibrium, which features special cancellations due to detailed balance, do the coefficients of the power-law terms vanish, thereby destroying LRSCs (except at criticality, which is dominated by a different cooperative phenomenon). Therefore, it is the equal-time correlations in equilibrium that are special; even the equilibrium two-time correlation functions have the long-range effects known as *long time tails.* The existence of LRSCs is also consistent with hydrodynamic fluctuation theory (Kirkpatrick, Cohen, & Dorfman 1980, Garrido *et al.* 1990b), and it has been verified experimentally via light scattering from a fluid in a container whose walls are kept at different temperatures (see Law *et al.* 1990).

One might have expected LRSCs to impede reliable computer simulation of nonequilibrium anisotropic systems. However, although peculiar finite-size effects occur (§2.6), useful information may be extracted from MC data for these systems, as in equilibrium. This is to be related to two facts. Firstly, the effects of LRSCs do not seem to be more dramatic than the more familiar difficulties within the critical region for equilibrium systems. Secondly, one expects a crossover in the behavior of correlations between high T and the neighborhood of T_C, where correlations may not differ so essentially from those in equilibrium. In fact, it has been claimed (Leung & Cardy 1986, Janssen & Schmittmann 1986) that in the critical region correlations decay $\propto r^{2-d-\eta}$. A prediction for a continuum model (whose relation to the DLG is not clear, however; see §2.6) is that for $d = 2$, $\eta_\parallel \simeq \frac{2}{3}$, where $\eta_\parallel$ is the exponent associated with longitudinal correlations. This is quite different from the $2-d$ Ising value of $\frac{1}{4}$, or the mean-field value of zero. In three dimensions, field theory predicts $\eta_\parallel \simeq 0$, while the Ising value is about 0.04.

The existence of LRSCs for the DLG is suggested by a high-T expansion of $G(\mathbf{r})$, which follows from (2.2) if three spin correlations are zero; the latter has been shown to occur exactly in a related system by Zhang *et al.*

(1988); see below. The model introduced by Spohn (1991) in which a steady current of particles is maintained by appropriate boundary conditions with fixed, unequal densities may be shown rigorously to have correlations of this kind. As an extension of the MC studies reported above, Zhang *et al.* (1988) have studied numerically lattices with one dimension larger than the other(s). This aims to account for the underlying anisotropy, and for the expectation (based on a field theoretic model; see §2.6) of two independent correlation lengths near T_C. These authors report that in three dimensions, as expected, the behavior of (2.21) along $\hat{\mathbf{x}}$ is $G_{\hat{\mathbf{x}}}(r) \sim r^{-a}$ with $a \approx 3$ at high T, and with $a = 1.0 \pm 0.3$ near T_C. A correlation length is then defined phenomenologically at each temperature such that $G_{\hat{\mathbf{x}}}(r) \sim [r/\xi_{\hat{\mathbf{x}}}(T)]^{-3}$ for $r > \xi_{\hat{\mathbf{x}}}(T)$. On this assumption, one observes that $\xi_{\hat{\mathbf{x}}}$ increases sharply (*diverges*) as $T \to T_C$ ($\approx 1.06T_C^0$) with $\nu_{\hat{\mathbf{x}}} \approx 0.8$ (which is smaller than the field-theoretic value $\nu_{\hat{\mathbf{x}}} = \frac{4}{3}$). For $d = 2$, the correlation length is introduced via

$$G_{\hat{\mathbf{x}}}(r) \sim \left[1 + \left(\frac{r}{\xi_{\hat{\mathbf{x}}}}\right)^2\right]^{-1} \quad \text{for large } r. \tag{2.23}$$

Numerical results (see below) indicate that $\nu_{\hat{\mathbf{x}}} \approx 0.7$. It should be noted, however, that statistical errors affecting the computation of correlation lengths by means of relations such as (2.23) are large. Most studies have failed to yield quantitative conclusions regarding the transverse correlation function, $G_{\hat{\mathbf{y}}}(r)$, which is observed to decay very rapidly, becoming negative at short distances for high enough temperature.

Achahbar & Marro (1995) have analyzed in detail spatial correlations in Λ_∞. Their primary conclusion is that, for $n = \frac{1}{2}$, $S(\mathbf{k})$ exhibits a pronounced peak at $\mathbf{k} =(0,0,1)$ for $T < 0.95T_C^0$, which corresponds to having phase segregation in only one of the planes, and the peak moves towards $\mathbf{k} =(1,0,0)$ as T is increased. This corresponds to the formation of strips. The situation is similar for $n < \frac{1}{2}$. The study of correlations also provides further support for the result $\hat{n}_\infty(T) \neq n_0(T)$ described in table 2.3. One observes that $G_{\hat{\mathbf{x}}}(r)$ differs depending on whether $E = 0$ or $E > 0$, and that $G_{\hat{\mathbf{x}}}(r) \neq G_{\hat{\mathbf{y}}}(r)$ for $r \leqslant 2$ in a saturating field. One observes, on the other hand, large differences in $G_{\hat{\mathbf{y}}}(r)$ between the equilibrium and nonequilibrium cases for large r at phase points for which the interface exists.[15]

The possibility that Λ_∞ exhibits LRSCs is intriguing. As for λ_∞, one observes in Λ_∞ anisotropic clustering in the one-phase region — see

[15] We refer the reader to the original work Vallés & Marro (1987) and Achahbar & Marro (1995) for a detailed phenomenological description of off-critical correlations in λ_∞ and Λ_∞, respectively.

figures 2.10(a) and 2.10(d) — but this is not necessarily the hallmark of slow decay of correlations. One might argue that correlations should not decay as slowly here as for λ_∞, given that particles can hop to the other plane. In fact, the decay of correlations in Λ_∞ is (at high enough temperatures) similar to that in λ_∞; we give below a proof of this for a closely related system. As a consequence, assuming the existence of the two correlation lengths, along $\hat{\mathbf{x}}$ and $\hat{\mathbf{y}}$, respectively, one may estimate them, for large enough r, using an expression such as (2.23). The divergence of $\xi_{\hat{\mathbf{x}}}$ and $\xi_{\hat{\mathbf{y}}}$ as $T \to T_C$ may then be characterized by the exponents $v_{\hat{\mathbf{x}}}$ and $v_{\hat{\mathbf{y}}}$, respectively. *On these assumptions*, the data provide some evidence that $v_{\hat{\mathbf{x}}} \approx 2v_{\hat{\mathbf{y}}}$ over a wide range of temperatures, and it has been estimated that $v_{\hat{\mathbf{x}}} = 0.7 \pm 0.3$, which is consistent with independent estimates (see above, and §2.6) but differs from the field-theoretic prediction. In any case, the observation that $v_{\hat{\mathbf{x}}} \approx 2v_{\hat{\mathbf{y}}}$ may just indicate that the temperature dependencies of $\xi_{\hat{\mathbf{x}}}$ and $\xi_{\hat{\mathbf{y}}}$ are interrelated, i.e., that a unique, relevant correlation length exists, as discussed in §2.6.

High-T expansions

The nature of correlations in anisotropic nonequilibrium systems may be investigated further by considering Λ and the master equation (2.2) with the rate

$$c(\vec{\sigma};\mathbf{r},\mathbf{s}) = \begin{cases} 1 & \text{for} \quad \hat{\mathbf{x}}\cdot(\mathbf{r}-\mathbf{s}) = \pm 1 \\ \phi(\beta\delta H) & \text{for} \quad \hat{\mathbf{x}}\cdot(\mathbf{r}-\mathbf{s}) = 0\,, \end{cases} \tag{2.24}$$

where $\delta H = H_\Lambda\left(\vec{\sigma}^{\mathbf{rs}}\right) - H_\Lambda(\vec{\sigma})$, instead of (2.3). Here particle–hole exchanges in directions other than $\hat{\mathbf{x}}$ take their usual equilibrium rate at temperature T, while those along $\pm\hat{\mathbf{x}}$ are completely random, as if induced by a heat bath at *infinite* temperature. This system, to be denoted Λ^T, is closely related to Λ_∞; however, Λ^T corresponds to a more symmetric case and, consequently, is more convenient for computations.

Let us follow Garrido *et al.* (1990b) and Alonso *et al.* (1995), and write from (2.2) that

$$\frac{\partial G(\mathbf{r})}{\partial t} = \sum_{\substack{\mathbf{r}',\mathbf{s}',\ |\mathbf{r}'-\mathbf{s}'|=1 \\ \mathbf{r}'\in(\mathbf{s},\mathbf{r}+\mathbf{s}),\mathbf{s}'\notin(\mathbf{s},\mathbf{r}+\mathbf{s})}} \left\langle \sigma_{\mathbf{s}}\sigma_{\mathbf{s}+\mathbf{r}}\ (\sigma_{\mathbf{r}'}\sigma_{\mathbf{s}'} - 1)\ c(\vec{\sigma};\mathbf{r}',\mathbf{s}')\right\rangle\,. \tag{2.25}$$

Here $\mathbf{r} = (\mathbf{x}, \ell)$, where $\ell = 1, 2$ refers to the plane and $\mathbf{x} = (x, y)$ is the location within the given plane. Then one may expand around $\beta = 0$, for example, $\phi(\beta\delta H) = 1 - \beta\delta H + \ldots$ and $\langle\cdots\rangle \simeq \langle\cdots\rangle_{\beta=0} + \beta\langle\cdots\rangle^{(1)}$ at high enough temperature. Thus one has that

$$G(\mathbf{r}) = \langle\sigma_{(\mathbf{0},1)}\sigma_{(\mathbf{x},\ell)}\rangle = m^2 + \beta g^{(1)}(\mathbf{0},1;\mathbf{x},\ell) + \mathcal{O}(\beta^2)\,, \quad (\mathbf{x},\ell) \neq (\mathbf{0},1)\,, \tag{2.26}$$

where $m = \langle\sigma_{\mathbf{r}}\rangle$. Let us define the Fourier transform of the first-order correlation, i.e.,

$$\hat{g}_{\ell',\ell}(\mathbf{k}) = \sum_{\mathbf{x}} e^{i\mathbf{k}\cdot\mathbf{x}} g^{(1)}(\mathbf{0},\ell';\mathbf{x},\ell)\,, \quad \ell',\ell = 1,2. \tag{2.27}$$

One obtains that

$$\hat{g}_{1,1}(\mathbf{k}) = \frac{1}{6-\alpha}\left[(5-\alpha)\ \omega(\mathbf{k};\mathbf{q}) + g^{(1)}(\mathbf{0},1;\mathbf{0},2) - \left(1-m^2\right)^2\alpha\right], \tag{2.28}$$

$$\hat{g}_{1,2}(\mathbf{k}) = \frac{1}{6-\alpha}\left[\omega(\mathbf{k};\mathbf{q}) - g^{(1)}(\mathbf{0},1;\mathbf{0},2) + \left(1-m^2\right)^2\alpha\right]. \tag{2.29}$$

Here $\alpha = \alpha(\mathbf{k}) \equiv 2\left(\cos k_x + \cos k_y\right)$ and

$$\omega(\mathbf{k};\mathbf{q}) \equiv \frac{1}{\alpha-2}\Big\{q_1 - q_2\cos k_x - q_3\cos k_y + \left(1-m^2\right)^2\left[\cos\left(k_x+k_y\right)\right.$$
$$\left.+\cos\left(k_x-k_y\right) + \cos 2k_y\right]\Big\}, \tag{2.30}$$

with $q_1 \equiv 2g^{(1)}(\mathbf{0},1;\hat{\mathbf{x}},1) + 2g^{(1)}(\mathbf{0},1;\hat{\mathbf{y}},1)$, $q_2 \equiv 2\left(1-m^2\right)^2 + g^{(1)}(\mathbf{0},1;\hat{\mathbf{x}},1)$, and $q_3 \equiv (1-m^2)^2 + g^{(1)}(\mathbf{0},1;\hat{\mathbf{y}},1)$. Therefore, assuming good convergence of (2.26), at high enough temperature one has that the dominant contribution to spatial correlations is

$$g^{(1)}(\mathbf{0},1;\mathbf{x},\ell) \simeq \frac{a_\ell x^2 - b_\ell y^2}{x^2+y^2}\,, \tag{2.31}$$

where the coefficients a_ℓ and b_ℓ depend on ℓ.

This result implies that differences between intraplane and interplane correlations are quantitative only (in spite of the fact that interplane bonds are broken). On the other hand, (2.31) also holds for the unique plane of λ^T, for which (2.28) and (2.29) reduce to (2.30), simply by fixing ℓ. Thus, although the density is not conserved separately in each plane of Λ^T, correlations in this system show a power-law decay, just as in λ^T. The fact that correlations such as (2.31) are established between any pair of points in different planes of Λ^T at high temperature gives a hint to understanding why Λ_∞ and λ_∞ (unlike Λ and λ) behave differently from each other.

As a further application of this method, consider a d-dimensional lattice gas whose dynamics are a combination of particle–hole exchanges with probability $1-p$ and creation/annihilation processes or *spin flips*, i.e., $\sigma_{\mathbf{r}} \to -\sigma_{\mathbf{r}}$, with probability p. As in (2.24), the exchanges are random along $\pm\hat{\mathbf{x}}$, and at temperature T otherwise. The flips occur at a rate

$p\,\phi\left(\beta\left[H\left(\vec{\sigma}^{\mathbf{r}}\right)-H\left(\vec{\sigma}\right)\right]\right)$, where $\vec{\sigma}^{\mathbf{r}}$ denotes $\vec{\sigma}$ after a flip at $\mathbf{r}$. One finds that at a high enough temperature,

$$G(\mathbf{r}) \propto \frac{1}{\sqrt{r^{\,d-1}}} \exp\left(-r\sqrt{\frac{2p}{1-p}}\right)\left[a_0+\frac{1}{r}\left(a_1 x^2 - a_2\,|\mathbf{y}|^2\right)\right], \tag{2.32}$$

where a_0, a_1, and a_2 are constants, $r = |\mathbf{r}|$, $\mathbf{r} = (x, \mathbf{y})$, x denotes the coordinate along $\hat{\mathbf{x}}$, and $\mathbf{y} = (y_1, y_2, \ldots, y_{d-1})$. That is, the power-law correlations described above are exponentially modulated for any $p \neq 0$.

It is likely that this interesting result holds for the version of DLG studied by Binder & Wang (1989), who added spin flips at a small rate to accelerate relaxation. There is a related result (Grinstein 1991, Grinstein, Jayaprakash, & Socolar 1993) for a continuum system. This consists of two Langevin equations with a *drift term* (see the next section) corresponding to anisotropic systems with conserved and nonconserved order parameters, respectively. It is concluded that a linear coupling between these subsystems preserves power-law behavior, while a nonlinear coupling, which might correspond to (a continuum version of) the Binder–Wang DLG variation, leads to exponential decay.

Finally, we mention that Hill *et al.* (1996) have performed a preliminary (MC) study of a Binder–Wang DLG variation in which two planes are coupled by NN bonds of strength $|J'| \geq 0$ (a case studied for $E = 0$ by Hansen *et al.* (1993)). Assuming that $J' \equiv 0$ corresponds to the Λ_∞ system, which is questionable given the result (2.32), this confirms, and extends to $J' \neq 0$ the results in figure 2.13 and table 2.2 for $n = \frac{1}{2}$.

2.6 Critical and scaling properties

According to symmetry arguments in the preceding section, one may expect, as for $E = 0$, Λ_∞ to have the same critical behavior as λ_∞; this is supported by simulation data. Some of the raw data for Λ_∞ are presented in figure 2.15. This illustrates the size dependence of $m(T)$, the extrapolation to $L \to \infty$, and a comparison with a similar result for the $2-d$ Ising model, and with the mean-field behavior. Figures 2.16 and 2.17 illustrate the relation between order parameters m, m', and φ, and provide evidence of power-law scaling, as in (2.13), with $\beta \simeq 0.3$. (The same result is reported to follow from independent sets of data, obtained in different simulations of Λ_∞, for $\Delta n(T)$, $\delta n(T)$, $m(T)$, $m'(T)$, and $\varphi(T)$ using several different methods; the same methods give $\beta \approx \frac{1}{8}$ for both $\Delta n(T)$ and $\delta n(T)$ when $E = 0$; see Marro *et al.* (1996).) The best present estimates are $T_\Lambda^*(\infty, n = \frac{1}{2}) = 1.30 \pm 0.02$, and

$$\beta = 0.3 \pm 0.05, \quad \mathscr{B} = 1.25 \pm 0.03. \tag{2.33}$$

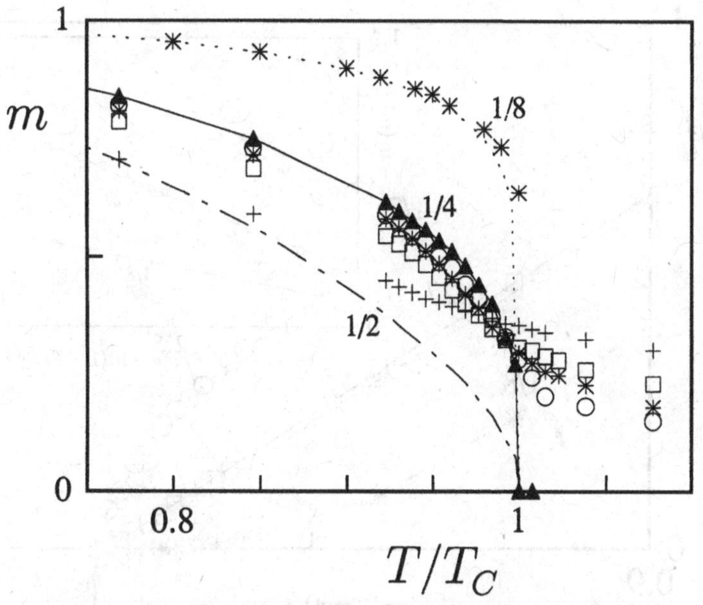

Fig. 2.15. The temperature variation of the parameter m defined in (2.7) and (2.20) for $n = \frac{1}{2}$ and $2 \times L \times L$ lattices with $L = 16$ (+), 32 (□), 64 (∗), and 128 (o). The solid line fits the case $L \to \infty$ (▲) obtained as an extrapolation from the data for the finite lattices. The Onsager solution (······), and corresponding MC results for $L = 128$ (∗) are shown for comparison and further information on finite-size effects. The mean-field result (– –––) is also shown. The three lines are also identified by the corresponding slope near criticality. Temperature is in units of the corresponding critical temperature.

These should be compared with $\beta \simeq 0.3$, $\mathscr{B} \simeq 1.2$, and $T_\lambda^*(\infty, n = \frac{1}{2}) \simeq 1.4$ for λ_∞. That is, β (and perhaps also the critical amplitude $\mathscr{B}$) is likely to be the same for Λ_∞ and for λ_∞, but not the critical temperature; this is confirmed analytically in chapter 3.

The fact that $\beta \neq \frac{1}{2}$ is also indicated by study of

$$\theta = e^{-2}\left[\frac{1}{4}(2-e)^2 - m^2\right], \tag{2.34}$$

where e represents the steady-state density of particle–hole pairs averaged over the planes of the infinite lattice. It has been shown (Marro *et al.* 1989) that θ is a measure of short-range order, and that

$$\theta \sim \alpha_1 |T - T_C|^{1-\alpha} - \alpha_2 |T - T_C|^{2\beta} \quad \text{as} \quad |T - T_C| \to 0\,, \tag{2.35}$$

where α_1 and α_2 are positive constants. The situation depicted in figure 2.18

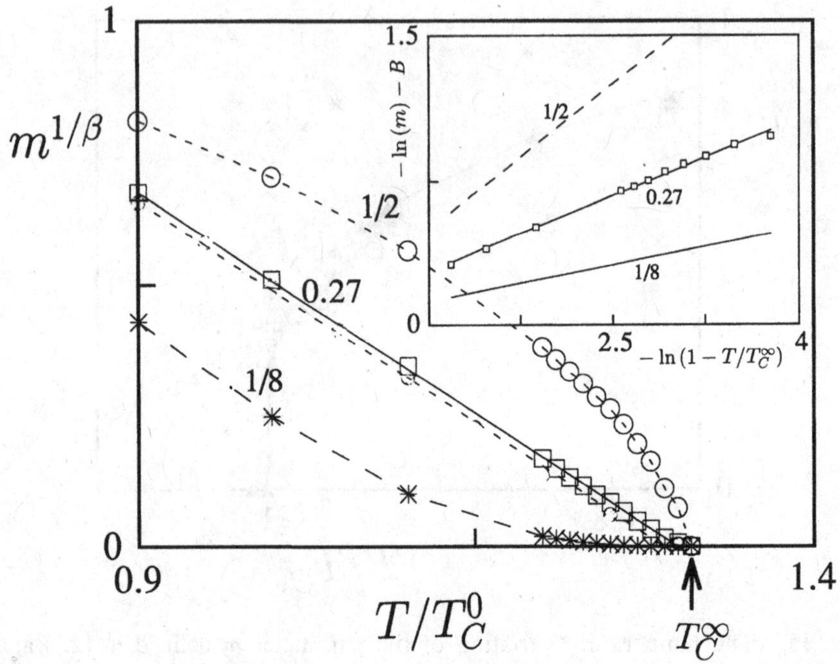

Fig. 2.16. Some graphic evidence that $m(T)$ for Λ_∞ is characterized by $\beta \simeq 0.3$ for the infinite system. The inset is a log–log plot of the data assuming $T_\infty^* = 1.30 \pm 0.02$ which gives $\beta \simeq 0.3$; lines of slope $\frac{1}{8}$ and $\frac{1}{2}$ are also shown. The main graph shows plots of $m^{1/\beta}$ for different values of β (as indicated) versus T; this gives a straight line only for β close to 0.3(□).

excludes the case $\beta = \frac{1}{2}$ and $\alpha = 0$: according to (2.35), the latter would lead to continuous behavior with T around the critical point (see the solid line in figure 3.10(b) for $\beta = \frac{1}{2}$ and $\alpha = 0$), while in fact a well-defined peak is observed.

On the other hand, analysis of the peak in $\partial e/\partial T$ (see figure 2.14) suggests $\alpha > 0$ rather than $\alpha = 0$, which would correspond to a logarithmic divergence or to a finite jump. However, the data are not sufficient to permit a definite conclusion in this regard.

Summing up, the data for both λ_∞ and Λ_∞ seem inconsistent with both Ising and mean-field critical behavior. (The latter has sometimes been associated with nonequilibrium phenomena, and is predicted by a related continuum model, as described below.) Instead, (2.13) and (2.33) seem to hold for both λ_∞ and Λ_∞, and there is also some evidence that $\alpha > 0$ (albeit small) and that $\nu_{\mathbf{x}} \approx 0.7$.

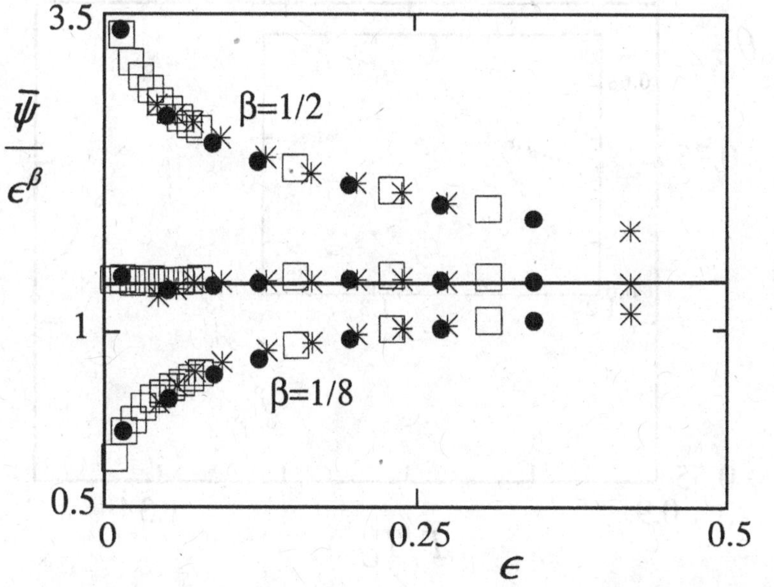

Fig. 2.17. The result shown in figure 2.16 is emphasized here by plotting independent data for ψ, which corresponds to m for λ_∞ (*) and Λ_∞ (□) as extrapolations to infinite size obtained from square lattices by the method in equation (2.12), and to φ for λ_∞ (•) for the 40×160 lattice. The only set of symbols that exhibits the expected independence of temperature *near* $\varepsilon = 0$ (fitted by the solid line $\mathscr{B} = 1.24$) is for $\beta = 0.27$, and $T^* = 1.38$ (*), $T^*_\Lambda = 1.30$ (□) and $T^*(L) = 1.37$ (•). (From Marro *et al.* (1996).)

Universality

Critical exponents consistent with $\beta \simeq 0.3$ and $\nu_{\hat{\mathbf{x}}} \simeq 0.7$ (more precisely, $\beta = 0.3 \pm 0.05$ and $\nu_{\hat{\mathbf{x}}} = 0.7 \pm 0.3$) have been reported in a series of independent simulations of related two-dimensional systems. In addition to λ_∞ and Λ_∞, these studies examined the following cases: the DLG with anisotropic couplings, namely, broken bonds between site pairs oriented parallel to the field (Vallés 1992); the DLG variation mentioned above in which spin flips at a small rate are added to exchanges (Binder & Wang 1989, Wang *et al.* 1989); a lattice gas or binary mixture with shear flow (Chan & Lin 1990); a lattice gas in which there is no field but exchanges along $\pm\hat{\mathbf{x}}$ occur completely at random (Cheng *et al.* 1991, Praestgaard, Larsen, & Zia 1994); and a cellular automaton representation of dissipative flow that simulates phase segregation in a quenched binary mixture (Alexander *et al.* 1992). Incidentally, $\beta \approx 0.3$ seems also to ensue from estimates for the relevant exponent for the growth of surfaces by hexagonal packing of discs described in §2.2; β then characterizes the

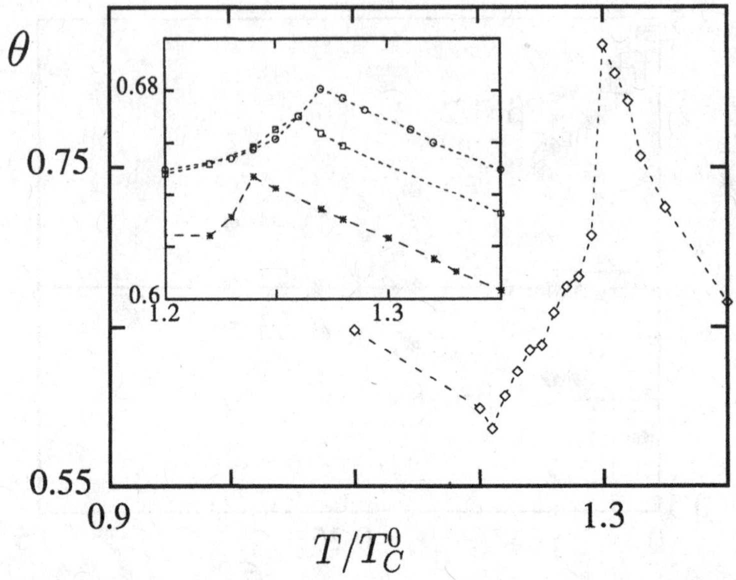

Fig. 2.18. Temperature variation of the short-range order parameter θ, as defined in equation (2.34), for the infinite volume limit of Λ_∞. The inset shows data for finite lattices: the peak locating the transition temperature increases and shifts towards larger temperatures as the size is increased. (From Achahbar & Marro (1995).)

scaling of surface with time, and is given by the ratio $\beta = \zeta/z$ between the roughness or wandering and dynamic exponents (Krug & Spohn 1992). A similar value was found for the order-parameter exponent in a field theory of the DLG with *random E* (Schmittmann & Zia 1991). The continuum model proposed for a constant driving field, by contrast, yields mean-field critical behavior.

Leaving aside this discrepancy, to which we return below, universality for a class of nonequilibrium systems is suggested. In addition to the ordinary Ising symmetries, modified here by the anisotropy of the process, the key features of this class are LRSCs, as described in §2.5, and a critical point characterized by the exponent values given above. Two absorbing questions arise. Firstly, the precise relation between all the models mentioned in the previous paragraph merits further study to determine the extent of this class. Secondly, it remains to confirm this in natural driven or anisotropic systems. With this aim, one may study FICs and other materials under nonequilibrium conditions. This study should focus on the effects of both the driving field and the anisotropic interface on critical behavior. The most convenient experimental situation is likely to be that of materials of low effective dimension (§2.2) under small electric fields, which may or may not saturate, with perhaps a ring geometry to realize a constant, nonconservative field (this requires increasing the magnetic flux for short periods of time). The only related experimental study of criticality known to us yielded *classical* behavior for (natural three-dimensional) fluids under shear (Beysens & Gbadamassi 1980),[16] in accordance with field theory (Onuki, Yamazaki, & Kawasaki 1981). Simulating the DLG via molecular dynamics may also add to our understanding; see Gillan (1985) and Ray & Vashishta (1989).

Field theory

Langevin equations combined with renormalization group arguments have been another main approach to understanding the properties of the DLG, including critical behavior. Some features of a field-theoretic model proposed as a continuum version of the DLG are briefly described next.[17] Let $\psi = \psi(\mathbf{r}, t)$ represent the bare order parameter, i.e., its average, $\bar{\psi}$, may correspond to m, m', or φ, etc., and consider *model B* (Hohenberg & Halperin 1977) with the familiar Hamiltonian

$$H = \int d\mathbf{r} \left[(\nabla\psi)^2 + \varepsilon\psi^2 + u\psi^4 \right], \quad u > 0\,. \tag{2.36}$$

The nonequilibrium nature of the system is taken into account by adding a drift, $j(\psi)\,\hat{\mathbf{x}}$, to the Langevin equation with Gaussian noise, ζ. The evolution is therefore governed by

$$\frac{\partial\psi(\mathbf{r},t)}{\partial t} = -\nabla\cdot\left[\eta\frac{\delta H}{\delta\psi} + j(\psi)\,\hat{\mathbf{x}}\right] + \nabla\zeta\,, \tag{2.37}$$

where η is a parameter. Assuming analyticity and symmetry under $\psi \to -\psi$, one writes $j(\psi) = a_0 + a_1\psi^2 + \cdots$. Then $a_1 = -a_0$ is implied by the fact that $\psi \to 1$ and $j \to 0$ as $T \to 0$, and $a_0 = j(T \to \infty)$. Therefore, the simplest assumption is that

$$j(\psi) = a_0\left(1 - \psi^2\right), \tag{2.38}$$

where a_0 is independent of T (but may depend on E).

System (2.36)–(2.38) reproduces the critical behavior of the Gaussian ψ^2 model with only an indirect influence of ψ^4. More explicitly, one obtains $u \sim \mu^{-2+\delta}$ and $E \sim \mu^{\delta/2}$ by standard *power counting* (μ is a momentum

[16] We refer the reader to the review by Dorfman, Kirkpatrick, & Sengers (1994) for related nonequilibrium fluids not covered here.

[17] We refer the reader to the specific review by Schmittmann & Zia (1995) for related field theory and for the original bibliography; this includes Gawedzki & Kupiainen (1986), Leung & Cardy (1986), Janssen & Schmittmann (1986), and Leung (1991, 1992).

scale and $d = 5 - \delta$, $\delta > 0$). That is, the relevant operator as $\mu \to \infty$ corresponds to the field term in the Lagrangian, while the u term (needed to stabilize properly below criticality) is *dangerously* irrelevant. For an upper critical dimension of 5, mean-field critical exponents, for example, $\beta = \frac{1}{2}$, ensue *exactly* for $2 < d \leqslant 5$, and weak logarithmic corrections to classical behavior are predicted for the marginal case $d = 2$. The role of the field in this picture suggests one should assume the existence of two correlation lengths, $\xi_{\hat{x}}$ and $\xi_{\hat{y}}$, which diverge with distinct exponents; one then obtains $\nu_{\hat{x}} = 1+(5-d)/6$ and $\nu_{\hat{y}} = \frac{1}{2}$, respectively. This implies that the essential DLG anisotropy can be removed simply by modifying the lattice sides accordingly.

An important issue is whether (2.36)–(2.38) contains the essential ingredients for DLG behavior. Simulations indicate that the DLG (as well as other closely related models mentioned above) are described by $\beta < \frac{1}{2}$. Moreover, assuming the existence of two correlation lengths, one finds they are not independent, and the sort of *effective isotropy* which has been predicted does not seem to be characteristic of the DLG (these two assertions are discussed further below). In other words, the main predictions of the continuum model conflict with the simulation data, casting serious doubt on the theory as a continuum representation of the DLG. This is not surprising, because the situation here is more complicated than in equilibrium, where the influence of dynamics and symmetries on the steady state is rather well established. A better understanding of relevant symmetries, including those associated with the *microscopic* dynamical rule,[18] would help in applying field theory to the DLG with confidence. In particular, we believe that a continuum version of the DLG should pay more attention to the detailed dependence of the mesoscopic coefficients in the Langevin-like equation on the microscopic parameter E, and perhaps focus on the role of the rare interface in this problem.

Finite-size scaling behavior

The behavior of the DLG with lattice size, for example, the one described in equation (2.12), is not the familiar one for the Ising model with periodic boundaries. Field theory has suggested that (anisotropic) rectangular, $L_{\hat{x}} \times L_{\hat{y}}$ lattices should be comparable to each other, exhibiting isotropic behavior if one fixes the ratio $s = L_{\hat{x}}^{\nu_{\hat{x}}/\nu_{\hat{y}}} L_{\hat{y}}^{-1}$; see Schmittmann & Zia (1995). Consequently, both s and $L_{\hat{x}}^{1/\nu_{\hat{y}}}(T - T_C)$ enter as independent variables in scaling formulas. This leads to simple relations at T_C which

[18] See Garrido & Muñoz (1995) and Garrido *et al.* (1997) for field-theoretic efforts in that direction. The influence of microscopic rules on emergent properties is a main concern throughout this book.

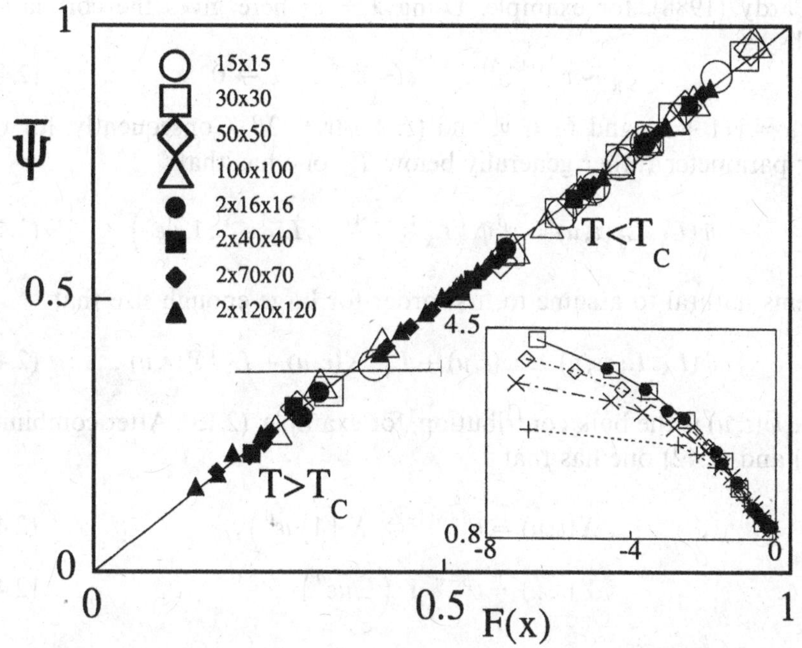

Fig. 2.19. The order parameter versus the r.h.s. of equation (2.47) for λ_∞ (open symbols) and Λ_∞ (filled symbols); all the available data within the ranges $4 < \kappa < 100$, $\kappa' > 1.5$ and $7 < \kappa < 200$, $\kappa' > 7$, respectively, have been included. The parameter values used in this graph are the ones in Vallés & Marro (1987) and Marro *et al.* (1996), which are consistent within statistical errors with the (more accurate) ones reported above. The inset is a plot $\varphi L_{\hat{\mathbf{x}}}^{1/3}$ versus $\varepsilon L_{\hat{\mathbf{x}}}^{2/3}$ which is suggested by field theory, equation (2.39).

were first checked against MC data by Binder & Wang (1989). Clear departures from the scaling prediction are found; see also Achahbar & Marro (1995). On the basis of the same scaling assumption, Leung (1991) proposed that

$$\bar{\psi} = L_{\hat{\mathbf{x}}}^{-1/3} \tilde{\psi} \left[(T - T_C) L_{\hat{\mathbf{x}}}^{2/3} \right] . \tag{2.39}$$

That this scaling is also not supported by the data was documented by Achahbar & Marro (1995), Wang (1996), and Marro *et al.* (1996); see inset in figure 2.19. We describe next simple arguments (Marro *et al.* 1996) leading to scaling formulas that have been reported to describe better the available data.

Let us assume (against expectations) that two relevant lengths exist, and consider the Hamiltonian (2.36), for instance. The pair correlation function for a rectangle $L_{\hat{\mathbf{x}}} \times L_{\hat{\mathbf{y}}}$ is expected to follow

$$G(x, y; \varepsilon, u) = \lambda^{\Delta+\eta} G(\lambda^{1+\Delta} x, \lambda y; \varepsilon \lambda^{-1/\nu}, u\lambda^{\theta/\nu}) \; ; \tag{2.40}$$

see Cardy (1988), for example. Using $\lambda = \varepsilon^{\nu}$ here gives the correlation lengths as

$$\xi_{\hat{\mathbf{x}}} \sim \varepsilon^{-\nu(1+\Delta)}\,, \quad \xi_{\hat{\mathbf{y}}} \sim \varepsilon^{-\nu}\,, \quad \varepsilon \to 0\,, \tag{2.41}$$

i.e., $\nu_{\hat{\mathbf{x}}} = \nu(1+\Delta)$ and $\nu_{\hat{\mathbf{y}}} = \nu$, and $(\Delta + \eta)\nu = 2\beta$. Consequently, for the order parameter rather generally below T_C one has that

$$\bar{\psi}(L_{\hat{\mathbf{x}}}, L_{\hat{\mathbf{y}}}; \varepsilon, u) = \varepsilon^{\beta}\,\tilde{\psi}\left(L_{\hat{\mathbf{x}}}^{-1}\varepsilon^{-\nu(1+\Delta)}, L_{\hat{\mathbf{y}}}^{-1}\varepsilon^{-\nu}; 1, u\varepsilon^{\theta}\right). \tag{2.42}$$

It seems natural to assume to first order for large enough size that

$$\bar{\psi}(L_{\hat{\mathbf{x}}}, L_{\hat{\mathbf{y}}}; \varepsilon, u) \simeq \psi(\varepsilon, u) + L_{\hat{\mathbf{x}}}^{-1} X(\varepsilon, u) + L_{\hat{\mathbf{y}}}^{-1} Y(\varepsilon, u)\,, \tag{2.43}$$

where $\psi(\varepsilon, u)$ is the bulk contribution, for example, (2.13). After combining (2.43) and (2.42) one has that

$$X(\varepsilon, u) = \varepsilon^{\beta-\nu(1+\Delta)}\, X\left(1, u\varepsilon^{\theta}\right), \tag{2.44}$$

$$Y(\varepsilon, u) = \varepsilon^{\beta-\nu}\, Y\left(1, u\varepsilon^{\theta}\right). \tag{2.45}$$

This implies that

$$\frac{X(\varepsilon, u)}{Y(\varepsilon, u)} = \varepsilon^{-\Delta\nu} f(1, u\varepsilon^{\theta})\,. \tag{2.46}$$

At issue is whether $\Delta = 0$ or not. These equations provide a method of checking this; getting $\Delta \neq 0$ would confirm the assumption that two independent lengths exist at criticality. Marro *et al.* (1996) fitted different sets of data to obtain the temperature variation of X and Y using equation (2.43); this fit suggests a bound of the form $|X/Y| < 1$, and a linear temperature dependence, namely, $X/Y \approx 0.26\varepsilon - 0.975$ over the physically relevant region, $1 < \xi < L$. Therefore, it seems that no singularity occurs as $\varepsilon \to 0^{-}$, and that $\Delta \approx 0$ in the formulas above. A more definite conclusion on this matter would require more (high-precision) data, for example, $\bar{\psi}$ for *many* different values of $L_{\hat{\mathbf{x}}}$ for given $L_{\hat{\mathbf{y}}}$, and vice-versa.

The assumption that a unique correlation length is relevant sufficiently near T_C is also contained in the proposal:

$$\bar{\psi} \sim L^{-\beta/\nu}\left(\mathscr{B}\,\kappa^{\beta} - \mathscr{B}_s\,\kappa^{\beta-\nu}\right) \quad \text{for} \quad T < T_C, \tag{2.47}$$

where $\kappa \equiv \varepsilon L^{1/\nu}$ and $\varepsilon \equiv 1 - T/T_C$. That is, this behavior, which has been shown to be a good description of the data for both λ_{∞} (Vallés & Marro 1987) and Λ_{∞} (Marro *et al.* 1996), as illustrated in figure 2.19, follows immediately from (2.43)–(2.45) for $L_{\hat{\mathbf{x}}} = L_{\hat{\mathbf{y}}}$ if Δ is zero; one has in this case

$$\mathscr{B}_s = -X(1, u\varepsilon^{\theta}) - Y(1, u\varepsilon^{\theta})\,. \tag{2.48}$$

One may generalize (2.47) to rectangular lattices. For $L_{\hat{x}} \neq L_{\hat{y}}$ and $\Delta \approx 0$, the prediction from (2.43)–(2.45) below T_C is that

$$\bar{\psi}(L_{\hat{x}}, L_{\hat{y}}; \varepsilon, u) \simeq \varepsilon^{\beta} \left\{ \mathscr{B} + \varepsilon^{-\nu} \left[\frac{X\left(1, u\varepsilon^{\theta}\right)}{L_{\hat{x}}} + \frac{Y\left(1, u\varepsilon^{\theta}\right)}{L_{\hat{y}}} \right] \right\}. \quad (2.49)$$

A very stringent test of this is to study the size and temperature dependence of the surface contributions, namely, $\bar{\psi}\varepsilon^{-\beta}$. This is performed for square geometries in Marro *et al.* (1996). The conclusion is that the data for $L_{\hat{x}} = L_{\hat{y}} = L$ supports (2.49) with $\beta \simeq 0.3$ and $\nu \simeq 0.7$ (but not with $\beta = \frac{1}{2}$ and $\nu = \frac{3}{2}$, for instance). One finds that the linear behavior near the origin extrapolates to $\mathscr{B} \simeq 1.24$, in agreement with the values used in figure 2.19. Furthermore, it is observed that $X\left(1, u\varepsilon^{\theta}\right) + Y\left(1, u\varepsilon^{\theta}\right)$ is constant, namely $-\mathscr{B}_s \simeq -1.04$ for $1 < \xi \sim \varepsilon^{-\nu} < L$ (i.e., departure from scaling occurs only for the *unphysical* region $\xi > L$). It is also worth mentioning that the corrections in (2.49) turn out to have different signs, namely, $X\left(1, u\varepsilon^{\theta}\right) > 0$, $Y\left(1, u\varepsilon^{\theta}\right) < 0$.

Correlation length

Properties (2.47)–(2.49) are consistent with the existence of two lengths but only as long as they depend on each other. Such a dependence seems to correspond to another interesting feature of the DLG which we describe next.

Anisotropic clusters have been observed both above T_C for any E (Vallés & Marro 1987, Achahbar & Marro 1995) and below T_C for small E (Yeung *et al.* 1992, Alexander *et al.* 1996). Therefore, it is natural to introduce two lengths, $\ell_{\hat{x}}$ and $\ell_{\hat{y}}$, to characterize these clusters. One may interpret $\ell_{\hat{x}}$ and $\ell_{\hat{y}}$ rather generally as mean displacements along $\hat{\mathbf{x}}$ and $\hat{\mathbf{y}}$, respectively, during a time interval Δt, for example, for a particle (hole) moving through the gas (liquid) phase for large E below T_C. Assuming the processes along the two principal directions are independent of each other, one may imagine a pure random walk along $\hat{\mathbf{y}}$, i.e., $\ell_{\hat{y}}^2 \sim D\Delta t$, where D is the diffusion coefficient, whilst one should expect that $\ell_{\hat{x}} \sim \upsilon\Delta t$, where υ is the (terminal) velocity for the driving process. That is, the prediction would be that $\ell_{\hat{x}} \sim \ell_{\hat{y}}^{\varsigma}$ with $\varsigma \approx 2$. This is precisely the observation described in §2.5, namely, that one obtains $a \sim b^{\varsigma}$ with $\varsigma \approx 2$ when $G(x, y) = \left(ax^2 - by^2\right)\left(x^2 + y^2\right)^{-2}$ is assumed for large r at high T.[19] One may interpret this as indicating that the shape of the (anisotropic) clusters

[19] The fact that $\ell_{\hat{x}} \sim \ell_{\hat{y}}^{\varsigma}$ has also been observed at lower T during the early time evolution of the DLG and some related continuum systems where ς is reported to be somewhat larger than 2; see Alexander *et al.* (1996) for a review.

is determined for each given value of E, and the relation between the two characteristic lengths is then essentially maintained as T is varied within some range (see §3.5). Consequently, a relation between these lengths exists, and one should probably expect a (unique) correlation length ξ, for example $\xi \sim \sqrt{\ell_{\hat{\mathbf{x}}}\ell_{\hat{\mathbf{y}}}} \sim \ell_{\hat{\mathbf{y}}}^{(\varsigma+1)/2}$ (as in self-affine interface phenomena; see Kértesz & Vicsek (1995)). This explains the observation above that only one length — which is influenced significantly by an interface along $\hat{\mathbf{x}}$ — is relevant; it is also consistent with the logarithmic behavior which has been reported for the size fluctuations of the interface of the DLG model (Schmittmann & Zia 1995).

Summing up, anisotropy is an intrinsic property of the DLG, closely related to the existence (for some values of the density, field, and temperature) of a linear interface along a principal direction. In particular, no support has appeared thus far from simulations for the assertion that the DLG has effective isotropy, i.e., that one may take care of anisotropic effects in a finite lattice by simply adjusting the sides; cf. Schmittmann & Zia (1995). Instead, simulations indicate that the interface is the cause of important surface effects apparently well described by equations (2.47) and (2.49). This is reinforced by the fact that the other second-order transitions in Λ_E for small E (see Marro *et al.* (1996) and §3.7), which lack such an interface, do not show dramatic finite-size effects and are characterized by the usual Ising critical behavior. It is worth mentioning that results on the lattice gas automaton defined in §1.1 confirm that the interface (and not the field) is relevant for the observed criticality (Marro & Achahbar 1998). The two-dimensional version of this system undergoes, among other things, a series of second-order transitions at phase points $b_c(p) > \frac{1}{2}$. The corresponding condensed phases exhibit a linear interface for any value of the field parameter p (see figure 1.2), and the transitions happen to be characterized by the same critical behavior, namely, $\beta \simeq 0.3$, independent of p, for example, for $p = 1$ (corresponding to a saturating field) and $p = \frac{1}{2}$ (corresponding to random field). This is an indication of the fact (of which we give further examples in subsequent chapters) that the Ising critical point is rather robust; for the DLG, only the presence of the peculiar linear interface seems strong enough to induce a measurable critical *anomaly*. We hope this issue will attract study in the near future.

3
Driven lattice gases: theory

Chapter 2 contains an essentially phenomenological description of the DLG. We now turn to theoretical descriptions, a series of mean-field approximations, which yield analytical solutions for arbitrary values of the driving field E and the jump ratio Γ.[1] We find that mean-field approximations are more reliable in the present context than in equilibrium. For example, a DLG model exhibits a *classical* critical point for $\Gamma, E \to \infty$ (§3.4), and simulations provide some indication of crossover towards mean-field behavior with increasing Γ (§2.3). On the other hand, the methods developed here can be used to treat the entire range of interest: attractive or repulsive interactions, and any choice of rate as well as any E and Γ value. As illustrated in subsequent chapters, these methods may be generalized to many other problems.

The present chapter is organized as follows. §3.1 contains a description of the method and of the approximations involved. The one-dimensional case is solved in §3.2 for arbitrary values of n, E, and Γ, and for various rates. A hydrodynamic-like equation and transport coefficients are derived in §3.3.[2] In §§3.4–3.6 we deal with two- (and, eventually, three-) dimensional systems. In particular, the limiting case for $\Gamma, E \to \infty$ studied in van Beijeren & Schulman (1984) and in Krug *et al.* (1986) is generalized in §3.4 by combining the one-dimensional solution of §3.2 with an Ω-*expansion,* to obtain explicit equations for finite fields and for the two

[1] We mainly follow Garrido *et al.* (1990a), and Alonso *et al.* (1995) for this chapter. We refer the reader to chapter 4 of Spohn (1991) for further theory for the DLG and related systems; see also Derrida, Domany & Mukamel (1992) for some aspects not covered here.

[2] We refer the reader to chapter 4 for a more detailed and systematic consideration of the problem of deriving macroscopic, hydrodynamic-like equations from microscopic kinetic models.

limits $\Gamma \to \infty$ and $\Gamma \to 0$. A two-dimensional model is solved in §3.5. This incorporates the assumption that an interface exists at low temperature as observed in MC simulations. §3.6 gives a method of finding the critical temperature and other properties by studying the stability of the high-temperature phase. The layered DLG introduced in §2.4 is further studied in §3.7. Some of the results in the present chapter are compared with the results obtained by other approaches, including MC results, and with some typical observations in real materials.

3.1 Macroscopic evolution

The aim here is to devise a method of studying kinetic lattice models, i.e., systems that are governed by the master equation (2.2) in which dynamics proceeds by exchange. The result is a description of the evolution of mean local quantities that neglects some fluctuations. This section describes the basics of the method and how the resulting kinetic equations may be derived from the original master equation under well-defined hypotheses of a mean-field character.

The microscopic dynamical function, say $A(\vec{\sigma})$, associated with an observable, typically depends on the occupation variables in a small, compact set of sites $D \subset \mathbf{Z}^d$. Thus, $A(\vec{\sigma}) = A(\vec{\sigma}_D)$, where $\vec{\sigma}_D$ represents the configuration within D, and the relevant macroscopic averages are

$$\langle A \rangle_t = \sum_{\vec{\sigma}} P(\vec{\sigma};t)\, A(\vec{\sigma}) = \sum_{\vec{\sigma}_D} Q(\vec{\sigma}_D;t)\, A(\vec{\sigma}_D)\,. \tag{3.1}$$

$Q(\vec{\sigma}_D;t) \equiv \sum_{\vec{\sigma}-\vec{\sigma}_D} P(\vec{\sigma};t)$ involves a sum over degrees of freedom outside D. The evolution of $\langle A \rangle_t$ follows from (2.2) as

$$\frac{\partial \langle A \rangle_t}{\partial t} = \sum_{|\mathbf{r}-\mathbf{s}|=1} \sum_{\vec{\sigma}} \delta A(\vec{\sigma};\mathbf{r},\mathbf{s})\, c(\vec{\sigma};\mathbf{r},\mathbf{s})\, P(\vec{\sigma};t)\,, \tag{3.2}$$

with $\delta A(\vec{\sigma};\mathbf{r},\mathbf{s}) \equiv A(\vec{\sigma}^{\mathbf{rs}}) - A(\vec{\sigma})$. This equation may be transformed into a closed set of equations for probabilities concerning D only. With this aim, let us define the *interior*, I, and the *surface*, S, of D; i.e., $D = I \cup S$ and $I \cap S = \emptyset$, where I is the set of lattice sites in D having all NNs as elements of D, while any $\mathbf{r} \in S$ has at least a NN outside D. We assume that, for any $\mathbf{r},\mathbf{s} \in I$, every site involved in $c(\vec{\sigma};\mathbf{r},\mathbf{s})$ belongs to D; we also require that

$$\sum_{|\mathbf{r}-\mathbf{s}|=1;\ \mathbf{r}\text{ and/or }\mathbf{s}\in S} \sum_{\vec{\sigma}} \delta A(\vec{\sigma}_D;\mathbf{r},\mathbf{s})\, c(\vec{\sigma};\mathbf{r},\mathbf{s})\, P(\vec{\sigma};t) = 0\,. \tag{3.3}$$

The former hypothesis is fulfilled by any of the familiar rates mentioned in §2.1, while, in general, the error incurred in the approximation (3.3) becomes negligible for D large enough.[3] Under these assumptions, (3.2) reduces to a closed equation for D, namely,

$$\frac{\partial \langle A(\vec{\sigma}_D)\rangle_t}{\partial t} = \sum_{|\mathbf{r}-\mathbf{s}|=1;\ \mathbf{r},\mathbf{s}\in I} \sum_{\vec{\sigma}_D} \delta A(\vec{\sigma}_D;\mathbf{r},\mathbf{s})\, c(\vec{\sigma}_D;\mathbf{r},\mathbf{s})\, Q(\vec{\sigma}_D;t)\,. \qquad (3.4)$$

Varying the size of D leads, in principle, to different levels of description — a key feature of the method, which is reminiscent of Kikuchi's (1951) *cluster variation* approach.

Condition (3.3) amounts to neglecting any coupling between the occupation variables within D and those in the rest of the system. Further insight into the nature of this approximation follows from the observation that

$$A(\vec{\sigma}_D) = a(\vec{\sigma}'_D) + \sigma_\mathbf{r} b(\vec{\sigma}'_D) + \sigma_\mathbf{s} c(\vec{\sigma}'_D) + \sigma_\mathbf{r}\sigma_\mathbf{s} d(\vec{\sigma}'_D)\,, \qquad (3.5)$$

for any $\mathbf{r},\mathbf{s} \in D$, for the functions of interest; here, $\vec{\sigma}'_D$ represents the configuration $\vec{\sigma}_D$ excluding the variables at sites $\mathbf{r}$ and $\mathbf{s}$. Thus (3.3) reduces to

$$\sum_{|\mathbf{r}-\mathbf{s}|=1;\ \mathbf{r}\text{ and/or }\mathbf{s}\in S} \left\langle (\sigma_\mathbf{s} - \sigma_\mathbf{r}) \left[b(\vec{\sigma}'_D) - c(\vec{\sigma}'_D)\right] c(\vec{\sigma};\mathbf{r},\mathbf{s}) \right\rangle_t = 0\,. \qquad (3.6)$$

For arbitrary $b(\vec{\sigma}'_D)$ and $c(\vec{\sigma}'_D)$ or, equivalently, for arbitrary $A(\vec{\sigma}_D)$, equation (3.6) asserts the cancellation of local currents between $\mathbf{s}$ and $\mathbf{r}$, namely, $(\sigma_\mathbf{s} - \sigma_\mathbf{r})\, c(\vec{\sigma};\mathbf{r},\mathbf{s})$. Consequently, (3.3) is equivalent to neglecting net currents through the surface S of the domain and, therefore, some local fluctuations. This (nonequilibrium) approximation generalizes the familiar Bethe–Peierls one in equilibrium if D has the appropriate *minimum size*, as defined below.

In practise (3.4) requires further approximations concerning $Q(\vec{\sigma}_D;t)$. An exact expression for the probability distribution on D is (Glauber 1963)

$$Q(\vec{\sigma}_D;t) = 1 + \langle s_\mathbf{r}\rangle_t \sum_{\mathbf{r}\in D} s_\mathbf{r} + \langle s_\mathbf{r} s_\mathbf{s}\rangle_t \sum_{\mathbf{r},\mathbf{s}\in D} s_\mathbf{r} s_\mathbf{s} + \cdots + \left\langle \prod_{\mathbf{r}\in D} s_\mathbf{r} \right\rangle_t \prod_{\mathbf{r}\in D} s_\mathbf{r}\,,$$

where $s_\mathbf{r} \equiv 2\sigma_\mathbf{r} - 1$ and averages are defined in (3.1), but one needs to write the coefficients in terms of a few low-order correlation functions. Of course, one should introduce an approximation for $Q(\vec{\sigma}_D;t)$ which is consistent with the size of D; otherwise, it is likely that the results will

---

[3] However, this error may not decrease monotonically to zero as D tends to the entire lattice, due to the sensitivity of interphases and boundaries of D at low temperature, the lack of consistency of the different conditions involved, etc., as noted in the appropriate places throughout the book.

not improve monotonically with increasing D. For instance, the average $\langle \prod_{\mathbf{r}\in D} s_{\mathbf{r}} \rangle_t$ may be approximated by a function of both $\langle s_{\mathbf{r}} \rangle_t$ and $\langle s_{\mathbf{r}} s_{\mathbf{s}} \rangle_t$, where $|\mathbf{r}-\mathbf{s}| = 1$, in order to remain within a first-order description. This means in practice that $Q(\vec{\sigma}_D;t) = Q(\vec{\sigma}_D; \langle s \rangle, \langle ss \rangle_{NN}; t)$ is a function of (only) the densities of NN pairs ••, •∘, and ∘∘ (where • represents an occupied site, and ∘ a hole). This situation corresponds to a *pair approximation* which suffices to provide an interesting description of the DLG.[4] Hereafter we denote by $p(\bullet) = n$ the particle density, and other relevant quantities as

$$p(\bullet,\bullet) = u,$$
$$p(\circ,\circ) = z \ = 1 - 2n + u;$$

then $p(\bullet,\circ) = p(\circ,\bullet) = n - u$, $p(\bullet,\bullet) + p(\bullet,\circ) = n$, and $p(\circ,\circ) + p(\circ,\bullet) = p(\circ) = 1 - n$.

In the present case, the *minimum domain D* is

$$+\ \mathbf{1}\ \ \mathbf{2}\ + \qquad\qquad \begin{matrix} & + & + & \\ + & \mathbf{1} & \mathbf{2} & + \\ & + & + & \end{matrix}$$

for $d = 1$ and 2, respectively; + denotes a surface site of S, and **1 2** stands for the only NN pair in I. In general, the minimum domain D for a d-dimensional system may be characterized by the occupation variables at the two sites in I, say σ_1 and σ_2, and by the number of particles in D_1 and D_2, say N_1 and N_2, where D_i consists of the $2d-1$ sites around the interior site i $(= 1,2)$; $D_1 \cap D_2 = \emptyset$, $D_1 \cup D_2 = S$. Thus, the probability of the occurrence of a given domain configuration is, within the pair approximation:

$$Q_{\sigma_1,\sigma_2,N_1,N_2}(u,n) = p(\sigma_1,\sigma_2) \prod_{j=1}^{2} \binom{2d-1}{N_j} p(\bullet \mid \sigma_j)^{N_j}\, p(\circ \mid \sigma_j)^{2d-1-N_j}. \tag{3.7}$$

Here, $p(\sigma,\sigma')$ is the joint probability for the NN pair of variables (σ,σ'), and $p(\sigma \mid \sigma') = p(\sigma,\sigma')\, p(\sigma')^{-1}$ is a conditional probability, with $p(\sigma)$ the probability of σ.

The master equation (2.2) involves exchanges only. Therefore, n is constant during the evolution, and the variable of interest is the density of NN pairs of particles, u, which is a measure of the energy, i.e., $p(\bullet,\circ) = n - u$. On the other hand, the exchanges involved in (2.2) are between the

[4] See Pathria (1972) and Ziman (1979), for instance, for a discussion of the first-order mean-field and pair approximations in equilibrium problems.

two sites (**1** and **2** as defined above) of I. Thus the evolution of u follows from equations (3.4) and (3.7) as

$$\frac{\mathrm{d}u}{\mathrm{d}t} = 2 \sum_{N_1=0}^{2d-1} \sum_{N_2=0}^{2d-1} (N_2 - N_1)\, \phi\left[\delta H\left(N_1, N_2; \bullet, \circ\right)\right] Q_{\bullet,\circ,N_1,N_2}(u) \,. \qquad (3.8)$$

We consider rates that depend on a small number of parameters. More precisely, ϕ depends on T, n, and E, and on the change of energy associated with the exchange, $\delta H\left(N_1, N_2; \sigma_1, \sigma_2\right) = -4J\left(N_1 - N_2\right)\left(\sigma_1 - \sigma_2\right)$. In addition to the evolution (3.8), we are interested in the stationary solution, i.e., $F(u_{st}) = 0$, where $F\left(u(t)\right)$ represents the r.h.s. of equation (3.8). The stability of these solutions requires that $\left(\partial F/\partial u\right)_{u=u_{st}} < 0$. The critical condition $\left(\partial F/\partial u\right)_{u=u_{st}} = 0$ may therefore correspond to a phase transition.

3.2 One-dimensional conduction

The method described in the preceding section is applied now to a one-dimensional lattice under the action of a uniform field, $\mathbf{E} = E\hat{\mathbf{x}}$ with $E > 0$; $\hat{\mathbf{x}}$ is a unit vector in the positive direction. This simple example has some relevance to surface growth, and to diffusion and ionic conduction in certain low-dimensional materials (§2.2).

For $d = 1$, the dynamics are assumed to proceed according to the rate

$$\phi(\delta H) = \frac{1}{2}\left(\phi_- + \phi_+\right) , \qquad (3.9)$$

where $\phi_\pm$ is for the hopping processes along the $\pm\hat{\mathbf{x}}$ directions, respectively, with, for example,

$$\phi_\pm = \min\left\{1, \exp\left(-\beta\delta H \pm E\right)\right\} , \qquad (3.10)$$

where we have absorbed a factor of β into E for simplicity. This means that one considers displacements of $\pm\hat{\mathbf{x}}$ with equal *a priori* probability, and favors jumps along the field direction more, the larger E is. For comparison purposes, we also refer to

$$\phi_\pm = \left\{1 + \exp\left(\beta\delta H \mp E\right)\right\}^{-1} . \qquad (3.11)$$

It follows from (3.8) that evolution is governed by

$$\frac{\mathrm{d}u}{\mathrm{d}t} = F(u; n, \beta, E, J) \qquad (3.12)$$

with

$$F(u; n, \beta, E, J) \equiv \frac{2\,(n-u)}{n\,(1-n)} \left[(n-u)^2\, \phi\,(-4J) - u\,(1 - 2n + u)\, \phi\,(4J)\right] ,$$

where $J > 0$ for attractive interactions, and $J < 0$ for repulsive ones. Here,

$$\left.\begin{aligned} \phi(4|J|) &= \begin{cases} \frac{1}{2}\left[1+\exp(-4\beta|J|-E)\right] & \text{for } E \geqslant 4\beta|J| \\ \exp(-4\beta|J|)\cosh E & \text{for } E \leqslant 4\beta|J| \end{cases} \\ \phi(-4|J|) &= \begin{cases} \frac{1}{2}\left[1+\exp(4\beta|J|-E)\right] & \text{for } E \geqslant 4\beta|J| \\ 1 & \text{for } E \leqslant 4\beta|J| \end{cases} \end{aligned}\right\} \tag{3.13}$$

for the choice (3.10), and

$$\phi(\pm 4|J|) = \frac{1}{2\left[1+\exp(\pm 4\beta|J|-E)\right]} + \frac{1}{2\left[1+\exp(\pm 4\beta|J|+E)\right]} \tag{3.14}$$

for the choice (3.11).

Equation (3.12) leads to a transcendental equation for u :

$$\frac{\left[2\delta(n-u)+\phi(4J)-\alpha\right]^{\alpha^+}}{\left[2\delta(n-u)+\phi(4J)+\alpha\right]^{\alpha^-}} = \left[2n(1-n)(n-u)^2\delta\right] e^{(-4t+C_0)} \tag{3.15}$$

with $\delta \equiv \phi(-4J) - \phi(4J)$, $\alpha \equiv \sqrt{\phi(4J)\left[4n(1-n)\delta + \phi(4J)\right]}$ and $\alpha^{\pm} \equiv \phi(4J)/\alpha \pm 1$; C_0 is a constant related to the initial condition. One also has from (3.15) the special solutions:

$$u = n - \frac{n-u_0}{\sqrt{\dfrac{4(n-u_0)^2}{n(1-n)}\phi(-4J)\,t+1}}, \tag{3.16}$$

where $u_0 \equiv u(t=0)$, for $\phi(4J) = 0$ and $\phi(-4J) > 0$, and

$$u = n - \frac{n(1-n)}{1-\dfrac{n^2-u_0}{n-u_0}\exp\left[-2\phi(4J)t\right]} \tag{3.17}$$

for $\phi(4J) = \phi(-4J) > 0$, for example. Therefore, the system may exhibit exponential decay as in (3.17), or slower relaxation as in (3.16); the latter corresponds to the existence of a critical point at $\beta = \infty$ (i.e., $T_C = 0$), as shown more explicitly below.

The steady solutions that follow from (3.15)–(3.17) as $t \to \infty$ are:

$u_{st} =$	conditions	from
$n + \frac{1}{2\delta}\left[\phi(4J) \mp \alpha\right]$	general solution	(3.15)
n	$J > 0$, $\beta \to \infty$, finite E	(3.16)
$0, 2n-1, n \geqslant \frac{1}{2}$	$J < 0$, $\beta \to \infty$, finite E	(3.15)
n^2	any J , $\beta \to 0$, finite E or else $E \to \infty$ for any β, J	(3.17)

(3.18)

A conclusion one may draw immediately is that both dynamics (3.10) and (3.11) tend to destroy correlations for both $\beta \to 0$ and $E \to \infty$. Furthermore, the state for $\beta \to \infty$ and $E > 0$ is practically indistinguishable from the equilibrium state $E = 0$ for both $J > 0$ and $J < 0$ as long as $n = \frac{1}{2}$ (while the behavior strongly depends on n otherwise). The stability of these solutions is related to the sign of $(\partial F/\partial u)_{st}$, namely, it is required that

$$3\delta\,(n-u)^2 + \phi(4J)\,[n\,(1+n) - 2u] > 0\,. \tag{3.19}$$

The field $E\hat{\mathbf{x}}$ induces a net current that may be written as

$$j\,(t;\beta,E,n) = \lim_{\delta t\to 0}\frac{1}{N\delta t}\,[N_{+\hat{\mathbf{x}}}(t,\delta t) - N_{-\hat{\mathbf{x}}}(t,\delta t)] \tag{3.20}$$

or, alternatively,

$$j\,(t;\beta,E,n) = \sum_{N_1=0,1;\;N_2=0,1}\{\phi_+\,[-4J\,(N_1-N_2)] - \phi_-\,[-4J\,(N_1-N_2)]\}\,Q_{\bullet,\circ,N_1,N_2}\,. \tag{3.21}$$

Here, N is the total number of particles, and $N_{\pm\hat{\mathbf{x}}}(t,\delta t)$ represents the number of particles hopping to NN empty sites in the $\pm\hat{\mathbf{x}}$ direction between times t and $t+\delta t$. Equation (3.21) yields qualitatively different behavior for different rates, values of the field, and interactions. That is, the current is variously given as follows:

$j\,(t;\beta,E,n)\,\dfrac{n\,(1-n)}{n-u} =$	conditions	rate
$\tilde{m} + u\,(1-2n+u)\left(1-e^{-4\beta J-E}\right) + (n-u)^2\left(1-e^{4\beta J-E}\right)$	$E \geq 4\beta\,\lvert J\rvert$	(3.10)
$\tilde{m} + 2u\,(1-2n+u)\,e^{-4\beta J}\sinh E$	$E \leqslant 4\beta J,\ J>0$	(3.10)
$\tilde{m} + 2\,(n-u)^2\,e^{-4\beta\lvert J\rvert}\sinh E$	$E \leqslant 4\beta\,\lvert J\rvert,\, J<0$	(3.10)
$2\sinh E\left\{\frac{1}{2}(n-u)\dfrac{(1-2n+2u)}{1+\cosh E} + \dfrac{u\,(1-2n+u)+(n-u)^2}{\left(e^{-4\beta\lvert J\rvert}+e^{-E}\right)\left(1+e^{4\beta\lvert J\rvert+E}\right)}\right\}$	any E, any J	(3.11)

(3.22)

where $\tilde{m} \equiv (n-u)\,(1-2n+2u)\left(1-e^{-E}\right)$, with $u = u(t)$. The infinite field limit for both rates is

$$j\,(t;\beta,E\to\infty,n) = n - u(t)\,, \tag{3.23}$$

i.e., a simple relation exists in this case between the current and the energy, as described already in §2.3. Several other limiting conditions may be worked out from these expressions.

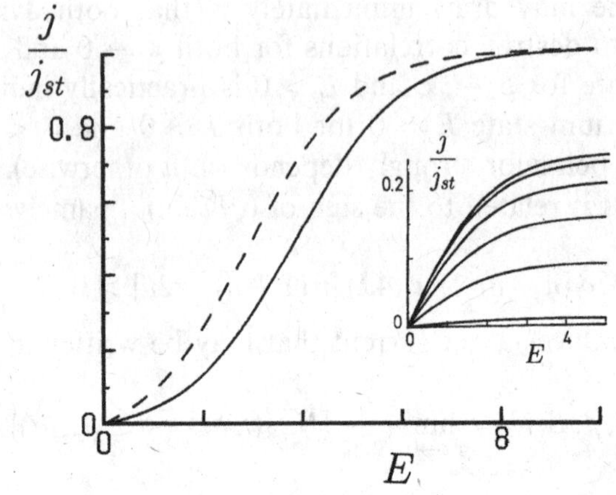

Fig. 3.1. The stationary current (normalized to the saturation value for large fields) versus the field strength for the one-dimensional DLG for $J > 0$ (attractive interactions), $T = 1$ (units of J/k_B), and the Kawasaki rate (3.11). The solid line is for $n = \frac{1}{2}$, and the dashed line is for $n = 0.1$. The inset is for $\beta \to 0$ and different densities, $n = 0.5, 0.4, 0.3, 0.2$, and 0.1, respectively, from top to bottom. (Adapted from Garrido *et al.* (1990a).)

The stationary current, j_{st}, follows by combining the equations in (3.22), or the corresponding one for a limiting condition with the equations in (3.18) for u_{st}. The resulting behavior may better be illustrated graphically. Figure 3.1 depicts $j_{st}(E)$ for $J > 0$; the repulsive case is qualitatively similar, except that j_{st} saturates then for smaller fields, i.e., around $E = 5$, and the increase for small fields is more abrupt, especially for small n where $\partial^2 j_{st}/\partial E^2$ is negative (unlike the case shown in figure 3.1). The use of different dynamics produces no qualitative changes, except that (3.10) induces a discontinuity of $\partial j_{st}/\partial E$ at $E = 4$ (which is an artifact related to the discontinuous nature of the Metropolis rate). Figure 3.2 represents $j_{st}(T)$ for the repulsive case; the attractive case has, independently of n, a behavior qualitatively similar to the one shown by the graph for $n = \frac{1}{2}$ in figure 3.2.

These graphs reveal that the one-dimensional system exhibits two qualitatively different kinds of behavior. The first, for $J < 0$ and $n \neq \frac{1}{2}$, is illustrated for $E = 10$ by the two upper curves in figure 3.2. Qualitatively similar conducting states are found at both low and high temperatures, with a sudden increase of the current near $T = 4/E$. This is related to the degeneracy of ground *antiferromagnetic-like* states for $n < \frac{1}{2}$ that favors large steady currents at low temperature. The second kind of behavior is illustrated by the bottom graph in figure 3.2 ($J < 0$ and $n = \frac{1}{2}$), and by any of the graphs in figure 3.1 ($J > 0$ at any density). This is characterized

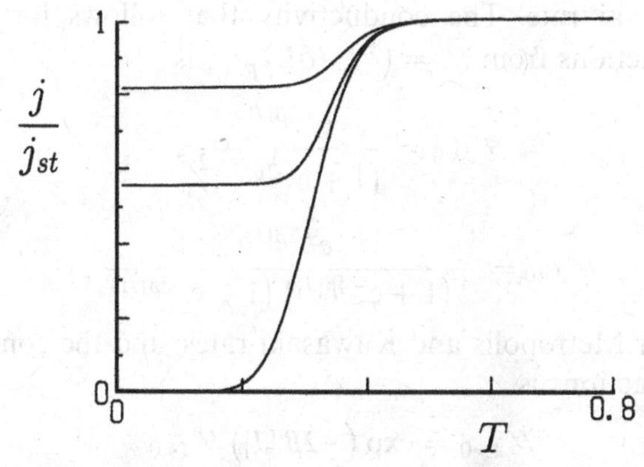

Fig. 3.2. The stationary current, as in figure 3.1, versus temperature, for $E = 10$, $J < 0$ (repulsive interactions), and different densities: $n = 0.3, 0.4$, and 0.5 from top to bottom. (Adapted from Garrido *et al.* (1990a).)

by a continuous changeover from an insulating state for $T \ll 4/E$ to a conducting state for $T > 4/E$. Similar changes are expected as one varies n (and, if possible, the sign of J) in *quasi* one-dimensional FIC materials. It follows that a comparison of the above with the experimental situation for a given substance might provide information about the sign of the corresponding *effective* ionic interactions (which is of interest in the light of our comment in page 17). Furthermore, a material exhibiting these properties is potentially of great practical value. It would also be interesting to consider the implications of this on surface growth, following the analogy depicted in figure 2.1.

Small fields

One may develop the above expressions for $t \to \infty$ around $E = 0$ to obtain the steady current for small fields. The result is

$$j_{st}(\beta, E, n) = \frac{n - u_{st}^0}{n(1-n)} \left[\left(n - u_{st}^0\right)\left(u_{st}^0 + z_{st}^0\right) + \psi_{st}^0 \right] E + \mathcal{O}\left(E^2\right) . \quad (3.24)$$

The superscript 0 indicates that the quantity corresponds to zero field, and

$$\psi_{st}^0 = -2u_{st}^0 z_{st}^0 \exp(-4\beta J) \quad \text{for } J > 0 ,$$

$$\psi_{st}^0 = 2\left(n - u_{st}^0\right)^2 \exp\left(-4\beta |J|\right) \quad \text{for } J < 0 ,$$

for the Metropolis rate, and

$$\psi_{st}^0 = 2\left[u_{st}^0 z_{st}^0 + \left(n - u_{st}^0\right)^2\right] \frac{e^{4\beta|J|}}{\left(1 + e^{4\beta|J|}\right)^2} - \frac{1}{2}\left(n - u_{st}^0\right)\left(u_{st}^0 + z_{st}^0\right)$$

for the Kawasaki rate. The conductivity that follows for $n = \frac{1}{2}$ and repulsive interactions from $\mathcal{S} = \left(\partial j_{st}/\partial E\right)_{E\to 0}$ is

$$\mathcal{S}_{J<0} = \frac{e^{-2\beta|J|}}{\left(1+e^{-2\beta|J|}\right)^2}, \tag{3.25}$$

$$\mathcal{S}_{J<0} = \frac{e^{-2\beta|J|}}{2\left(1+e^{-4\beta|J|}\right)\left(1+e^{-2\beta|J|}\right)}, \tag{3.26}$$

respectively, for Metropolis and Kawasaki rates, and the conductivity for attractive interactions is

$$\mathcal{S}_{J>0} = \exp\left(-2\beta\,|J|\right)\mathcal{S}_{J<0} \tag{3.27}$$

for both dynamics.

This indicates that, for a given temperature, repulsive interactions favor the conductivity more than attractive interactions do. Figure 3.3 illustrates the influence of dynamics on the steady state. That is, while the maxima of $\mathcal{S}$ always occur at high temperature (given that thermal motion favors breaking of clustering), and $\mathcal{S}(K) \to 0$ as $|K| \equiv |J|/k_B T \to \infty$ (as the system tends to be more ordered), a maximum of $\mathcal{S}(K)$ occurs for $K \leqslant 0$ *only* for Kawasaki dynamics. This may be related to the fact, familiar from MC simulations, that the Kawasaki rate has a lower efficiency (in the sense that it allows more transitions per unit time that increase the energy and, consequently, larger fluctuations) than the Metropolis rate. The extra fluctuations seem to favor the presence of a maximum conductivity state for the *antiferromagnetic* system before the infinite temperature limit is reached.

Figure 3.3 compares the analytical results for the Metropolis rate with some MC data. The agreement is excellent; for example, both reveal a discontinuity of $\partial\mathcal{S}(K)/\partial K$ at $K = 0$, which is another distinct feature of the Metropolis rate. There is also a comparison with a model by Dieterich *et al.* (1980) for weak fields, using a rate (different from those considered here) that permits an exact evaluation of $\mathcal{S}(K)$. As depicted by the inset in figure 3.3, this model behaves quite similarly to the mean-field DLG with Kawasaki rates; furthermore, both models satisfy $\mathcal{S}(K \to \infty)/\mathcal{S}(K \to 0) = (2n-1)\,n^{-2}$.

Fluctuation–dissipation relations

The (configurational) energy per lattice site is $\langle e\rangle = \langle H/N\rangle = -4Ju$, where N represents the volume. One may define a (nonequilibrium) *specific heat* as

$$C_E \equiv \frac{\partial\langle e\rangle}{\partial T} = -\frac{4J}{T^2}\frac{\partial u}{\partial \beta}. \tag{3.28}$$

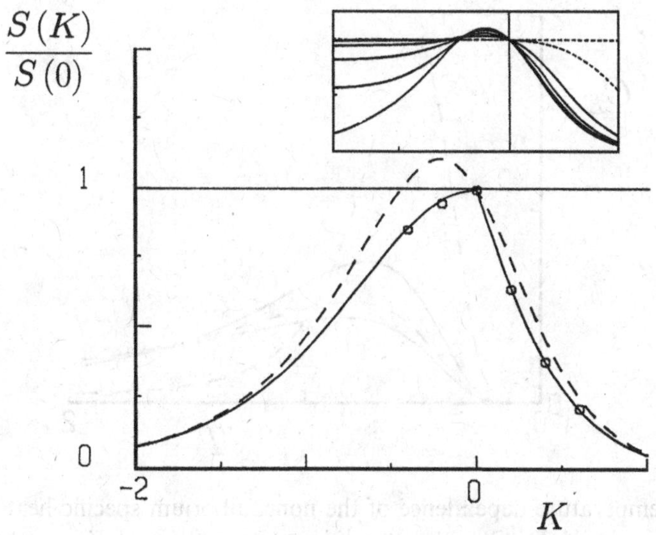

Fig. 3.3. The electrical conductivity at zero field, equations (3.25)–(3.27), normalized to the infinite temperature value, as a function of $K = J/k_B T$ for Metropolis (solid line) and Kawasaki (dashed line) rates for $n = \frac{1}{2}$. The circles correspond to MC data for the one-dimensional DLG with Metropolis rates. For comparison purposes, the inset represents exact results for different values of n in a related model by Dieterich *et al.* (1980).

The behavior of C_E is illustrated in figure 3.4 for attractive interactions and Kawasaki rates; a different choice for dynamics and/or interactions produces similar qualitative behavior. This reveals that: (i) the maximum that characterizes the system at equilibrium ($E = 0$) increases and shifts toward lower temperatures as E is increased; (ii) the location of the maximum is at $\beta_m \approx \frac{1}{4}E$ for large enough fields, i.e., around the temperature of the conductor–insulator transition described above; and (iii) C_E tends to a Dirac delta function centered at $T = 0$ as $E \to \infty$.

As mentioned in chapter 2, a specific question concerns the existence of a fluctuation–dissipation relation. That is, the relation that exists between C_E and the fluctuations of the energy, for example, $\delta^2 e \equiv \langle e^2 \rangle - \langle e \rangle^2$, where the average is given by equation (3.1) with (3.7). It follows after some algebra that

$$\delta^2 e = u \left(\frac{4J}{3n} \right)^2 \left(2u^2 + 4un - 9un^2 + 3n^2 \right). \qquad (3.29)$$

If one uses here and in equation (3.28) the general solution given in (3.18) for u_{st}, one obtains $\delta^2 e / C_E = f(n, T; \phi)$. That is, the relation between the two relevant quantities is a functional that, in general, depends

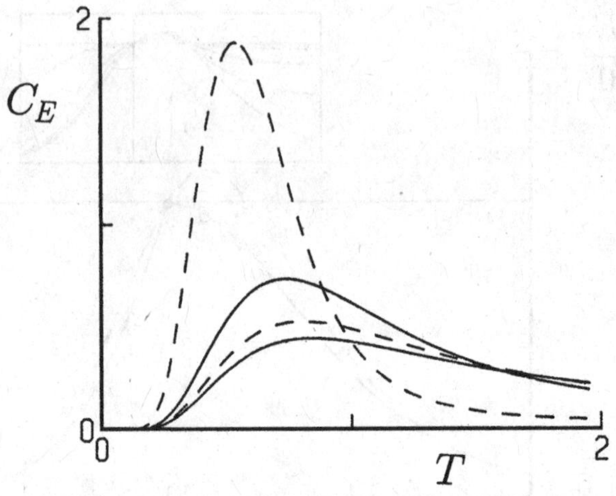

Fig. 3.4. Temperature dependence of the nonequilibrium specific heat (3.28) for the one-dimensional DLG with attractive interactions and Kawasaki rate for different fields: $E = 0, 1, 2,$ and 5 correspond, respectively, to the curves with increasing maxima. (From Marro *et al.* (1991).)

on the rate. As expected, assuming detailed balance, in the sense that $\phi(4J) = e^{-4J}\phi(-4J)$, the relation $\delta^2 e / C_E$ becomes independent of ϕ and $f\left(n = \frac{1}{2}, T; \phi\right) = k_B T^2$, which corresponds to the equilibrium case. A generalized fluctuation–dissipation relation has been derived for the DLG and other systems without local equilibrium (Price 1965, Eyink *et al.* 1996).

Correlations

The present (mean-field) method allows one to compute correlation functions such as $\langle \sigma_{r_1} \cdots \sigma_{r_k} \rangle$, $\langle \sigma_{r_1}(t) \sigma_{r_2}(t + t') \rangle$, etc., where $(r_1, r_2, \ldots, r_k)$ is any set of k sites whose elements are ordered along the line, i.e., $r_1 < r_2 < \cdots < r_k$. With this aim — see also Szabó, Szolnoki, & Bodócs (1991) — one may write (with the philosophy of the method) that

$$P(\vec{\sigma}) = p(\sigma_1) \prod_{r=1}^{N-1} \langle \sigma_r \mid \mathbf{p} \mid \sigma_{r+1} \rangle \ , \quad \sum_{\vec{\sigma}} P(\vec{\sigma}) = 1 \ , \tag{3.30}$$

where $\mathbf{p}$ is the *transfer matrix* (Kramers & Wannier 1941) of elements

$$\mathbf{p} = \begin{pmatrix} \dfrac{u}{n} & \dfrac{n-u}{n} \\ \dfrac{n-u}{1-n} & \dfrac{1-2n+u}{1-n} \end{pmatrix} \tag{3.31}$$

whose eigenvalues are 1 and $\Delta \equiv \left(u - n^2\right)/n(1-n) < 1$. It follows for spatial correlations that

$$\langle \sigma_{r_1} \cdots \sigma_{r_k} \rangle = n \prod_{j=1}^{k-1} \left[n + (1+n)\, \Delta^{r_{j+1}-r_j} \right] . \tag{3.32}$$

Therefore, $\langle \sigma_0 \rangle = n$, and $G(r) = \langle \sigma_0 \sigma_r \rangle - n^2 = n(1-n)\left[(u-n^2)/n(1-n)\right]^r$ with $r > 0$. This implies exponential behavior, namely,

$$G(r) \propto \exp\left(r/\xi\right) , \quad \xi^{-1} = -\ln \frac{u-n^2}{n\,(1-n)} , \tag{3.33}$$

and

$$S(k) \equiv \sum_{r=-N/2}^{r=N/2} \mathrm{e}^{ikr} G(r) = n\,(1-n) \frac{1-\Delta^2}{1+\Delta^2 - 2\Delta \cos k} \tag{3.34}$$

for the infinite, $N \to \infty$, system. The long-range order limit, i.e., $S(k=0)$ for $J > 0$ and $S(k=\pi)$ for $J < 0$, diverges as $\Delta \to \pm 1$, respectively. That is, there is a critical point at $T = 0$ for attractive interactions if $u = n$, and also for repulsive interactions if $n = \frac{1}{2}$ (the latter case exhibits the conductor–insulator transition); otherwise, the ground state degeneracy precludes long-range order at any temperature, excluding the trivial cases $n = 1, 0$.

3.3 Hydrodynamics in one dimension

The old problem of deriving macroscopic, hydrodynamic-like equations for microscopic systems of interacting particles has recently benefited from a series of successful efforts.[5] In order to illustrate both this approach and the versatility of the DLG in producing relevant macroscopic information, a derivation of a specific hydrodynamic equation is sketched here. A similar equation for a related Ginzburg–Landau model was obtained in Fritz & Maes (1988). Eyink *et al.* (1996, and references therein) have presented a fully microscopic derivation of more general hydrodynamic equations, including explicit expressions for all the quantities involved and an H-theorem describing the approach to homogeneous steady states.

Consider the model master equation:

$$\frac{\partial P_{\mathrm{E}}\left(\vec{\sigma};t\right)}{\partial t} = L_{E} P_{\mathrm{E}}\left(\vec{\sigma};t\right) + \epsilon L P_{\mathrm{E}}\left(\vec{\sigma};t\right) . \tag{3.35}$$

[5] See chapter 4 for some relevant bibliography, and for further details and concepts pertinent to this section; see also Derrida *et al.* (1992).

Here, L_E represents the r.h.s. of equation (2.2), and $L = L_{E=0}$; i.e., L corresponds to the action of a rate, say ϕ_0, that satisfies detailed balance. For $E = 0$, both (3.35) and the master equation (2.2) that characterizes the original DLG may be written as $\partial P(\vec{\sigma};t)/\partial t = LP(\vec{\sigma};t)$ whose steady solution is the equilibrium state (2.4). However, (3.35) involves for any $E \neq 0$ an additional term, $LP_E(\vec{\sigma};t)$, which is not contained in (2.2). This term is necessary for the argument below but it only represents a sort of *canonical perturbation* to the original DLG. To obtain a hydrodynamic description from (3.35), one also needs to proceed from the microscopic lattice to the macroscopic continuum and, simultaneously, to describe the system on a time scale where the effect of individual, microscopic events becomes negligible (§4.2). With this aim, one scales lengths, x, according to $r = \epsilon x$ and times according to $\tau = \epsilon^2 t$, and considers the *hydrodynamic limit*, i.e., $\epsilon \to 0$, $x \to \infty$, $t \to \infty$ (with both r and τ remaining finite), of (3.35). Under these conditions, the effect of L_E in (3.35) becomes *microscopically* negligible as compared to the canonical term, while it remains *macroscopically* measurable. The system is then expected to be described by a sort of local equilibrium with respect to the equilibrium distribution (2.4). The principal effect of this limit is to transform the quantity $\langle\sigma_x\rangle_t \equiv \sum_{\vec{\sigma}} \sigma_x P_E(\vec{\sigma};t)$ into a deterministic variable, say $m(r;\tau)$; moreover, for a one-dimensional system, (3.35) transforms into a hydrodynamic-like equation for $m(r;\tau)$, namely,

$$\frac{\partial m(r;\tau)}{\partial \tau} + \frac{\partial j(r;\tau)}{\partial r} = 0, \quad j(r;\tau) = j_o(r;\tau) + j_E(r;\tau). \tag{3.36}$$

Here,

$$j_0(r;\tau) = -\left(\frac{\partial m}{\partial r}\right)\left\{1 + \frac{1}{2}\left[1 - m\left(\frac{\partial \alpha}{\partial m}\right) + \alpha\right][\phi_0(\gamma) - \phi_0(0)]\right\}, \tag{3.37}$$

and

$$\begin{aligned} j_E(r;\tau) = {} & \tfrac{1}{2}m^2[\phi_E(\gamma - E) - \phi_E(\gamma + E)] - \tfrac{1}{4}\alpha^2[\phi_E(E) - \phi_E(-E)] \\ & + \tfrac{1}{4}\left(\alpha + \tfrac{1}{2}\alpha^2\right)[\phi_E(\gamma + E) + \phi_E(-\gamma - E) \\ & - \phi_E(\gamma - E) - \phi_E(-\gamma + E)] \end{aligned} \tag{3.38}$$

with $\gamma \equiv 4\beta J$, and $\alpha = \alpha(m) \equiv [e^{-\frac{1}{2}\gamma}(2m^2 - 1) + \theta]/(e^{-\frac{1}{2}\gamma} + \theta)$, where $\theta \equiv \sqrt{1 - m^2(1 - e^{-\gamma})}$, and $m = m(r;\tau)$.

These equations lead to explicit expressions for the coefficients of bulk diffusion,

$$\mathscr{D}(m) = 1 + (1 - m^2)\frac{\phi_0(\gamma) - 1}{1 - m^2(1 - e^{-\gamma}) + e^{-\frac{1}{2}\gamma}}, \tag{3.39}$$

and conductivity,

$$\mathscr{S}(m) = -\left(1-m^2\right)\Phi_+ + \frac{1}{2}(1-\alpha)(\Phi_+ - \Phi_-) + \frac{1}{4}\left(1-\alpha^2\right)(\Phi_+ + \Phi_- + 1)\,, \tag{3.40}$$

where $\Phi_\pm \equiv \left[\partial\phi_E(u)/\partial u\right]_{u=\pm\gamma}$. Then $m = 0$ reproduces the Einstein relation

$$\mathscr{S}(0) = \mathscr{D}(0)/\left(1+\mathrm{e}^{-2\beta J}\right) \tag{3.41}$$

for any dynamics such that $\phi_0 \equiv \phi_E$. This leads, moreover, to the specific prediction

$$\mathscr{D}(0)_{J>0} = \mathrm{e}^{2\beta|J|}\,\mathscr{D}(0)_{J<0} \tag{3.42}$$

(for $d = 1$); see also the mean-field result (3.27).

On the other hand, we remark that (3.36)–(3.38) transforms, in the limit $E \to \infty$, $\beta = 0$, into Burger's equation (Burger 1974) for the simple asymmetric exclusion process:

$$\frac{\partial m}{\partial \tau} + \frac{\partial}{\partial r}\left(\mathscr{K}m^2 - \frac{\partial m}{\partial r}\right) = 0\,, \quad \mathscr{K} = \frac{1}{2}\left[\phi_E(-\infty) - \phi_E(\infty)\right]\,. \tag{3.43}$$

For large enough fields, the approximate result $j_E(r;\tau) \approx \mathscr{K}m(r;\tau)$ follows; see Martin, Siggia, & Rose (1978), and de Dominicis & Peliti (1978), for example.

3.4 Conduction in two dimensions

The essential anisotropy of the dynamic rule is an important feature of the DLG. One may highlight this fact by writing

$$c_E(\vec{\sigma};\mathbf{r},\mathbf{s}) = \begin{cases} c_{\hat{\mathbf{x}}}(\vec{\sigma};\mathbf{r},\mathbf{s}) & \text{for } \mathbf{r}-\mathbf{s} = \pm\hat{\mathbf{x}} \\ c_{\hat{\mathbf{y}}}(\vec{\sigma};\mathbf{r},\mathbf{s}) & \text{for } \mathbf{r}-\mathbf{s} = \pm\hat{\mathbf{y}} \\ c_{\hat{\mathbf{z}}}(\vec{\sigma};\mathbf{r},\mathbf{s}) & \text{for } \mathbf{r}-\mathbf{s} = \pm\hat{\mathbf{z}} \\ 0 & \text{otherwise,} \end{cases} \tag{3.44}$$

where the rates $c_{\hat{\mathbf{x}}}$, $c_{\hat{\mathbf{y}}}$, and $c_{\hat{\mathbf{z}}}$ along the principal directions are defined in accordance with (2.3). This section is devoted to the case $d = 2$, i.e., $c_{\hat{\mathbf{z}}} \equiv 0$ in (3.44). It turns out to be interesting at this point to allow for different functions ϕ, say $\phi_\parallel$ and $\phi_\perp$, and different *a priori* frequencies, $\Gamma_\parallel$ and $\Gamma_\perp$, for *longitudinal* ($\hat{\mathbf{x}}$) and *transverse* jumps, respectively; we define $\Gamma \equiv \Gamma_\parallel/\Gamma_\perp$. The master equation splits into two classes of terms associated respectively with the two kinds of jumps.

The first solution to this problem was given by van Beijeren & Schulman (1984). They found (for $d = 2$) an exact expression for the stationary probability $P_E(\vec{\sigma})$ in the limit $\Gamma \to \infty$ and $E \to \infty$, when the rate function in (2.3) is $\phi(X) = \exp\left(-\frac{1}{2}X\right)$. This limit was further considered by Krug *et al.* (1986) by using different techniques to confirm and extend the previous results. We describe below a different solution that includes, still in the limit $\Gamma \to \infty$, arbitrary rates and finite values for E, and reproduces the known results as $E \to \infty$ (the case $\Gamma \to 0$ is also treated). This is based on the analytical description presented earlier in this chapter, and, to some extent, uses techniques similar to those employed by Krug *et al.* We describe below a different technique that can be applied to arbitrary Γ and E. This is based on the kinetic cluster mean-field method of §3.1 and on the observation (from simulations) that the low-T ordered phase is anisotropic with striped configurations.

Fast driving fields

In the limit $\Gamma \to \infty$, jumps along the field direction are so much more frequent than those in the perpendicular direction that one may consider the system as a collection of longitudinal columns that immediately *stabilize*, i.e., reach the new steady state with the current number of particles, after each jump between columns. One may analyze this situation on two different times scales. On a microscopic time scale, $\partial P_E(\vec{\sigma};t)/\partial t = 0$ holds, and $c_{\hat{\mathbf{y}}}(\vec{\sigma};\mathbf{r},\mathbf{s}) = 0$, i.e., only jumps along the column occur. Then

$$P_E(\vec{\sigma}) = \prod_i \pi_E(\vec{\sigma}_i;\eta_i;\vec{\eta}) \ , \tag{3.45}$$

where $\pi_E(\vec{\sigma}_i;\eta_i;\vec{\eta})$ represents the probability that the ith column with η_i particles has the configuration $\vec{\sigma}_i$; π_E may depend on the occupation of the surrounding columns, denoted $\vec{\eta}$. On this time scale one has

$$\sum_{\mathbf{r}-\mathbf{s}=\pm\hat{\mathbf{x}}} \left[c_{\hat{\mathbf{x}}}(\vec{\sigma}^{\mathbf{rs}};\mathbf{r},\mathbf{s})\,\pi_E(\vec{\sigma}_i^{\mathbf{rs}};\eta_i;\vec{\eta}) - c_{\hat{\mathbf{x}}}(\vec{\sigma};\mathbf{r},\mathbf{s})\,\pi_E(\vec{\sigma}_i;\eta_i;\vec{\eta})\right] = 0 \ , \tag{3.46}$$

where $\mathbf{r}$ is a site in the ith column. On the other hand, one observes on a coarser time scale that

$$P_E(\vec{\sigma};t) = p_E(\vec{\eta};t)\prod_i \pi_E(\vec{\sigma}_i;\eta_i;\vec{\eta}) \ , \tag{3.47}$$

where $p_E(\vec{\eta};t)$ is the probability that the columns have η_i, $i = 1, 2, \ldots, M$, particles at time t. This satisfies the master equation:

$$\frac{\partial p_E(\vec{\eta};t)}{\partial t} = \sum_{\vec{\eta}'}\left[c(\vec{\eta}',\vec{\eta})\,p_E(\vec{\eta}';t) - c(\vec{\eta},\vec{\eta}')\,p_E(\vec{\eta};t)\right] \tag{3.48}$$

with

$$c(\vec{\eta}',\vec{\eta}) \equiv N\left\langle \phi_\perp\left[\beta H(\vec{\sigma}^{\mathbf{rs}}) - \beta H(\vec{\sigma})\right]\right\rangle_{\vec{\eta}}, \quad \mathbf{r}-\mathbf{s} = \pm\hat{\mathbf{y}} \ , \tag{3.49}$$

where N is the number of lattice sites within each column, and the average is with respect to the probability distribution (3.45).

The above implies that the original two-dimensional problem reduces, in the limit $\Gamma \to \infty$, to two one-dimensional problems, governed, respectively, by equations (3.46) and (3.48). This was first realized by van Beijeren & Schulman who restricted themselves to the situation in which jumps against **E** are forbidden (while one performs all jumps along **E** permitted by the hard core exclusion). In this *strong field limit* all longitudinal correlations vanish, so that one has a random distribution of particles; this eliminates the problem of computing $\langle \cdots \rangle_{\vec{\eta}}$.

Alternatively, one may compute $c(\vec{\eta}', \vec{\eta})$ as given by equation (3.49) for arbitrary values of E. This may be done via the following steps: (i) Solve equation (3.46) by the kinetic mean-field method. This provides values for the density $p(\bullet, \bullet) = u$ of pairs along the columns; then define $\mathbf{u}$ such that $(\mathbf{u})_i \equiv u_i = u_i(\mathbf{n})$, with $i = 1, 2, \ldots, M$, where $(\mathbf{n})_i \equiv n_i$ and $n_i = \eta_i / N$ is the density of particles in column i. (ii) Find an explicit expression for (3.49) by using both the solution $u_i = u_i(\mathbf{n})$ from the previous step and the method set forth in the preceding sections. (iii) Solve equation (3.48) for large N by means of the Ω-expansion technique (see van Kampen 1981, Krug *et al.* 1986). We now describe each step in greater detail.

(i) Consider three neighboring columns, $i = 1, 2, 3$; they are characterized by densities n_i and by the unknowns u_i. The probability of a local configuration:

$$
\begin{array}{lcccc}
\text{column:} & i = & 1 & 2 & 3 \\
 & & & & \\
 & & & \sigma_4 & \\
\text{occupation variables:} & & \sigma_2 & 0 & \sigma_6 \\
 & & \sigma_1 & 1 & \sigma_5 \\
 & & & \sigma_3 &
\end{array}
$$

is given by

$$
Q_{1,0}(\sigma_1, \ldots, \sigma_6) = \frac{u_2^{\sigma_3} z_2^{1-\sigma_4}}{n_2(1-n_2)} (n_2 - u_2)^{2-\sigma_3+\sigma_4} q_1(\sigma_1, \sigma_2) q_3(\sigma_5, \sigma_6), \quad (3.50)
$$

where

$$
q_i(\sigma, \sigma') = \begin{cases} u_i, & \text{for } \sigma = \sigma' = 1 \\ n_i - u_i, & \text{for } \sigma = 1, \sigma' = 0 \text{ or } \sigma = 0, \sigma' = 1 \\ z_i, & \text{for } \sigma = \sigma' = 0 \end{cases} \quad (3.51)
$$

with $z_i = 1 - 2n_i + u_i$. Thus equation (3.8) reads

$$
\frac{\partial u_2}{\partial t} = \sum_{\sigma_1, \ldots, \sigma_6} (\sigma_4 - \sigma_3) \phi_{\parallel}(\delta H) Q_{1,0}(\sigma_1, \ldots, \sigma_6), \quad (3.52)
$$

where $\delta H = -4J(\sigma_2 + \sigma_4 + \sigma_6 - \sigma_1 - \sigma_3 - \sigma_5)$ and $\phi_\parallel$ is the rate acting along columns. The corresponding stationary solution is

$$\begin{aligned}[\Psi_2(0) + \Psi_2(8J)]\,[(n_1 - u_1)(u_3 + z_3) + (n_3 - u_3)(u_1 + z_1)] \\ + \Psi_2(4J)\,[2(n_1 - u_1)(n_3 - u_3) + (u_1 + z_1)(u_3 + z_3)] \\ + [\Psi_2(-4J) + \Psi_2(12J)]\,(n_1 - u_1)(n_3 - u_3) = 0\ , \quad (3.53)\end{aligned}$$

where $\Psi_2(X) \equiv (n_2 - u_2)^2\,\phi_\parallel(-X) - n_2 z_2 \phi_\parallel(X)$. There are similar equations for u_1 and u_3, i.e., for $\Psi_1(X)$ and $\Psi_3(X)$ that are defined similarly to $\Psi_2(X)$, so that one needs to solve M nonlinear coupled equations to obtain u_i with $i = 1, 2, \ldots, M$. One may check that, for $\phi_\parallel$ =const., this system of equations has the solution of van Beijeren & Schulman, namely,

$$u_i = n_i^2\ , \quad i = 1, \ldots, M\ ; \qquad (3.54)$$

this corresponds to the case discussed before in which there are no correlations along the columns, i.e., to the limit $E \to \infty$.

(ii) Next, consider jumps from column i to column $i+1$, i.e., the starting configuration is now

column:	$j =$	$i-1$	i	$i+1$	$i+2$
			σ_1	σ_2	
occupation variables:		σ_3	1	0	σ_4
			σ_5	σ_6	

According to (3.49) and (3.45), the probability of these jumps is

$$\begin{aligned}R_i^+(n_{i-1}, n_i, n_{i+1}, n_{i+2}) &= \sum_{\vec{\sigma}} \prod_j \pi_E(\vec{\sigma}_i; \eta_i; \vec{\eta})\,\phi_\perp(\beta\delta H) \\ &= \sum_{\substack{\vec{\sigma}_j \\ \{j=i-1,i,i+1,i+2\}}} \pi_E(\vec{\sigma}_{i-1}; \eta_{i-1}; \vec{\eta})\,p(\vec{\sigma}_i)\,p(\vec{\sigma}_{i+1})\,p(\vec{\sigma}_{i+2})\,\phi_\perp(\beta\delta H)\ ,\end{aligned} \qquad (3.55)$$

where the second equality follows because δH depends only on $\vec{\sigma}_j$, with $j = i-1, i, i+1, i+2$, and $\pi_E(\vec{\sigma}_i; \eta_i; \vec{\eta})$ is normalized to unity. Moreover, only the occupation variables around the pair affected by the exchange enter δH in practise. One may then define local probabilities as

$$\pi_i(\{\sigma_{ik}; k = \ell, \ell+1, \ldots\}) = \sum_{\vec{\sigma}_i - \{\sigma_{ik}\}} \pi_E(\vec{\sigma}_i; \eta_i; \vec{\eta})\ , \qquad (3.56)$$

where the sum is over all column configurations $\vec{\sigma}_i$ excluding the variables σ_{ik}, $k = \ell, \ell+1, \ldots$ Consequently, one has for the situation of interest simply that

$$R_i^+ = \sum_{\sigma_1, \ldots, \sigma_6} \pi_{i-1}(\sigma_3)\,\pi_i(\sigma_1, 1, \sigma_5)\,\pi_{i+1}(\sigma_2, 0, \sigma_6)\,\pi_{i+2}(\sigma_4)\,\phi_\perp(\beta\delta H)\ . \qquad (3.57)$$

The πs here may be computed as above, i.e.,

$$\left.\begin{aligned}
\pi_i(\sigma_\ell) &= \begin{cases} n_i & \text{if} \quad \sigma_\ell = 1 \\ 1 - n_i & \text{if} \quad \sigma_\ell = 0\,, \quad \text{for } \ell = 3 \text{ or } 4\,, \end{cases} \\
\pi_i(\sigma_1, 1, \sigma_5) &= \frac{u_i^{\sigma_1+\sigma_5}\,(n_i - u_i)^{2-\sigma_i-\sigma_5}}{n_1^3}\,, \\
\pi_{i+1}(\sigma_2, 0, \sigma_6) &= \frac{z_{i+1}^{2-\sigma_2-\sigma_6}\,(n_{i+1} - u_{i+1})^{\sigma_2+\sigma_6}}{(1-n_i)^3}\,.
\end{aligned}\right\} \tag{3.58}$$

Equations (3.57)–(3.58), together with (3.53) for $u_i = u_i(\mathbf{n})$ and $u_{i+1} = u_{i+1}(\mathbf{n})$, are the solution here. When one assumes that $E \to \infty$, so that (3.54) holds, and uses the appropriate function for ϕ in (3.57), then

$$\begin{aligned}
R_i^+ &= \dot{n}_i v_{i+1}\left(\frac{n_{i-1}}{\alpha} + \alpha v_{i-1}\right)\left(\alpha n_{i+2} + \frac{v_{i+2}}{\alpha}\right) \\
&\quad + \left(\frac{n_i^2}{\alpha^2} + 2n_i v_i + \alpha^2 v_i^2\right)\left(\alpha^2 n_{i+1}^2 + 2n_{i+1}v_{i+1} + \frac{v_{i+1}^2}{\alpha^2}\right)\,, \quad \alpha \equiv e^{\beta J}\,,
\end{aligned} \tag{3.59}$$

which is the result of van Beijeren & Schulman.

(iii) The densities n_i undergo N changes of size N^{-1} during a unit time interval. One may then use the column height $N \equiv \Omega$ as a system size parameter to perform an Ω-expansion of the master equation (3.48) for $E \to \infty$.

The mean local densities are governed for large N by conservation-like deterministic equations, i.e.,

$$\frac{\partial n_i}{\partial t} = -\left[J_i(\mathbf{n}) - J_{i-1}(\mathbf{n})\right]\,, \tag{3.60}$$

where, $J_i = R_i^+ - R_{i+1}^-$ is the current between columns i and $i+1$; $R_{i+1}^- = R_{i+1}^-(n_{i+2}, n_{i+1}, n_i, n_{i-1})$ represents the number of jumps per unit time from $i+1$ to i (cf. equation (3.55)). The fluctuations around the means are governed by Langevin-type equations, namely,

$$\frac{d\psi_j}{dt} = \sum_{i=1}^{M} L_{ji}(\mathbf{n})\,\psi_j + \xi_j(t)\,. \tag{3.61}$$

Here, ξ_j is a Gaussian force of zero mean and covariance

$$\begin{aligned}
g_{ij} = \sum_\ell \big[&(\delta_{i,\ell+1} - \delta_{i,\ell})(\delta_{j,\ell+1} - \delta_{j,\ell})\,R_\ell^+ \\
&+ (\delta_{i,\ell} - \delta_{i,\ell+1})(\delta_{j,\ell} - \delta_{j,\ell+1})\,R_\ell^-\big]\,,
\end{aligned}$$

and one obtains that $L_{ij} = -(\delta/\delta n_j)(J_i - J_{i-1})$ by linearizing (3.60) around the solution $\mathbf{n}(t)$.

Spatial correlations behave differently here than for $E \to \infty$, as expected. One may write

$$S(\mathbf{k}) = \sum_{\mathbf{r}} e^{i\mathbf{k}\cdot\mathbf{r}} \left[\langle \sigma_{0,0}\sigma_{j,\ell}\rangle - \langle \sigma_{0,0}\rangle \langle \sigma_{j,\ell}\rangle\right], \tag{3.62}$$

where $\mathbf{k}$ becomes continuous for $N, M \to \infty$; $\mathbf{k} \equiv (k_{\hat{\mathbf{x}}}, k_{\hat{\mathbf{y}}})$, $\mathbf{r} \equiv (j, \ell)$, and

$$\left\langle \sigma_{j,\ell}\, \sigma_{j',\ell'} \right\rangle = \sum_{\vec{\eta}} \pi_E(\vec{\eta}; t) \sum_{\vec{\sigma}_\ell, \vec{\sigma}_{\ell'}} \sigma_{j,\ell}\, \sigma_{j',\ell'}\, p_E(\vec{\sigma}_\ell, \eta_\ell; \vec{\eta})\, p_E(\vec{\sigma}_{\ell'}, \eta_{\ell'}; \vec{\eta}) \,. \tag{3.63}$$

It follows immediately that

$$\langle \sigma_{j,\ell}\rangle = \sum_{\vec{\eta}} \pi_E(\vec{\eta}; t)\, n_\ell \equiv \langle n_\ell\rangle_p \,, \tag{3.64}$$

and

$$\left\langle \sigma_{j,\ell}\, \sigma_{j',\ell'} \right\rangle = \begin{cases} \left\langle \left\langle \sigma_{j,\ell}\, \sigma_{j',\ell'} \right\rangle_n \right\rangle_p & \text{within each column} \\ \langle n_\ell n_{\ell'}\rangle_p & \text{between columns,} \end{cases} \tag{3.65}$$

where $\langle \cdots \rangle_n \equiv \sum_{\vec{\sigma}_j} P_E(\vec{\sigma}_j, \eta_j; \vec{\eta}) \cdots$, and $\langle \cdots \rangle_p$ is defined by (3.64). After some algebra one finds

$$S(\mathbf{k}) = S_\perp(\mathbf{k}) + \left\langle S'(k_{\hat{\mathbf{x}}})\right\rangle_p \,, \tag{3.66}$$

where

$$\left.\begin{aligned} S_\perp(\mathbf{k}) &\equiv \textstyle\sum_{\mathbf{r}} e^{i\mathbf{k}\cdot\mathbf{r}} \left[\langle n_0 n_j\rangle_p - \langle n_0\rangle_p \langle n_j\rangle_p\right], \\ S'(k_{\hat{\mathbf{x}}}) &= \textstyle\sum_{r} e^{ik_{\hat{\mathbf{x}}}r} \left[\langle \sigma_{0,0}\sigma_{r,0}\rangle_p - n_0^2\right]. \end{aligned}\right\} \tag{3.67}$$

The behavior of $S_\perp(\mathbf{k})$ is given by

$$S_\perp(k_{\hat{\mathbf{y}}}) = -\frac{G(k_{\hat{\mathbf{y}}})}{2\mathscr{L}(k_{\hat{\mathbf{y}}})} = \frac{D(n)}{R_{23}(n) + R_{14}(n)\left(1 + 2\cos k_{\hat{\mathbf{y}}}\right)}, \tag{3.68}$$

where $G(k_{\hat{\mathbf{y}}})$ and $\mathscr{L}(k_{\hat{\mathbf{y}}})$ are one-dimensional Fourier transforms of g_{ij} and L_{ij}, respectively, and

$$R_{j\ell}(n) = \left[\delta R_i^+/\delta x_j - \delta R_i^+/\delta x_\ell\right]_{x_i = n}$$

and $D(n) = R_2^+(x_1, x_2, x_3, x_4)$ for $x_i = n$, $i = 1, 2, 3, 4$. Equation (3.68) is formally identical to the expression by Krug *et al.* for the whole structure function; the second term in (3.66) may produce non analyticities associated with LRSCs (§2.5).

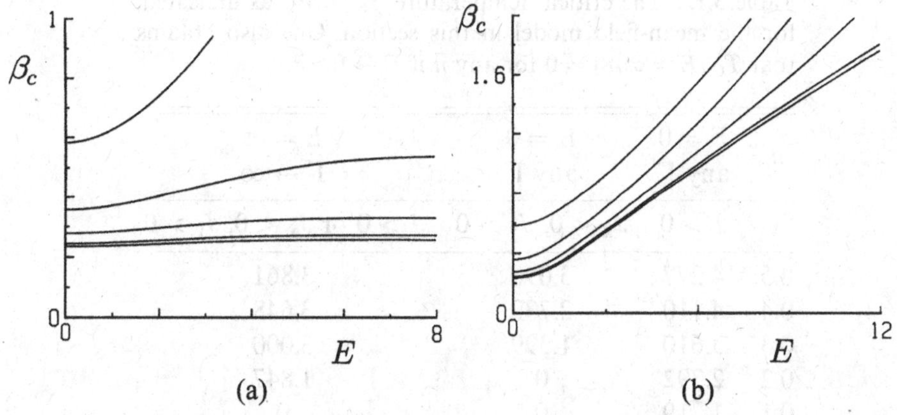

Fig. 3.5. The effect of Γ and n on the transition temperature, $\beta_C = [k_B T_C(n, E, \Gamma)]^{-1}$, of the two-dimensional system for attractive interactions, and different densities, $n = 0.1, 0.2, 0.3, 0.4,$ and 0.5 from top to bottom. (a) The fast-rate limit, $\Gamma \to \infty$. (b) The slow-rate limit, $\Gamma \to 0$. (Adapted from Garrido *et al.* (1990a).)

Critical temperature

Many of the properties of the two-dimensional system follow from equations (3.53) and (3.57)–(3.58) for which finding a solution is a matter of algebra. For example,[6] the critical properties for attractive interactions and $\Gamma \to \infty$ may be obtained from the condition $R_{23} + 3R_{14} = 0$, which makes $S(\mathbf{k})$ diverge. The critical temperature $T_C(E, n)$ shown in figures 3.5(a) and 3.6(a) follows from this condition. This reveals the existence of a critical density, namely, $n_C = 0.152$, that separates two kinds of behavior. For $\frac{1}{2} \geq n > n_C$, $T_C(E)$ for given n has a shallow maximum for small E — which is not quite evident on the scale of figure 3.5; it is more pronounced the smaller n (larger than n_C) is — and decreases asymptotically towards a finite value, T_C^{∞}, as E is increased. For a given field, $T_C(n)$ increases with n. As E is increased, by contrast, $T_C(E)$ decreases monotonically and goes to zero as $E \to \infty$, for $n < n_C$. That is, phase segregation is suppressed in the infinite field limit, and the field tends to destroy (transverse) correlations in such a way that $T_C(E) < T_C^0$ in the limit $\Gamma \to \infty$; see §3.5 for the case $\Gamma \approx 1$. In table 3.1 we list some illustrative values for $T_C(E, n)$.

The critical behavior for repulsive ($J < 0$) interactions follows from the condition that $R_{23} = R_{14}$. This reveals that no phase transition exists for any values of E and n.

[6] Some more details are given in §3.5 where the case of arbitrary Γ is considered.

Table 3.1. The critical temperature $T_C(E,n)$, as indicated, for the mean-field model in this section. One also obtains that $T_C(E=\infty,n)=0$ for any n if $\Gamma \to 0$.

n	$E=0$ any Γ	$E=0$ any Γ	$E=\infty$ $\Gamma \to \infty$
	$J>0$	$J_{\hat{x}}<0,\ J_{\hat{y}}>0$	$J>0$ or $J_{\hat{x}}<0,\ J_{\hat{y}}>0$
0.5	4.277	3.090	3.861
0.4	4.110	2.732	3.648
0.3	3.610	1.399	3.000
0.2	2.792	0	1.847
0.1	1.719	0	0

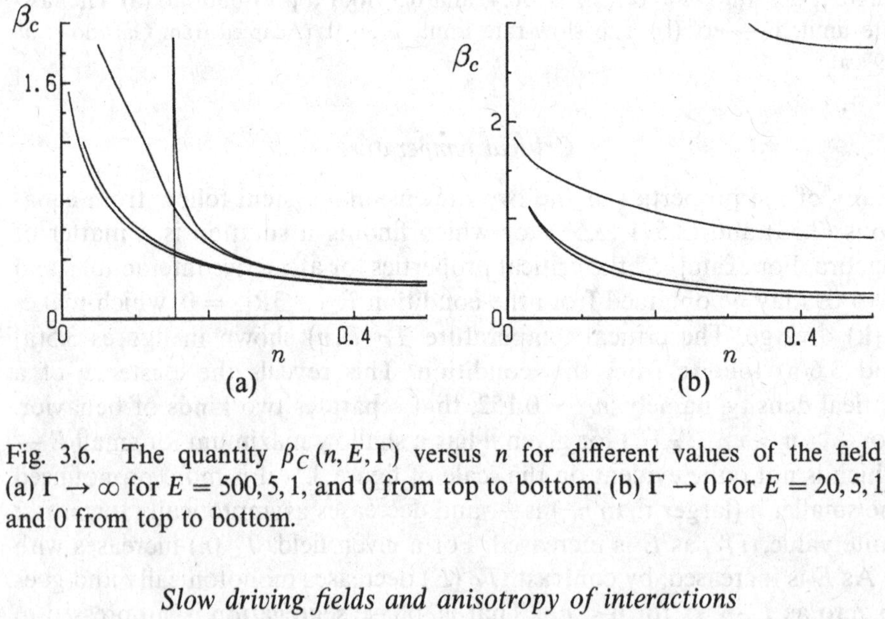

Fig. 3.6. The quantity $\beta_C(n,E,\Gamma)$ versus n for different values of the field: (a) $\Gamma \to \infty$ for $E=500, 5, 1$, and 0 from top to bottom. (b) $\Gamma \to 0$ for $E=20, 5, 1$, and 0 from top to bottom.

Slow driving fields and anisotropy of interactions

The method presented above allows one to consider arbitrary rates for the two principal directions of the lattice. As an illustration, we consider briefly: (a) the case $\Gamma \to 0$ in which jumps transverse to the field are more frequent than longitudinal ones, and (b) the anisotropic case in which the interactions are attractive in one direction and repulsive in the other.

The slow-rate limit $\Gamma \to 0$ simply follows when one replaces $\phi_\parallel$ in (3.52) and $\phi_\perp$ in (3.55) by $\phi_\perp = \phi(\delta H)$ and $\phi_\parallel = \phi(\delta H, E)$, respectively. The resulting behavior for $T_C(E)$ is illustrated in figures 3.5(b) and 3.6(b) (see also table 3.1) for attractive interactions. One observes that $T_C(E)$

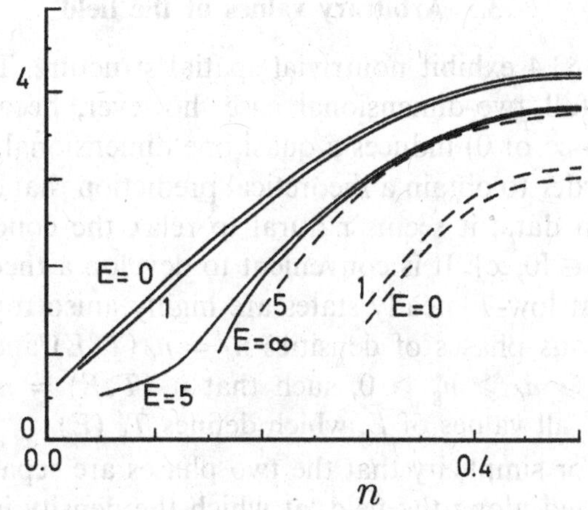

Fig. 3.7. The transition temperature as a function of density, n, for representative values of the field (as indicated), within the fast-rate limit $\Gamma \to \infty$, for attractive interactions, $J_{\hat{x}} = J_{\hat{y}} > 0$ (solid lines) and for the case in which the interactions along the field are repulsive, $J_{\hat{x}} < 0$, and the transverse ones are attractive, $J_{\hat{y}} > 0$ (dashed lines).

decreases monotonically with increasing E, so that $T_C^{\infty} = 0$, for all values of n; moreover, $\partial^2 T_C(E)/\partial E^2$ apparently changes sign around $E = 5$ for large enough n reflecting that this has for $E \to 0$ the same steady state (the equilibrium state) as $\Gamma \to \infty$. The system with $E \to \infty$ and $\Gamma \to 0$ may be interpreted, consequently, as a collection of rows, each in a steady state with respect to the equilibrium transition probability at temperature T. The rows are independent except for the fact that they exchange particles at random, i.e., independently of temperature, which prevents any long-range order.

The case $\Gamma \to 0$ with repulsive interactions shows no phase segregation for any value of E and n, as for $\Gamma \to \infty$. It thus seems interesting to consider further anisotropies in the model by allowing for different signs of the interactions along the two directions of the lattice, $J_{\hat{x}}$ and $J_{\hat{y}}$.

A principal result is that the phase segregation is suppressed in the limit $\Gamma \to \infty$ for any values of E and n if attractions are in the direction of the field (while transverse interactions are repulsive). On the contrary, long-range order may occur if repulsions are in the direction of the field, and the particles attract each other along the transverse direction. The latter case, $J_{\hat{x}} < 0$ and $J_{\hat{y}} > 0$, is illustrated in figure 3.7 which also contains a comparison with the case $J_{\hat{x}} = J_{\hat{y}} > 0$. The two cases are observed to approach a common limit as $E \to \infty$. Figure 3.7 underscores the importance of both n and the sign of interactions in determining the steady state.

3.5 Arbitrary values of the field

The models in §3.4 exhibit nontrivial spatial structure. They do not correspond to a full two-dimensional case, however, because the limiting condition ($\Gamma \to \infty$ or 0) induces a quasi one-dimensional situation as E is increased. In order to obtain a theoretical prediction that can be compared with simulation data, it seems natural to relax the condition on Γ, and allow any $E, \Gamma \in [0, \infty]$. It is convenient to develop a theory based on the assumption that low-T steady states are highly anisotropic, consisting of two homogeneous phases of densities $n_\ell = n_\ell(T,E)$ and $n_g = n_g(T,E)$, respectively, $1 \geqslant n_\ell \geqslant n_g \geqslant 0$, such that $n_\ell(T,E) = n_g(T,E) = n$ for $T \geqslant T_C(E)$ for all values of E, which defines $T_C(E)$.

We assume for simplicity that the two phases are separated by a sharp interface, oriented along the field, at which the density jumps from n_ℓ to n_g, and that a current of particles occurs from one side (phase) to the other depending only on the global properties of the phases. The assumption of a uniform density within each phase means, in effect, that following each transfer across the interface, the phases immediately adjust to the new value of the density. One may then describe the DLG by means of two connected kinetic cluster equations, one for each (homogeneous) phase.

The model

Consider a homogeneous phase of density n and temperature T under the action of a field $E\hat{\mathbf{x}}$. Transitions are governed by the rate (3.44) with $c_{\hat{\mathbf{x}}} = \phi_\parallel(\delta H, E)$, $c_{\hat{\mathbf{y}}} = \phi_\perp(\delta H)$, $c_{\hat{\mathbf{z}}} = 0$. The densities of NN pairs of particles along the two directions, $u_\parallel$ and $u_\perp$, satisfy kinetic equations that are similar to (3.8), except that jumps inside the domain D depend on the orientation of the latter relative to the field. (We refer to the domains shown in §3.4.) The number of particles in D_i (see §3.1) is $N_i = N_i' + N_i''$, where $i = 1, 2$, $N_1' = \sigma_3$, $N_2' = \sigma_4$, $N_1'' = \sigma_1 + \sigma_5$ and $N_2'' = \sigma_2 + \sigma_6$. Therefore,

$$\frac{\partial u_\parallel}{\partial t} = \sum_{N_1', N_2' = 0}^{1} \sum_{N_1'', N_2'' = 0}^{2} \Big[\left(N_2' - N_1'\right) \phi_\perp(\delta H) Q^\perp_{+-N_1'' N_2'' N_1' N_2'}(u_\perp, u_\parallel) + \left(N_2'' - N_1''\right) \phi_\parallel(\delta H, E) Q^\parallel_{+-N_1'' N_2'' N_1' N_2'}(u_\perp, u_\parallel) \Big] . \qquad (3.69)$$

Here, $\delta H = 4J(N_1 - N_2)$, and $Q^\parallel$ and $Q^\perp$ are defined as in (3.7) but correspond to the cluster configuration shown just above equations (3.50) (in which case the jump is *vertical*, i.e., along the field), and (3.55) (in which case the jump is *horizontal*), respectively. Thus

$$Q^{\perp}_{+-N_1''N_2''N_1'N_2'}\left(u_\perp,u_\parallel\right)=\binom{2}{N_1''}\binom{2}{N_2''}\frac{u_\perp^{N_1'}z_\perp^{1-N_2'}u_\parallel^{N_1''}z_\parallel^{2-N_2''}}{[\tilde{n}\,(1-\tilde{n})]^3}$$

$$\times\left[\tfrac{1}{2}\,(1-u_\perp-z_\perp)\right]^{2-N_1'+N_2'}\left[\tfrac{1}{2}\,(1-u_\parallel-z_\parallel)\right]^{2-N_1''+N_2''}, \quad (3.70)$$

with $z_{\perp(\parallel)}=1-2\tilde{n}+u_{\perp(\parallel)}$, and

$$\left.\begin{aligned}Q^{\perp}_{-+N_2''N_1''N_2'N_1'}\left(u_\perp,u_\parallel\right)&=Q^{\perp}_{+-N_1''N_2''N_1'N_2'}\left(u_\perp,u_\parallel\right),\\ Q^{\parallel}_{+-N_1''N_2''N_1'N_2'}\left(u_\perp,u_\parallel\right)&=Q^{\perp}_{+-N_1''N_2''N_1'N_2'}\left(u_\parallel,u_\perp\right).\end{aligned}\right\} \quad (3.71)$$

Equations (3.69)–(3.71) for $\partial u_{\perp(\parallel)}/\partial t$ are a basic description for any homogeneous (anisotropic) phase. The liquid corresponds to $\tilde{n}=n_\ell$, $u_\parallel=u_{\ell,\parallel}$, and $u_\perp=u_{\ell,\perp}$, and the gas at the same temperature corresponds to $\tilde{n}=n_g$, $u_\parallel=u_{g,\parallel}$, and $u_\perp=u_{g,\perp}$. Then one may relate the phases by means of

$$\frac{\partial u_\ell}{\partial t}=j_{g\to\ell}-j_{\ell\to g}\,. \quad (3.72)$$

One further assumes that the number of particles crossing the interface per unit time from the liquid (gas) to the gas (liquid) equals the probability for the local configuration at the interface times the probability per unit time for the appropriate interchange, namely, $j_{\ell\to g}=\mathcal{Q}_{\ell g}\left(n_\ell,n_g\right)\phi_\perp\left(\delta H_{\ell g}\right)$, and similarly for $j_{g\to\ell}$. Here, $\delta H_{\ell g}=12J\left(n_\ell-n_g\right)$,

$$\mathcal{Q}_{\ell g}\left(n_\ell,n_g\right)=\frac{\left(u_{\ell,\perp}\right)^{n_\ell}\left(u_{\ell,\parallel}\right)^{2n_\ell}}{\left[n_\ell\left(1-n_g\right)\right]^2}\left[\tfrac{1}{2}\left(1-2n_\ell+u_{\ell,\perp}\right)\right]^{1-n_\ell}$$

$$\times\left[\tfrac{1}{2}\left(1-2n_\ell+u_{\ell,\parallel}\right)\right]^{2(1-n_\ell)}\left[\tfrac{1}{2}z_{g,\parallel}\left(1-2n_g+u_{g,\parallel}\right)\right]^{2(1-n_g)}$$

$$\times\left[\tfrac{1}{2}z_{g,\perp}\left(1-2n_g+u_{g,\perp}\right)\right]^{1-n_g}, \quad (3.73)$$

and $\mathcal{Q}_{g\ell}\left(n_g,n_\ell\right)=\mathcal{Q}_{\ell g}\left(n_g,n_\ell\right)$. This refers to the following local (average) configuration at both sides of the interface:

$$\begin{array}{cc|cc} & \text{liquid} & \text{gas} & \\ & n_\ell & n_g & \\ & + & + & \\ + & \sigma_1 & \sigma_2 & + \\ & + & + & \end{array}$$

One assumes here that σ_1 is surrounded, on the average, by $3n_\ell$ particles (within the liquid) and, similarly, that σ_2 has an average of $3n_g$ occupied neighboring sites (in the gas); $\sigma_1=1$, $\sigma_2=0$ for $j_{\ell\to g}$.

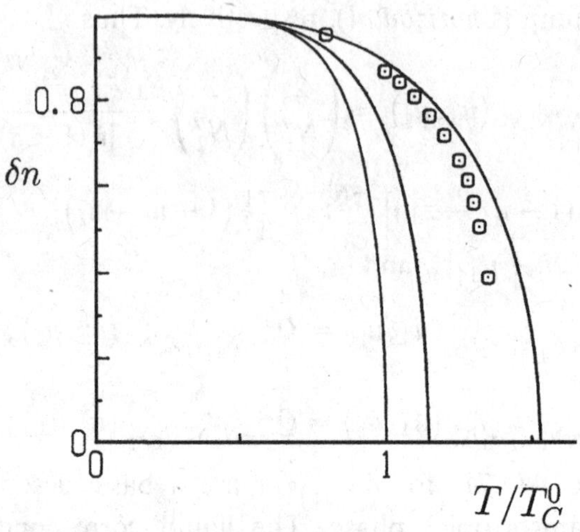

Fig. 3.8. The order parameter $\delta n \equiv n_\ell - n_g$ for $n = \frac{1}{2}$, $\Gamma = 1$, and $E = 0, 1$, and ∞ (from left to right). The circles represent MC data for the *infinite* lattice for $n = \frac{1}{2}, \Gamma = 1$, and $E = \infty$. (Adapted from Garrido *et al.* (1990a).)

Mean-field ordering

The approach described above is now applied to the case of attractive interactions with a Metropolis rate. Then $\phi_\parallel = \frac{1}{2}(1-p)(\phi_\parallel^+ + \phi_\parallel^-)$, with $p \equiv (1+\Gamma)^{-1}$ and $\phi_\parallel^\pm = \exp(-\beta\delta H \pm E)$ for $-\beta\delta H \pm E \leqslant 0$ and $\phi_\parallel^\pm = 1$ otherwise, and $\phi_\perp = p\exp(-\beta\delta H)$ for $\delta H \geq 0$ and $\phi_\perp = p$ otherwise. This case is interesting because it corresponds to most of the simulations described in chapter 2.

Figure 3.8 illustrates the behavior of the order parameter $\delta n = n_\ell - n_g$ for $n = \frac{1}{2}$ and $p = \frac{1}{2}$ (i.e., $\Gamma = 1$). As expected for a mean-field approximation, one finds that $\delta n \approx [1 - T/T_C(E)]^{1/2}$. The quantitative differences observed in the critical temperature and other properties between the mean-field and MC results (for example, figure 3.8; see also table 3.3) are related to the fact that mean-field models overestimate long-range correlations and, consequently, describe an order which is more stable against thermal fluctuations than that observed in simulations.

It is also observed that $T_C(E)$ is an increasing function of E. This is in accordance with MC observations (§2.3) where it may be understood as the result of competition between two effects of the field:

(i) a tendency to produce strip-like geometries, i.e., to enhance long-range correlations (that would increase $T_C(E)$), and

(ii) a tendency to destroy some local correlations within the strip, thus

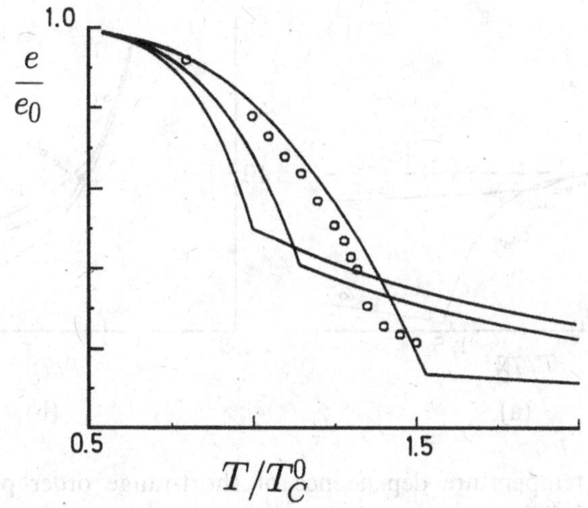

Fig. 3.9. The energy (normalized to $e_0 \equiv -2J$) versus temperature for $n = \frac{1}{2}$, $\Gamma = 1$, and $E = 0, 1$, and ∞ (from left to right). The circles represent MC data for the *infinite* lattice for $n = \frac{1}{2}, \Gamma = 1$, and $E = \infty$.

reducing effective cohesion at short distances (that would decrease $T_C(E)$).

The (anisotropic) effect (i) dominates over the more local one (ii) in the present case of attractive interactions; one should expect, however, that (ii), which effectively destroys short-range order, will dominate over (i) for repulsive interactions, as has been reported in MC simulations. On the other hand, decreasing n sufficiently prevents the formation of strips, which makes (ii) dominant at low density, in accordance with the phase diagrams in figures 2.2. and 2.13. (The situation is, in fact, more subtle because mean-field theory shows that suppressing all correlations, as in the single-site approximation, produces the highest T_C, and T_C decreases as one considers further local correlations, i.e., as one goes to the pair, and then to the four-site approximation, etc.)

The energy density is $\langle e \rangle = -4Ju$, where $u = \frac{1}{2}(u_\ell + u_g)$, and $u_{\ell(g)} = \frac{1}{2}(u_{\ell(g),\perp} + u_{\ell(g),\parallel})$ is the density of pairs of particles within the liquid (gas), with $u_{\ell(g),\perp}$ and $u_{\ell(g),\parallel}$ the stationary solutions of (3.69)–(3.73). Figure 3.9 illustrates the behavior of $\langle e \rangle/(-2J)$. One observes that this decreases with increasing E for any given $T \geq T_C(E)$ but increases for any given $T < T_C(E)$. That is, the field effectively condenses the particles below $T_C(E)$, so that effect (i) described above is dominant, while it acts as a sort of *dispersing* agent for $T > T_C(E)$, so that (ii) is then dominant. Similar information follows from $u_{\ell(g)}$ for each phase.

It is interesting to consider two other measures of order. The first one

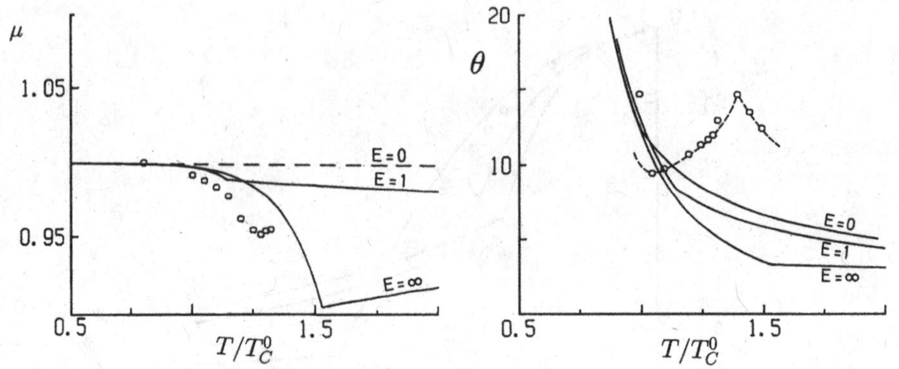

(a) (b)

Fig. 3.10. The temperature dependence of short-range order parameters for $n = \frac{1}{2}$, $\Gamma = 1$, and different values of E, as indicated: (a) the case μ in (3.74), and (b) the case θ in (3.75). The circles represent MC data for the *infinite* lattice for $n = \frac{1}{2}, \Gamma = 1$, and $E = \infty$.

is the parameter of local anisotropy defined as

$$\mu = \frac{u_{\perp}}{u_{\parallel}}, \tag{3.74}$$

i.e., the ratio between the number of NN pairs of particles along the transverse direction and those along the direction of the field. Figure 3.10(a) indicates that $\mu = 1$, independent of T, for the isotropic system, $E = 0$, but any field induces anisotropic clusters, and the anisotropy tends to increase with E. On the other hand, μ decreases monotonically with increasing T for $T < T_C(E)$ for any E, while the qualitative behavior above $T_C(E)$ depends on E. This may be understood as follows: at $T = 0$, the system is fully condensed and locally isotropic; any finite T allows for a particle current along the field, and, for sufficiently large $T < T_C(E)$ — in fact, rather near $T_C(E)$ — there are anisotropic clusters whose anisotropy increases with T. For $T > T_C(E)$, μ increases slowly but monotonically with T from its minimum value $\mu_C \equiv \mu[T_C(E)]$ if the field is large enough, i.e., clusters are relatively unstable against the action of both field and thermal fluctuations. If the field is small, for example, $E = 1$, it does not help thermal fluctuations as efficiently in destroying local order but rather tends to induce a higher anisotropy as the particle mobility increases due to higher T. As $T \to \infty$, μ goes to μ_∞, which decreases with increasing E, regardless of the value of E. This sort of local anisotropic order above $T_C(E)$ is an important nonequilibrium effect described in §2.5. The overall behavior is consistent with the description of cluster anisotropy in §2.6.

Further information about order in Ising-like systems with attractive

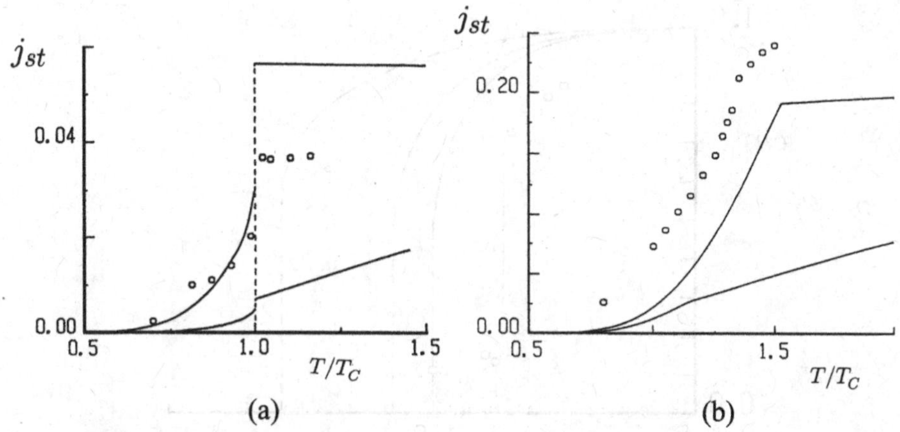

Fig. 3.11. The temperature dependence of the ionic current for $\Gamma = 1$, and (from top to bottom) $E = \infty$ and 1. (a) $n = \frac{1}{2}$; the circles are MC data for the *infinite* lattice for $E = \infty$. (b) $n = 0.1$; the circles are MC data for $E = \infty$ and $L = 50$.

interactions is provided by the short-range order parameter defined in §2.6, namely,

$$\theta \equiv \frac{\langle p(\bullet,\bullet)\, p(\circ,\circ)\rangle}{\langle p(\bullet,\circ)\rangle^2} \approx \left\langle \frac{p(\bullet,\bullet)\, p(\circ,\circ)}{p(\bullet,\circ)^2} \right\rangle . \tag{3.75}$$

One may write $\theta = [\frac{1}{4}(1+u)^2 - m^2](1-u)^2$ which leads to the critical behavior in equation (2.35). While θ develops, in general, a well-defined peak for finite lattices in simulations (which provides in practise an accurate method for estimating the size dependence of the critical temperature), the classical case $\beta = \frac{1}{2}$ and $\alpha = 0$ is unique in that it implies no singularity at T_C but a monotonic decrease with increasing T. These differences are illustrated in figure 3.10(b).

The field induces a particle current — see equation (3.21) — which is represented in figure 3.11. A main fact here is that the slope of the curves changes discontinuously at $T_C(E)$, as in both real and MC experiments for $d \geq 2$ (this does not occur for the one-dimensional system considered in §3.2).

It is also interesting to compare the simulation result that the phase transition is apparently continuous for any n within the range $0.5 \geq n \geq 0.35$ (while clear discontinuities occur for $n \leqslant 0.2$) with mean-field theory. The latter predicts a phase transition of first order for small values of n; see figure 3.12. This is indicated, for example, by the behavior of both the energy, $\langle e\rangle(-2J) = u_\ell + n^{-1}(1-n)\,u_g$, and the short-range order parameter; the temperature dependence of the particle current is

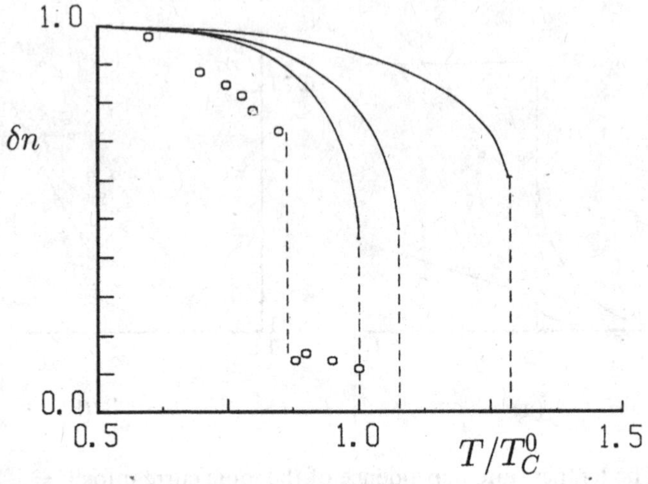

Fig. 3.12. The order parameter $\delta n \equiv n_\ell - n_g$ for $n = 0.1$, $\Gamma = 1$, and $E = 0, 1$ and ∞ (from left to right); the circles correspond to MC data for $E = \infty$ and $L = 50$.

illustrated in figure 3.11(b); we refer the reader to Garrido *et al.* (1990) for further details.

A variation of the present model allows one to study varying Γ. The above assumptions regarding the interface, expressed in equations (3.72) and (3.73), should be improved for this purpose. In particular, one needs to consider a domain around the interface larger than the one shown after equation (3.73), if one is to take proper account of the effect of varying p on the interface. However, how this should be done needs further study given that the interface details appear to exert a controlling influence on the system, particularly in the limits $E, \Gamma \to \infty$. A method of determining the transition temperature $T_C\,(n, E, \Gamma)$ that does not depend critically on assumptions about the interface is presented in §3.6.

Stability of strips

The method in this section is based on the assumption that the DLG condenses below $T_C\,(E)$ into a single liquid strip directed along the field. This is consistent with the observation (in simulations) that states with more than one strip do not seem to be stable in two dimensions. The present formalism allows for a demonstration of this, namely, one may prove that multi-strip states decay into a single strip (though the relaxation time may be large), as suggested by figure 2.7 (see also figure 2.3). A complete proof would use the method above to write a set of coupled kinetic cluster equations for a system with several strips, and then analyze these equations numerically; see the next section. The following argument leads to the same conclusion assuming that the steady state corresponds to minimum energy.

Consider a half-filled rectangular lattice, $L_{\hat{x}} \times L_{\hat{y}}$ with E along $\hat{\mathbf{x}}$. Suppose that the steady state at low enough temperature consists of s identical liquid strips separated from each other by gas strips; the width of each strip is $\lambda = L_{\hat{y}}/2s$. The total energy associated with such a configuration is

$$\mathscr{E} = -4JL_{\hat{x}} \sum_{i=1}^{s} \{\lambda (u_{2i} + u_{2i+1}) - a(T)\,[(1 - n_{2i-1}) + (1 - n_{2i+1})]\} \ , \quad (3.76)$$

where the index $2s + 1$ is understood to correspond to the first site. Here, n_j and u_j are, respectively, the particle density and the density of pairs within the jth strip; even j corresponds to liquid strips, odd j to gas. The first two terms in the r.h.s. of (3.76) correspond to the energy within the liquid and gas, respectively, and the rest refers to the interface. It is not necessary to write an explicit function for $a(T)$, but one knows that it is positive and $a(T \to \infty) = 1$. Then one may write that $n_g = s^{-1} \sum_i n_{2i-1}$ and $u_{\ell(g)} = (2s)^{-1} \sum_i u_{2i(2i-1)}$ and, consequently, the energy per site is

$$e = \frac{\mathscr{E}}{N} \simeq -4J \left[u_\ell + u_g - \frac{2a(T)\,(1 - n_g)\, s}{L_{\hat{y}}} \right] , \quad (3.77)$$

where $N = L_{\hat{x}} \times L_{\hat{y}}$. Assuming that u_ℓ, u_g, n_g, and $a(T)$ are all independent of s, which seems to be supported by the MC data, it follows that any departure from the state with s strips must satisfy

$$\frac{\delta e}{\delta t} = 8a(T)J\,(1 - n_g) \left(\frac{\delta s}{\delta t} \right) L_{\hat{y}}^{-1} . \quad (3.78)$$

Furthermore, one assumes here that observation is on a scale in which n_g and $u_{\ell(g)}$ are stationary in practise, but fluctuations around those mean values are large enough to allow for the coalescence of strips. Then assuming that the system searches for the lowest energy which is compatible with the existing constraints, (3.78) states that s tends to decrease, with no apparent lower bound other than $s = 1$, where the system becomes stationary with respect to all dynamical processes. On the other hand, it is also noticeable that the minimum interfacial energy — at zero temperature and field — for the finite two-dimensional system is given, depending on the density, by three different possible configurations: a square, and either a strip or a cross with bars connected by periodic boundary conditions; see Marro (1976) for a three-dimensional system.

3.6 Stability of the homogeneous phase

The mean-field description above may be complemented by a theory based on the vanishing of the diffusion coefficient at the critical point. The diffusion coefficient is determined by examining the current in response to a small density gradient, and the calculation proceeds in two steps (Dickman 1988). Firstly, the set (3.69)–(3.73) of kinetic equations for cluster densities reflecting nonequilibrium dynamics under the action of the driving field is integrated at the critical density $n = \frac{1}{2}$. The aim is to find the spatially homogeneous steady state for $T > T_C(E)$ characterized by $n_\ell(T,E) = n_g(T,E) = n$ for all E. Secondly, the stability of this state is tested by imposing a weak (typically, one part in 10^6) density gradient. Since MC simulations always show the interface oriented along the driving field, the gradient is imposed transverse to the field (see §3.7 for some details).

This approach is sensitive only to a local instability of the homogeneous phase and, consequently, cannot detect a first-order phase transition. This is not a problem in the present context, however, since both the mean-field models above and simulations always yield a continuous transition for $n = \frac{1}{2}$. The homogeneous state becomes unstable at the critical temperature $T_C(E)$. Applying this scheme for the case $\Gamma = 1$ and $E \to \infty$ in the two-dimensional lattice, one obtains the critical temperatures (units of J/k_B): $T_C = 3.206$ at the pair approximation level and $T_C = 3.134$ in the square approximation, as compared with the simulation result $T_C \simeq 3.13$ with the error in the last figure. For $d = 3$, the pair approximation yields a critical temperature which agrees with the simulation result, $T_C \simeq 4.83$, to within uncertainty.

The same approach may be used to obtain values for the transition temperature $T_C(n, E, \Gamma)$ as one varies Γ. The main findings are summarized in figure 3.13 depicting $T_C(E, \Gamma)$ versus E/T for several choices of Γ. One observes that while $T_C(E, 1)$ is a monotonically increasing function of E, for $\Gamma > 1$ the critical temperature exhibits a maximum, located near $E/T = 1.5$. For large driving fields, the effect of increasing Γ is to reduce the critical temperature. The same behavior is observed in simulations, as illustrated in table 3.2.

A pair mean-field theory for the DLG with *repulsive* interactions was devised by Dickman (1990a), yielding predictions in good agreement with simulations. The approach has been extended to larger clusters by Szabó and coworkers (1994, Szabó & Szolnoki 1996).

3.7 The layered system

This section describes an extension (Alonso *et al.* 1995) of the method in the preceding section. It allows one to study the layered system Λ_E

Table 3.2. Effect of the parameter Γ on the critical temperature $T_C(n, E, \Gamma)$, in units of the corresponding equilibrium value, for $d = 2$, $n = \frac{1}{2}$, and saturating values of the field, as observed in simulations (MC), and as obtained, using the mean-field theory described in this section (MF).

Γ	MC	MF
1	1.38±0.02	1.3806
5	~1.18	1.2775
20	~1.05	1.1302
80	~1.0	1.0546

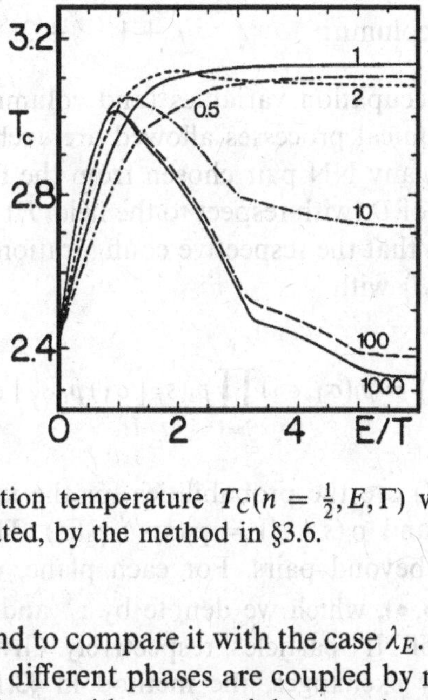

Fig. 3.13. The transition temperature $T_C(n = \frac{1}{2}, E, \Gamma)$ versus E/T for several choices of Γ, as indicated, by the method in §3.6.

introduced in §2.4, and to compare it with the case λ_E. Unlike the procedure in §3.5, in which two different phases are coupled by means of an equation for the current, this provides a partial, even incorrect description of the (inhomogeneous) states of Λ_E within the temperature range $T^* > T > T'$. That is, the present method predicts *homogeneous* states at low temperature with no indication of strips, for example, in figure 2.10 . However, a test of stability under low density gradients reveals that such homogeneity is unstable, while the same method predicts stable states for $T > T^*$. This, in fact, is the procedure used below to determine T^*. The other relevant transition temperature in Λ_E, i.e., T' (which is always below T^*), is estimated as the lowest temperature for which the densities within each plane, $n_1(T)$ and $n_2(T)$, remain equal. It should also be mentioned that the present method neglects some correlations between planes. This is not justified *a priori* for $T^* > T > T'$, when peculiar interplane correlations along the interface appear to determine the system behavior. Nevertheless, the method relies only on the description for $T > T^*$ and $T < T'$, when simulations suggest weak interplane correlations, i.e., a small amplitude of the (interplane) LRSCs studied in §2.5.

Consider the domain $D = D_1 \cup D_2$ with D_2 *on top* of D_1 in a different plane. Each subdomain D_i, $i = 1, 2$, is defined as

$$\begin{array}{lccc} & & s_2' & \\ & s_1' & \sigma_2 & s_3' \\ & s_1 & \sigma_1 & s_3 \\ & & s_2 & \\ \\ \text{column:} & \ell & \ell+1 & \ell+2 \end{array}$$

where the different occupation variables (and column indexes) are indicated. The only dynamical processes allowed are exchanges involving an *interior* NN pair (i.e., any NN pair chosen from the four interior sites of D). Both orientations of D (with respect to the field, E) must be considered. Further, it is assumed that the respective configurations have probabilities $p(\vec{\sigma}_D) = p(\vec{\sigma}_{D_1})\, p(\vec{\sigma}_{D_2})$ with

$$p(\vec{\sigma}_{D_i}) = p_i(\sigma_1, \sigma_2) \prod_{j=1}^{3} p_i(s_j \mid \sigma_1)\, p_i(s_j' \mid \sigma_2) \,. \tag{3.79}$$

Here, $p_i(s)$ and $p_i(s, s')$ are the probabilities for the indicated occupation variables within D_i, and $p_i(s \mid s') = p_i(s, s')/p_i(s')$. This amounts to neglecting correlations beyond pairs. For each plane, one is interested in $p_i(\bullet) = n_i$, and in $p_i(\bullet, \bullet)$, which we denote by $z_i^{\hat{x}}$ and $z_i^{\hat{y}}$ for longitudinal and transverse pairs of NN particles, respectively. Given the conservation of $n = \frac{1}{2}(n_1 + n_2)$ by exchanges, the method in §3.1 leads (after some computer algebra) to a set of five coupled nonlinear differential equations. The main conclusions are as follows.

The mean-field version of Λ_E exhibits, for any value of E, steady states such that

$$\begin{aligned} n_1 = n_2,\ z_1^{\hat{x}} = z_2^{\hat{x}},\ z_1^{\hat{y}} = z_2^{\hat{y}} \quad &\text{for} \quad T > T'(E), \\ n_1 \neq n_2,\ z_1^{\hat{x}} \neq z_2^{\hat{x}},\ z_1^{\hat{y}} \neq z_2^{\hat{y}} \quad &\text{for} \quad T < T'(E) \,. \end{aligned}$$

This corresponds to condensation in only one of the planes observed in simulations. Also in accord with simulations is the result that $T'(E)$ decreases with increasing E, with a well-defined limit for $E \to \infty$. $T'(0)$

Table 3.3. The two transition temperatures for the layered system (in units of the corresponding equilibrium value) as obtained from simulations (MC) and mean-field theory (MF) for the indicated rates, for $n = \frac{1}{2}$ and saturating fields, and a comparison with the two-dimensional system.

System	Method	Rate	T^*	T'
	MC	Metropolis	1.30	0.95
Λ_∞	MF	Metropolis	1.09	0.80
	MF	Kawasaki	1.23	0.99
	MC	Metropolis	1.38	—
λ_∞	MF	Metropolis	1.11	—
	MF	Kawasaki	1.32	—

is the Bethe critical temperature for the square Ising model. Interestingly, while this phase transition at $T'(E)$ is of second order for all values of E for the Kawasaki rate, the Metropolis rate induces a nonequilibrium *tricritical point* at $E_C = 3.5 \pm 0.1$. That is, the transition is of second order for $E < E_C$, and of first order for $E > E_C$. This has two important implications. On one hand, it provides theoretical support for the simulation result that discontinuities occur for $E \to \infty$ (for the Metropolis rate). On the other, it suggests that *continuous* phase transitions may occur for small fields in simulations, which have been indeed observed for $E < E_C \simeq 2$ (Marro *et al.* 1996). Transition temperatures are compared in table 3.3. The overall behavior is illustrated in figure 3.14; the discontinuous nature of the function ϕ for the Metropolis rate reflects discontinuities that are evident in figure 3.14(b).

The phase transition that occurs in Λ_∞ above $T'(E)$ has been studied for $n = \frac{1}{2}$ by the following method. One assumes that, as observed in simulations for $E = \infty$, the phase transition is of second order. Then one generates a mean-field homogeneous solution as described above. Next, the solution is modified by introducing a density gradient ∇n along $\hat{\mathbf{y}}$, slightly altering the state. This is performed in practice by considering the cluster depicted above in this section with the field acting vertically. For pairs such as (s_1, s_1') that are within the column ℓ along $\hat{\mathbf{x}}$, one transforms

$$p(\bullet,\bullet) \rightarrow z^{\hat{\mathbf{x}}} - (5-2\ell)\,\nabla n, \quad \text{and}$$
$$p(\circ,\circ) \rightarrow z^{\hat{\mathbf{x}}} + (5-2\ell)\,\nabla n,$$

while leaving $p(\bullet,\circ) = p(\circ,\bullet) = \frac{1}{2} - z^{\hat{\mathbf{x}}}$ unchanged; for pairs within the

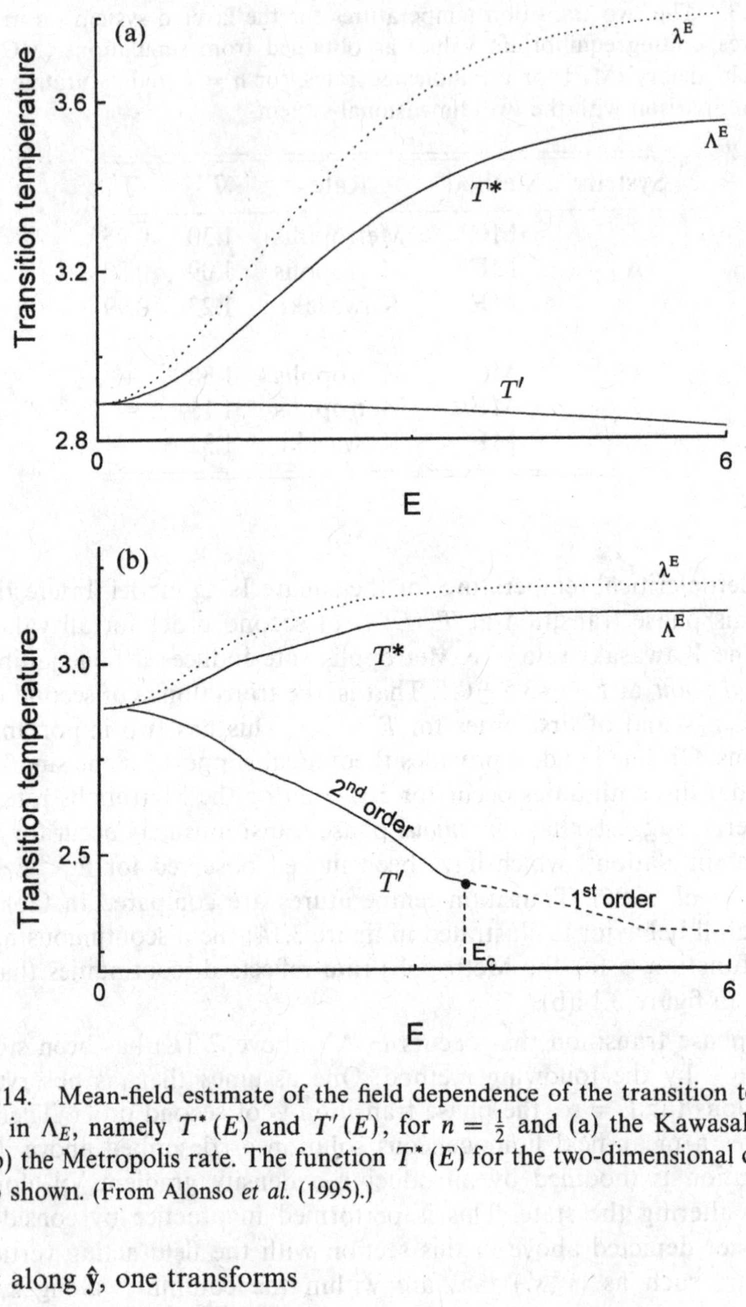

Fig. 3.14. Mean-field estimate of the field dependence of the transition temperatures in Λ_E, namely $T^*(E)$ and $T'(E)$, for $n = \frac{1}{2}$ and (a) the Kawasaki rate, and (b) the Metropolis rate. The function $T^*(E)$ for the two-dimensional case λ_E is also shown. (From Alonso *et al.* (1995).)

row k along $\hat{\mathbf{y}}$, one transforms

$$\begin{aligned} p(\bullet,\bullet) &\rightarrow z^{\hat{\mathbf{y}}} - 2(2-k)\nabla n, \\ p(\circ,\circ) &\rightarrow z^{\hat{\mathbf{y}}} + 2(2-k)\nabla n, \\ p(\circ,\bullet) &\rightarrow \tfrac{1}{2} - z^{\hat{\mathbf{y}}} - \nabla n, \quad \text{and} \\ p(\bullet,\circ) &\rightarrow \tfrac{1}{2} - z^{\hat{\mathbf{y}}} + \nabla n. \end{aligned}$$

∇n is typically taken to be 10^{-6}. Next, one computes the new values for $p(\vec{\sigma}_{D_i})$ using (3.79), and the corresponding transverse particle current, i.e.,

$$j_{\hat{\mathbf{y}}} = \sum_{\vec{\sigma}_D} (\sigma_{\mathbf{r}} - \sigma_{\mathbf{s}})\, c(\vec{\sigma};\mathbf{r},\mathbf{s})\, p(\vec{\sigma}_{D_i}), \quad \hat{\mathbf{x}}\cdot(\mathbf{r}-\mathbf{s}) = 0\,. \tag{3.80}$$

If $j_{\hat{\mathbf{y}}}$ compensates ∇n, then the original state is interpreted as stable; one obtains $T^*(E)$ from $j_{\hat{\mathbf{y}}}[T^*(E)] = 0$. The result is illustrated in figure 3.14. This indicates that $T^*(E)$ increases with E, with apparently a well-defined limit, $T_C^\infty = \lim_{E\to\infty} T^*\left(n = \frac{1}{2}, E\right)$, and depends importantly on the rate. Figure 3.15 depicts the behavior of the resulting $p(\bullet,\bullet)$ which is a measure of the energy. The anisotropy of the systems is reflected in the fact that $z^{\hat{\mathbf{x}}} > z^{\hat{\mathbf{y}}}$ quite generally, and one observes again the strong influence the nature of the rate may have on the steady state.

The comparison between the cases λ_E and Λ_E is interesting. Figure 3.14 (and the simulations in chapter 2 for a saturating field) seem to indicate that $T^*(E)$ might generally be smaller for the layered system. Furthermore, $z^{\hat{\mathbf{x}}} - z^{\hat{\mathbf{y}}}$ in figure 3.15 is always larger for λ_E, and the differences increase with E. That is, the possibility of particles hopping to the other plane seems to lessen anisotropy, and $T^*(E)$ is in turn decreased, the effect becoming larger as E is increased. This may be interpreted as a direct consequence of microscopic differences between the dynamical rules for the two systems. The two-dimensional system λ_E evolves as a consequence of the competition of two processes, along $\hat{\mathbf{x}}$ and $\hat{\mathbf{y}}$, that are induced, respectively, by the field and by the heat bath. The rule for Λ_E involves an extra thermal process (along $\hat{\mathbf{z}}$) that amounts to additional randomness and has the effect of compensating the anisotropic action of the field.

Concluding remark

The previous two chapters show that the effort devoted during the last decade or so to studying the DLG has been fruitful. The DLG is an important example of nonequilibrium ordering, one that has been studied in considerable detail. Furthermore, the DLG probably captures the essential features of anisotropic states such as those induced by an external drive (for example, in FIC materials). One may expect this to remain an active area of research for some years, focusing on the relation between the nonequilibrium phase transition and the peculiarities of the attendant interface. This is consistent with the observation above that two-dimensional mean-field models are extremely sensitive to hypotheses about the interface (Garrido *et al.* 1990a). Furthermore, the same is suggested by the fact that two other phase transitions occurring in the DLG, namely, for

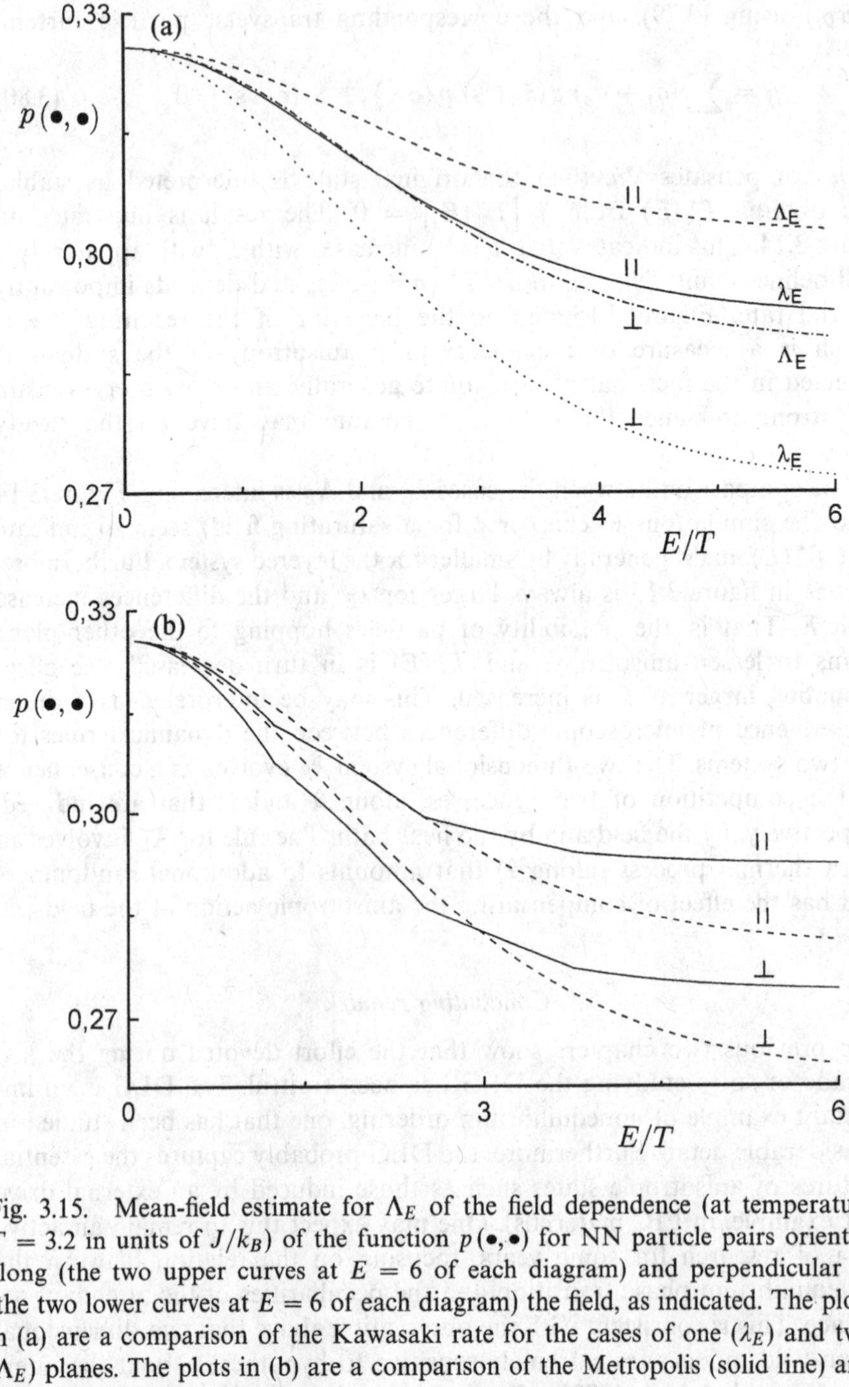

Fig. 3.15. Mean-field estimate for Λ_E of the field dependence (at temperature $T = 3.2$ in units of J/k_B) of the function $p(\bullet,\bullet)$ for NN particle pairs oriented along (the two upper curves at $E = 6$ of each diagram) and perpendicular to (the two lower curves at $E = 6$ of each diagram) the field, as indicated. The plots in (a) are a comparison of the Kawasaki rate for the cases of one (λ_E) and two (Λ_E) planes. The plots in (b) are a comparison of the Metropolis (solid line) and Kawasaki (dashed line) rates. (From Alonso *et al.* (1995).)

$J < 0$ in λ_∞ (§2.3) and the one at low temperature for $E < E_C$ in Λ_E (Marro *et al.* 1996), in which no anisotropic interface is developed, seem to belong to the Ising universality class, and no *peculiar* finite-size effects are then observed.

4
Lattice gases with reaction

A class of biochemical processes is modeled by an open system into which reactants are continuously fed, and out of which products are continuously removed to maintain a steady nonequilibrium state. Studying such idealized models helps in understanding the formation and stability of complex spatial and temporal structures in biological systems. In a model with this aim, one needs ultimately to assume that molecules also diffuse. Systems that are governed by an interplay of reaction and diffusion processes are relevant to many problems in physics and other disciplines (Turing 1952). This chapter describes kinetic lattice gases whose dynamics involve competition between reaction and diffusion. The pioneering study of such stochastic models was due both to mathematicians[1] and physicists[2]; the more recent activity described below was motivated by theoretical as well as practical interest in related nonequilibrium situations.

The chapter is organized as follows. In §4.1, the nonlinear reaction–diffusion equation, a hydrodynamic-like description in which mean values and fluctuations have both spatial and temporal variations, is introduced. §4.2 presents a lattice model with competition between creation/annihilation processes at temperature T, and diffusion at temperature T'. The relation between the hydrodynamic and microscopic levels of description is examined in §4.3, and the implications of the resulting macroscopic equations are discussed in §4.4. In particular, some exact results for homogeneous states and for some time-dependent inhomogeneous conditions are illustrated for the simplest case of random diffusion, i.e., $T' \to \infty$, and for finite T'. §4.5 reviews the main results of a series of MC simulations. A mean-field approach based on a variation of the

[1] See, for instance, Kurtz (1970, 1978), Aronson & Weinberger (1975), Fife (1979), Arnold & Theodosopulu (1980).

[2] See Nicolis & Prigogine (1977), Haken (1978, 1983), and chapters 5–9.

cluster method introduced in chapter 3 is presented in §4.6. The chapter ends with a summary of steady state properties in §4.7.

4.1 Macroscopic reaction–diffusion equations

Consider the simple autocatalytic reaction (Schlögl 1972, Grassberger & de la Torre 1979):

$$A + X \underset{k_1'}{\overset{k_1}{\rightleftarrows}} 2X\,. \tag{4.1}$$

This corresponds to a system with concentration c_A of molecules of type A, and concentration m of molecules of type X (the autocatalytic species). Molecules of A react with X at rate k_1 to form another X; the reverse reaction occurs with rate k_1'. It follows that $c_A m k_1 = m^2 k_1'$ in the stationary regime. Now suppose that molecules of B are also present, with concentration c_B, and that a new chemical C is formed by reaction of B with X:

$$B + X \underset{k_2'}{\overset{k_2}{\rightleftarrows}} C\,. \tag{4.2}$$

The steady regime is characterized by $c_B m k_2 = c_C k_2'$ provided this is independent of (4.1). If the two reactions are coupled, the net time variation of m may be written as

$$\frac{\mathrm{d}m}{\mathrm{d}t} = F(m) = \frac{\partial \Phi(m)}{\partial m}. \tag{4.3}$$

The second equality defines a potential $\Phi(m)$, and

$$F(m) \equiv c_A m k_1 - m^2 k_1' - c_B m k_2 + c_C k_2'\,. \tag{4.4}$$

The steady regime corresponds to $F(m) = 0$. This implies that m is proportional to $c_A k_1 - c_B k_2$ for $c_B k_2 < c_A k_1$; otherwise the catalyst is eventually depleted ($m = 0$) for $k_1' \neq 0$ and $c_C k_2' = 0$. Models similar to (4.1)–(4.4) have been devised in many disciplines, from physics to the social sciences.[3]

The model embodied in (4.3) is too simple to describe nonequilibrium ordering in nature, which often involves spatial inhomogeneities. This is

[3] Explicit details for a model for the development of two species, one of which is a food source for the other, may be found in Kurtz (1970), and for a model for several chemically reacting systems in Haken (1978, 1983). See also Kemeny & Snell (1962), Kurtz (1971), Aronson & Weinberger (1975), Nicolis & Prigogine (1977), Ebeling (1986), and Ohtsuki & Keyes (1987), for example.

the case if one considers, for example, the (spatial and temporal) oscillations in the famous Belousov–Zhabotinsky reaction — see Zhabotinsky & Zaikin (1973), for instance — or, more generally, the coexistence of several (nonequilibrium) phases. With this aim one needs to assume that m is a function of both t and $\mathbf{r}$. One also requires a diffusion term which induces spatial correlations. In fact, many different systems seem to be governed by reaction–diffusion equations of the form

$$\frac{\partial m(\mathbf{r},t)}{\partial t} = D\nabla^2 m(\mathbf{r},t) + F\left[m(\mathbf{r},t)\right], \tag{4.5}$$

where $F(m)$ derives from a potential $\Phi(m)$ as before (Smoller (1983), for instance). A rather natural generalization of (4.5) is to assume that the relevant quantity or order parameter (for example, the particle, momentum, or energy density of a fluid) satisfies

$$\frac{\partial m(\mathbf{r},t)}{\partial t} = \mathscr{F}(m, \nabla m, \ldots). \tag{4.6}$$

The dependence on $\mathbf{r}$ exhibited by (4.5) and (4.6) deserves comment. Consider the chemical problem from a more microscopic point of view. Let us partition the reactor into many small cubic boxes of side a, with N_x molecules in box x. Define the joint probability $P(\cdots, N_{x-1}, N_x, N_{x+1}, \cdots)$. Equation (4.5) amounts to a continuum description of the discrete system. This is not merely a formal transformation but involves scaling of space and time, which has an important physical motivation. It is necessary to sum over the many degrees of freedom that are irrelevant at the macroscopic level. For example, one sums over the many (typically of order 10^{13} per second) microscopic changes produced by collisions during a macroscopic instant $\mathrm{d}t$ between the many (of order 10^{24}) atoms in a macroscopic drop of fluid. The fluctuating microscopic behavior is thus averaged over in deterministic equations such as (4.6). The same method may be used, in principle, to obtain macroscopic equations not only for $m(\mathbf{r},t)$ but for correlation functions such as $C(\mathbf{r},\mathbf{r}';t) \equiv \langle N_x N_{x'}\rangle - \langle N_x\rangle\langle N_{x'}\rangle$. This is important because an understanding of fluctuations about the mean values is essential to understanding many phenomena. That is, restricting $m(\mathbf{r},t)$ to mean values is equivalent to a mean-field approximation in which one neglects correlations generated by the reaction process itself; see Kang & Redner (1985), for instance.

Reaction–diffusion equations, each characterized by its potential function Φ, are the relevant description at a macroscopic (or, more precisely, hydrodynamic) level for a large class of phenomena involving an optimization strategy. The underlying feature is that a fast reaction mechanism allows for the existence of many attractors that are 'visited' thanks to a coupled, relatively-slow diffusion mechanism. Moreover, two or more coupled reaction–diffusion equations may describe the behavior of some even more complex systems.[4] The field is also interesting because the critical behavior of reaction–diffusion systems may be varied; e.g., Ohtsuki & Keyes (1987b). The literature quoted in this section contains many specific examples of reaction–diffusion phenomena and a detailed description of their behavior. We focus next on a stochastic lattice-gas model that seems to contain some of the essential physics in a large class of phenomena apparently induced by competition between reaction and diffusion.

4.2 The microscopic model

The preceding section highlights the importance of determining the function $F(m)$ involved in *macroscopic* equations such as (4.5). A phenomenological determination of $F(m)$ may be of practical interest, as in many examples of (4.6), but a controlled derivation of these equations from basic principles leading to expressions for the corresponding potentials is desirable. Hydrodynamic-like equations have been derived from Hamilton's equations for very special cases (hard rods moving in one dimension, harmonic crystals, the mean-field limit of very long-range forces), but the problem is too difficult to be solved completely for realistic microscopic dynamics at present.[5] Some outstanding progress in this direction is described below: A stochastic lattice gas is defined in the present section that leads rigorously in §4.3 to an equation with the structure of (4.5), and to expressions for $F(m)$ that contain some important information.

The model is the ordinary lattice gas except that configurations, $\mathbf{s} = \{s_{\mathbf{x}} = \pm 1; \mathbf{x} \in \mathbf{Z}^d\}$, evolve in time due to the combination of two processes.[6] That is, $\mathbf{s}$ is changed stochastically by birth–death (*reaction*) processes activated by a heat bath at temperature T, and by *diffusion* (NN exchanges) driven by a heat bath at temperature T'. (The simplest case of a completely random diffusion corresponding to $T' \to \infty$ is most interesting — for example, it can be solved exactly —, but it only requires a little extra effort to define the model for finite T'.) The two processes are independent, with p the *a priori* probability of exchanges per bond and $1 - p$ that of *reactions*. Consequently, the configurational probability

---

[4] As an example, we mention that Chopard & Droz (1991) have studied a reaction–diffusion problem that involves a reaction front, in which two types of molecules diffuse and react to form a new species. See also Dab, Boon, & Li (1991).

[5] See Lebowitz (1986), for instance; see §3.2 for a hydrodynamic equation with the structure of (4.6) that describes some of the macroscopic behavior of the DLG; see also Eyink *et al.* (1996).

[6] We find it convenient to introduce here the *spin variable* $s_{\mathbf{x}}$ which is related to the *occupation variable* $\sigma_{\mathbf{x}}$ in previous chapters by $s_{\mathbf{x}} = 2\sigma_{\mathbf{x}} - 1$.

distribution, $\mu(\mathbf{s}, t)$, evolves in time according to the master equation

$$\frac{\partial \mu(\mathbf{s}, t)}{\partial t} = [p\mathscr{L}_K + (1-p)\mathscr{L}_G]\,\mu(\mathbf{s}, t). \tag{4.7}$$

- $\mathscr{L}_K$ is the Kawasaki (1966, 1972) generator that enters the definition of the original kinetic lattice gas. (It corresponds to the kinetic process in §2.1 for the DLG when the driving field is turned off.) That is, $\mathscr{L}_K$ performs stochastic exchanges $s_\mathbf{x} \rightleftarrows s_\mathbf{y}$ of the variables at NN sites $\mathbf{x}$ and $\mathbf{y}$ with a probability per unit time or rate of $c(\mathbf{s}; \mathbf{x}, \mathbf{y})$. More formally, one may write that

$$\mathscr{L}_K = \sum_{|\mathbf{x}-\mathbf{y}|=1} (\mathscr{K}_{\mathbf{xy}} - \mathbf{1})\, c(\mathbf{s}; \mathbf{x}, \mathbf{y}), \quad \mathscr{K}_{\mathbf{xy}} g(\mathbf{s}) = g(\mathbf{s}^{\mathbf{xy}}), \tag{4.8}$$

where $g(\mathbf{s})$ is an arbitrary function of the configuration, and $\mathbf{s}^{xy}$ is the configuration obtained from $\mathbf{s}$ after the exchange.

- $\mathscr{L}_G$ in (4.7) is a generator of the Glauber (1963) type, similar to the one that characterizes the ordinary kinetic version of the Ising model. That is, $\mathscr{L}_G$ produces stochastic changes of $s_\mathbf{x}$ to $-s_\mathbf{x}$ (*spin flips* in the magnetic language) at a rate $c(\mathbf{s}; \mathbf{x})$. One has formally that

$$\mathscr{L}_G = \sum_{\mathbf{x}} (\mathscr{G}_\mathbf{x} - \mathbf{1})\, c(\mathbf{s}; \mathbf{x}), \quad \mathscr{G}_\mathbf{x} g(\mathbf{s}) = g(\mathbf{s}^\mathbf{x}), \tag{4.9}$$

where $\mathbf{s}^x$ represents the configuration obtained from $\mathbf{s}$ after performing the indicated flip or *reaction*.

For simplicity, both rates are assumed to satisfy detailed balance,[7] but with respect to different temperatures, which implies the following familiar facts. On one hand,

$$\mathscr{L}_K \mu_{eq}^{T'}(\mathbf{s}; m) = 0, \tag{4.10}$$

and

$$\mu_{eq}^{T'}(\mathbf{s}; m) = \frac{1}{Z(\tilde{\mu})} \exp\left[-\frac{H(\mathbf{s})}{k_B T'} + \tilde{\mu} \sum_{\mathbf{x}} s_\mathbf{x}\right]. \tag{4.11}$$

Here $\tilde{\mu} = \tilde{\mu}(m)$ is an external field, $m \propto \partial \ln Z(\tilde{\mu})/\partial \tilde{\mu}$, and

$$Z(\tilde{\mu}) = \sum_{\mathbf{s}} \exp\left[-\frac{H(\mathbf{s})}{k_B T'} + \tilde{\mu} \sum_{\mathbf{x}} s_\mathbf{x}\right]. \tag{4.12}$$

$H(\mathbf{s})$ is the configurational energy that we assume for simplicity to be of the simplest Ising type, namely,

$$H(\mathbf{s}) = -J \sum_{|\mathbf{x}-\mathbf{y}|=1} s_\mathbf{x} s_\mathbf{y}. \tag{4.13}$$

---

[7] A more detailed discussion of rates is given in §§4.4 and 7.4.

On the other hand,

$$\mathscr{L}_G \mu_{eq}^T(\mathbf{s}) = 0, \tag{4.14}$$

and

$$\mu_{eq}^T(\mathbf{s}) = \left\{ \sum_{\mathbf{s}} \exp\left[-\frac{H(\mathbf{s})}{k_B T} \right] \right\}^{-1} \exp\left[-\frac{H(\mathbf{s})}{k_B T} \right]. \tag{4.15}$$

Properties (4.10) and (4.14) correspond to the two limiting canonical cases of the master equation (4.7). That is, the latter reduces to Glauber's master equation for the kinetic Ising model with a nonconserved order parameter, for either $p \equiv 0$ or $c(\mathbf{s}; \mathbf{x}, \mathbf{y}) \equiv 0$. In this case, any *reaction rate* $c(\mathbf{s}; \mathbf{x})$ in (4.9) that satisfies detailed balance drives the system to the same steady state: the Gibbs equilibrium state (4.15) corresponding to temperature T and energy $H(\mathbf{s})$. This undergoes (in the infinite volume limit which is considered throughout, except for MC models) a phase transition of second order at temperature $T_C^0 \geqslant 0$ for $d \geqslant 1$, where the equality holds for $d = 1$. On the other hand, one recovers the Kawasaki kinetic lattice-gas model, which corresponds to the Ising model with a conserved order parameter, for either $p \equiv 1$ or $c(\mathbf{s}; \mathbf{x}) \equiv 0$. Any *diffusion rate* $c(\mathbf{s}; \mathbf{x}, \mathbf{y})$ in (4.8) that satisfies detailed balance then drives the system to the Gibbs state (4.11) corresponding to energy $H(\mathbf{s})$, temperature T', and a fixed value m for the order parameter. As described in §2.1, the system exhibits in this case a critical point of the Ising class at $T' = T_C^0$ when $m = 0$.

In general, the steady state $\mu_{st}(\mathbf{s})$ which is implied for $t \to \infty$ by (4.7) under reaction–diffusion competition lies away from equilibrium. In a sense, the situation is similar to the one for the DLG. (The only steady current in the present case is the *local* energy current between the two baths in contact with each $s_{\mathbf{x}}$; see Droz, Rácz, & Schmidt (1989).) An interpretation of the model is that in addition to the familiar mechanism that induces magnetic ordering at low enough temperature, $\mathbf{s}$ is acted on by an external agent which induces diffusion, disturbing that order. The perturbing agent cannot be related to the heat bath at T nor included in the Hamiltonian. In fact, as for the DLG, the condition of detailed balance does not hold in (4.7) because there is no single Hamiltonian with suitable properties — for example, short range — that plays in the expression for $\mu_{st}(\mathbf{s})$ the role played by $H(\mathbf{s})$ in equations (4.11) or (4.15); this is discussed in detail for a related model in chapter 7. Therefore, the steady state is not unique but in general depends, even qualitatively, on the specific form of the reaction and diffusion rates. Under such conditions, the system may suffer instabilities and, consequently, exhibit nonequilibrium phase transitions. One may expect the latter to be related to the equilibrium case for $p = 0$ or 1. However, it seems reasonable to expect the states to depend on the values for the model parameters p and $T - T'$, in addition to the more familiar dependence on d, J, and T, and on the symmetries of the microscopic dynamical process. The nonequilibrium ordering exhibited by the reaction–diffusion lattice gas might be even more varied than for the DLG given the complex (competing) nature of the dynamical process in the former case.

4.3 Transition to hydrodynamics

The preceding discussion reveals one reason why theoreticians have been motivated to explore the implications of (4.7)–(4.15): this is a microscopic process with reaction and diffusion, and a well-defined mathematical system in nonequilibrium statistical physics. It has been shown by de Masi, Ferrari, & Lebowitz (1986a; see also Lebowitz (1986)) that the master equation (4.7) may be transformed *rigorously* within appropriate limits into a macroscopic equation (4.5). More explicitly, when diffusion is completely random, i.e., as $T' \to \infty$, and is infinitely fast compared to reactions, i.e., as $p \to 1$, the order parameter that one measures on a macroscopic spatial scale satisfies (4.5). The argument may be stated as follows.[8]

Consider first the master equation (4.7) with $T' \to \infty$, i.e., $c(\mathbf{s}; \mathbf{x}, \mathbf{y}) = 1$, so that exchanges occur completely at random. Denote by ϵ^{-2} the rate of exchanges relative to reactions; space is then to be scaled with ϵ^{-1}. An appropriate macroscopic order parameter or magnetization density may be obtained from the microscopic variable $s_{\mathbf{x}}$. With this aim, consider a partition of the (infinite) d-dimensional lattice into cubic boxes of side ϵ^{-1} centered on $\mathbf{r}$; let $\Omega_{\mathbf{r},\epsilon}$ denote the part of the original lattice lying inside this box. Define a new random variable on this scale,

$$m_\epsilon(\mathbf{r}; \mathbf{s}) = \epsilon^d \sum_{\mathbf{x} \in \Omega_{\mathbf{r},\epsilon}} s_{\mathbf{x}}(t), \tag{4.16}$$

whose probability distribution depends on the initial state. Assume that

$$\lim_{\epsilon \to 0} \frac{1}{\epsilon^d} \left| \langle s_{\mathbf{x}} \rangle_\epsilon - m_o(\epsilon \mathbf{x}) \right| \to 0, \tag{4.17}$$

where $m_o(\epsilon \mathbf{x})$ is a smooth function of $\mathbf{r}$, $|m_o(\epsilon \mathbf{x})| \leqslant 1$. Under these conditions, the following results hold in the limit $\epsilon \to 0$:

- The quantity $m_\epsilon(\mathbf{r}; \mathbf{s})$ at time t transforms into the integral over the unit box at $\mathbf{r}$ of a deterministic, nonfluctuating function, $m(\mathbf{r}, t)$.

[8] This section is based upon Lebowitz (1986), Garrido, Marro, & González-Miranda (1989), and Spohn (1989); see also Eyink *et al.* (1996).

- $m(\mathbf{r},t)$ satisfies the reaction–diffusion equation:

$$\frac{\partial m(\mathbf{r},t)}{\partial t} = \frac{1}{2}\nabla^2 m(\mathbf{r},t) + F\left[m(\mathbf{r},t)\right]. \tag{4.18}$$

The initial condition here is $m(\mathbf{r},0) = m_o(\mathbf{r})$, and

$$F\left[m(\mathbf{r},t)\right] = -2\left\langle s_{\mathbf{x}} c(\mathbf{s};\mathbf{x})\right\rangle_{m(\mathbf{r},t)} . \tag{4.19}$$

- The indicated average is with respect to the Bernoulli product distribution, in which $\langle s_{\mathbf{x}}\rangle_m = m$ for all $\mathbf{x}$. This corresponds to the *local equilibrium state* for a system at infinite temperature with macroscopic magnetization $m(\mathbf{r},t)$.

As indicated above, one is also interested in the (microscopic) fluctuations about the deterministic solutions $m(\mathbf{r},t)$ of (4.18). It follows that:

- The microscopic state, i.e., the probability distribution at time t deviates from local equilibrium as described by the Bernoulli product.
- The fluctuations form a random Gaussian field of amplitude $\epsilon^{d/2}$, and the correlation function $C(\mathbf{r},\mathbf{r}';t)$ satisfies a linear inhomogeneous equation whose coefficients depend on the solution of (4.18)–(4.19).

We refer the reader to de Masi *et al.* (1986a) for the proof of this important theorem. Here we describe a nonrigorous derivation of the same result, which also provides some information for finite T'. First, one multiplies (4.7) by $s_{\mathbf{x}}$ and sums over configurations to obtain:

$$\frac{\partial\left\langle s_{\mathbf{x}}\right\rangle}{\partial t} = -2\left\langle s_{\mathbf{x}} c(\mathbf{s};\mathbf{x})\right\rangle + \frac{1}{2\epsilon^2}\sum_{\mathbf{s}} s_{\mathbf{x}}\mathscr{L}_K \mu(\mathbf{s};t). \tag{4.20}$$

Here, $\frac{1}{2}\epsilon^{-2}$ is the relative *a priori* rate for diffusion, and $\langle\cdots\rangle$ indicates an average with $\mu(\mathbf{s};t)$. The last term in (4.20) may be transformed using (4.8) to write

$$\frac{\partial\left\langle s_{\mathbf{x}}\right\rangle}{\partial t} = -2\left\langle s_{\mathbf{x}} c(\mathbf{s};\mathbf{x})\right\rangle + \frac{1}{2\epsilon^2}\sum_{i=1}^{d}\sum_{\mathbf{z}=\mathbf{x}\pm\mathbf{i}}\left\langle\left(s_{\mathbf{z}} - s_{\mathbf{x}}\right) c(\mathbf{s};\mathbf{x},\mathbf{z})\right\rangle . \tag{4.21}$$

Here, $\mathbf{x}\pm\mathbf{i}$ are the NNs of $\mathbf{x}$ along the $\pm\hat{\mathbf{i}}$ directions, with $i = 1,2,\ldots,d$. As indicated above, it is convenient to have the system partitioned, and define the mean magnetization within the box $\Omega_{\mathbf{r},\epsilon}$, $m_\epsilon(\mathbf{r};\mathbf{s})$. This is determined by competition between the reactions occurring within the box and the exchanges with the neighboring boxes. The two mechanisms may be expected to produce variations of the same order of magnitude in m_ϵ. More precisely, the variation induced by reactions is of order of the volume of the box, ϵ^{-d}, and the one induced by exchanges is a surface effect of order $\epsilon^{-(d-1)}$ times the gradient of the magnetization involved at each exchange, $1/\epsilon^{-1}$, times the explicit factor $\frac{1}{2}\epsilon^{-2}$. Consequently, one obtains from (4.21) that

$$\frac{\partial m_\epsilon(\mathbf{r};t)}{\partial t} = -2\epsilon^d \sum_{\mathbf{x}\in\Omega_{\mathbf{r},\epsilon}} \langle s_{\mathbf{x}} c(\mathbf{s};\mathbf{x})\rangle + \frac{1}{2}\epsilon^{d-2} \sum_{i=1}^{d} \sum_{\mathbf{x}\in\Omega_{\mathbf{r},\epsilon}} \sum_{\mathbf{z}=\mathbf{x}\pm\mathbf{i}} \langle (s_{\mathbf{z}} - s_{\mathbf{x}})\, c(\mathbf{s};\mathbf{x},\mathbf{z})\rangle, \tag{4.22}$$

where $m_\epsilon(\mathbf{r};t) \equiv \langle m_\epsilon(\mathbf{r};\mathbf{s})\rangle$.

Next, one introduces new variables for both time and space, namely, $\tau = \epsilon^{-2}t$ for temporal observations, and an *observable lattice spacing* $a = \epsilon^{-1}a_0$, where a_0 is the original lattice spacing. Then one considers the limit $\epsilon \to 0$, $t \to 0$, and $a_0 \to 0$, with τ and a remaining finite (in fact, $a = 1$), which corresponds to the limiting process from discrete to continuum mentioned above. In this limit, the boxes $\Omega_{\mathbf{r},\epsilon}$ develop a macroscopic size, and $m_\epsilon(\mathbf{r};t) \equiv \langle m_\epsilon(\mathbf{r};\mathbf{s})\rangle$ transforms into the macroscopic deterministic variable $m(\mathbf{r},t)$. Simultaneously, $\epsilon \to 0$ makes the diffusion very fast as compared to the reaction process, i.e., an infinite number of exchanges occur in the box for each reaction, and the number of exchanges through the surface of the box becomes small. Under these circumstances, in time units of ϵ^{-2}, one expects local equilibrium within the box with respect to the fast diffusion process. More precisely, the original probability distribution satisfies (4.7) or, equivalently,

$$\frac{\partial \mu_\epsilon}{\partial t} = \mathscr{L}_G \mu_\epsilon + \left(2\epsilon^2\right)^{-1} \mathscr{L}_K \mu_\epsilon. \tag{4.23}$$

Let us assume that the solution of this is

$$\mu_\epsilon(\mathbf{s};t) = \mu_0(\mathbf{s};t) + \epsilon^\alpha \nu(\mathbf{s};t) + \mathscr{O}\left(\epsilon^\beta\right), \tag{4.24}$$

with $\nu(\mathbf{s};t)$ independent of ϵ, and $\beta > \alpha > 0$. One has to leading order that

$$\frac{\partial \mu_0}{\partial t} = \left(\mathscr{L}_G + \frac{1}{2\epsilon^2}\mathscr{L}_K\right)\mu_0 + \frac{1}{2}\epsilon^{\alpha-2}\mathscr{L}_K \nu + \mathscr{O}\left(\epsilon^{\beta-2}\right). \tag{4.25}$$

This reveals that, for small enough ϵ, there will only be a well-behaved solution if $\mathscr{L}_K \mu_0(\mathbf{s};t) = 0$; see (4.10). Therefore, (4.10) and the fact that $\partial\mu/\partial t = \mathscr{L}_K \mu$ has the stationary solution $\mu_{eq}^{T'}(\mathbf{s};m)$ suggest that the relevant probability distribution in the limit $\epsilon \to 0$, assuming that surface terms are negligible, is simply

$$\mu_0(\mathbf{s};t) = \lim_{\epsilon\to 0} \mu_\epsilon(\mathbf{s};t) = \prod_{\mathbf{r}} \mu_{eq}^{T'}(\mathbf{s}_{\mathbf{r}}; m(\mathbf{r};t)). \tag{4.26}$$

The product is over all system points $\mathbf{r}$ or boxes at $\mathbf{r}$, and the vector $\mathbf{s_r}$ represents the configuration within the latter. The result (4.26) corresponds to the existence of local equilibrium characterized by local magnetization $m(\mathbf{r};t)$.

The probability distributions (4.24) and (4.26) are the ones used to compute the averages involved in (4.22) when ϵ is small enough or zero, respectively. In order to recover the result of de Masi *et al.*, one assumes $\alpha > 1$ (see below). It follows *to leading order* in ϵ that

$$\frac{\partial m(\mathbf{r};t)}{\partial t} = -2\left\langle s_{\mathbf{x}} c(\mathbf{s};\mathbf{x})\right\rangle_{m(\mathbf{r};t)} + \frac{1}{2\epsilon^2}\sum_{i=1}^{d}\sum_{\mathbf{r}'=\mathbf{r}\pm a\mathbf{i}} C\left(m(\mathbf{r};t), m(\mathbf{r}';t)\right) + \mathcal{O}(\epsilon). \tag{4.27}$$

Here,

$$C\left(m(\mathbf{r};t), m(\mathbf{r}';t)\right) = \left\langle j\left(\mathbf{r},\mathbf{r}';\mathbf{s}\right)\right\rangle_0 \equiv \left\langle (s_{\mathbf{z}} - s_{\mathbf{x}})\, c(\mathbf{s};\mathbf{x},\mathbf{z})\right\rangle_0, \tag{4.28}$$

where $\mathbf{x} \in \Omega_{\mathbf{r},\epsilon}$ and $\mathbf{z} \in \Omega_{\mathbf{r}',\epsilon}$, represents the mean current between two macroscopic locations with local magnetizations $m(\mathbf{r};t)$ and $m(\mathbf{r}';t)$, respectively, and $\langle\cdots\rangle_m \equiv \sum_{\mathbf{s}}\cdots\mu_{eq}^{T'}(\mathbf{s};m)$, $\langle\cdots\rangle_0 \equiv \sum_{\mathbf{s}}\cdots\mu_0(\mathbf{s};t)$. Now, take $a = \epsilon^{-1}a_0 = 1$, and the lengths scaled by ϵ, i.e., $\mathbf{i}$ is replaced by $\epsilon\mathbf{i}$. Equation (4.27) leads in the limit $\epsilon \to 0$ (assuming that $\mathcal{O}(\epsilon)$ vanishes) to the macroscopic equation:

$$\begin{aligned}\frac{\partial m(\mathbf{r};\tau)}{\partial \tau} &= -2\left\langle s_{\mathbf{x}} c(\mathbf{s};\mathbf{x})\right\rangle_{m(\mathbf{r};\tau)} \\ &\quad + \tfrac{1}{2}\sum_{i=1}^{d}\frac{\partial}{\partial r_i}\left[\left.\frac{\partial C(u,v)}{\partial v}\right|_{\substack{u=v=\\ m(\mathbf{r};\tau)}}\frac{\partial m(\mathbf{r};\tau)}{\partial r_i}\right],\end{aligned} \tag{4.29}$$

where $\mathbf{r} = \sum_{\mathbf{i}} r_{\mathbf{i}}\hat{\mathbf{i}}$. The coefficients in this equation are evaluated in the local equilibrium ensemble (4.26); they depend, in general, on the rates of both reaction and diffusion mechanisms.

The preceding derivation requires a comment. We first summarize some important points, and state the range of validity of the assumptions involved. The system has been partitioned into cubes $\Omega_{\mathbf{r},\epsilon}$, of side $a_0\epsilon^{-1}$. The mean magnetization $m_\epsilon(\mathbf{r};t)$ at each cube is determined by competition between reactions inside the cube and exchanges with neighboring cubes; each produces variations of m_ϵ of the same order of magnitude. Therefore, $T' > T_C$ is required; otherwise, diverging correlations might invalidate some of the arguments (and inhomogeneities would occur that may not be described correctly by (4.29)). Also, as $\epsilon \to 0$, the sites $\mathbf{x}$ form a dense set on the scale of $\mathbf{r}$, m_ϵ transforms into the fully macroscopic variable $m(\mathbf{r};t) = \langle s_{\mathbf{x}}\rangle$, where $\mathbf{x} \in \Omega_{\mathbf{r},\epsilon}$, and the dominant probability distribution is (4.26). If no corrections to the latter were necessary, very fast exchanges would bring the system to a sort of local equilibrium at temperature T' in which the relevant quantities depend on $m(\mathbf{r};\tau)$, which satisfies (4.29). It is nevertheless likely that $\alpha \leqslant 1$, contrary to the trial assumption above. An extra term then arises in (4.29) that follows from the $\mathcal{O}(\epsilon)$ terms in (4.27). That is, while the reaction term in (4.29) is the correct one, the diffusion term involves an approximation. More explicitly, the precise structure of the resulting equation is

$$\frac{\partial m(\mathbf{r};\tau)}{\partial \tau} = F(m) + \frac{\partial}{\partial \mathbf{r}}\left[\Delta(m)\frac{\partial m}{\partial \mathbf{r}}\right], \tag{4.30}$$

where $\Delta(m)$ is the bulk diffusion coefficient given by the Green–Kubo formula; see Spohn (1982). A formal expression for this coefficient follows from the last term in (4.27) as

$$\Delta(m) = \left.\frac{\partial C(u,v)}{\partial v}\right|_{u=v=m(\mathbf{r};\tau)} + \left[\frac{\partial m(\mathbf{r};\tau)}{\partial r_i}\right]^{-1} \int dr_i \lim_{\epsilon\to 0} \frac{\langle j(\mathbf{r},\mathbf{r}+\epsilon\mathbf{i};\mathbf{s})\rangle_1 - \langle j(\mathbf{r}-\epsilon\mathbf{i},\mathbf{r};\mathbf{s})\rangle_1}{2\epsilon^{2-\alpha}}. \tag{4.31}$$

Here, $j(\mathbf{r},\mathbf{r}';\mathbf{s})$ is defined in (4.28), and $\langle\cdots\rangle_1 \equiv \sum_{\mathbf{s}}\cdots v(\mathbf{s};t)$; see equation (4.24).[9] The derivation provides no expression for $v(\mathbf{s};t)$, however, so that only the first part of $\Delta(m)$ in (4.31) is known explicitly. If the second part is considered, the macroscopic equation for $m(\mathbf{r};t)$ turns out to depend on the reaction and diffusion mechanisms, and on the correction $v(\mathbf{r};t)$ to local equilibrium. Some information may be obtained for the latter, i.e., for the second part of $\Delta(m)$ from the implicit equation

$$\mathscr{L}_K v(\mathbf{s};t) = -2\mathscr{L}_G \mu_0(\mathbf{s};t) \tag{4.32}$$

that follows from (4.25) by using a procedure similar to the one above. Alternatively, one may neglect the second part of (4.31); the result is a mean-field approximation of the kind described in chapter 3.

In any event, (4.29) has two interesting features. It can be used to obtain information about homogeneous states for finite T' as described in the next section. It also leads correctly to the result of de Masi *et al.* That is, in the limit $\epsilon \to 0$, the probability distribution (4.26) reduces as $T' \to \infty$ to the factorized one:

$$\mu_{eq}^0(\mathbf{s} : m(\mathbf{r};t)) = \frac{1}{Z}\exp\left(\tilde{\mu}\sum_{\mathbf{x}\in\Omega_{\mathbf{r},\epsilon}} s_{\mathbf{x}}\right), \tag{4.33}$$

[9] The expression for $\Delta(m)$ may depend on the specific relevant probability distribution.

where $\tilde{\mu} = \tilde{\mu}(m)$ is defined in (4.10). The probability (4.33) is precisely the Bernoulli distribution which has been shown rigorously to occur strictly for $T' \to \infty$, i.e., when the infinite temperature diffusion prevents any local correlation, as if the fluid were strongly stirred with a stick by the observer. Moreover, the last term in (4.31) vanishes for $T' \to \infty$, and one recovers the exact equation (4.18)–(4.19).

4.4 Nonequilibrium macroscopic states

This section illustrates the macroscopic behavior implied by reaction diffusion equations for lattice systems. Some exact results for both random diffusion and the case of finite T' are described. The focus is on mean densities m for steady homogeneous states, but correlations in some interesting inhomogeneous cases are also described.

Homogeneous states for random diffusion

Consider a homogeneous situation, $m(\mathbf{r};t) = m(t)$, in the random-diffusion limit, $T' \to \infty$, $(c(\mathbf{s};\mathbf{x},\mathbf{y}) \equiv 1)$. Equations (4.18)–(4.19) reduce to

$$\frac{\partial m(t)}{\partial t} = F(m) = -2\left\langle s_{\mathbf{x}} c(\mathbf{s};\mathbf{x})\right\rangle_m . \tag{4.34}$$

Consequently, the steady solution, m^*, corresponds to

$$F\left(m^*\right) = 0, \qquad \left.\frac{\partial F(m)}{\partial m}\right|_{m=m^*} \leqslant 0, \tag{4.35}$$

where the second condition guaranties local stability. First, we remark that some general results may be derived from (4.35). With this aim, it is convenient to transform the expression for $F(m)$ that holds for $T' \to \infty$.

The Hamiltonian (4.13) may be written as

$$H(\mathbf{s}) = -K \sum_{\{\mathbf{x},\mathbf{y};|\mathbf{x}-\mathbf{y}|=1\}} s_{\mathbf{x}} s_{\mathbf{y}} = H_0(\mathbf{s}) - K s_{\mathbf{x}} \sum_{\{\mathbf{y};|\mathbf{x}-\mathbf{y}|=1\}} s_{\mathbf{y}}, \tag{4.36}$$

where $K \equiv J/k_B T$; the second sum is over the q neighbors of site $\mathbf{x}$ (i.e., q is the *lattice coordination number*), and $H_0(\mathbf{s})$ is independent of $s_{\mathbf{x}}$. One may write quite generally that

$$c(\mathbf{s};\mathbf{x}) = f_0(\mathbf{s}) \exp\left[-K s_{\mathbf{x}} \sum_{\{\mathbf{y};|\mathbf{x}-\mathbf{y}|=1\}} s_{\mathbf{y}}\right]. \tag{4.37}$$

It is also convenient to write

$$c(\mathbf{s};\mathbf{x}) = f(\mathbf{s})\left[A_+(\mathbf{s}) + s_{\mathbf{x}} A_-(\mathbf{s})\right]. \tag{4.38}$$

For the lattices of interest,

$$f(\mathbf{s}) = f_0(\mathbf{s}) \prod_{i=1}^{d} \cosh\left[K\left(s_{\mathbf{x}+\mathbf{i}} + s_{\mathbf{x}-\mathbf{i}}\right)\right], \tag{4.39}$$

where $f(\mathbf{s})$ and $f_0(\mathbf{s})$ are arbitrary except that both are strictly positive, and invariant under the change $\mathbf{s} \to \mathbf{s}^{\mathbf{x}}$ if they need to satisfy detailed balance, namely, $c(\mathbf{s};\mathbf{x})/c(\mathbf{s}^{\mathbf{x}};\mathbf{x}) = \exp\{-[H(\mathbf{s}^{\mathbf{x}}) - H(\mathbf{s})]\}$. Furthermore,

$$A_{\pm}(\mathbf{s}) \equiv \frac{1}{2}\prod_{i=1}^{d}\left[1 - \frac{\alpha}{2}\left(s_{\mathbf{x}+\mathbf{i}} + s_{\mathbf{x}-\mathbf{i}}\right)\right] \pm \frac{1}{2}\prod_{i=1}^{d}\left[1 + \frac{\alpha}{2}\left(s_{\mathbf{x}+\mathbf{i}} + s_{\mathbf{x}-\mathbf{i}}\right)\right], \tag{4.40}$$

with $\alpha \equiv \tanh(2K)$, have no explicit dependence on $s_{\mathbf{x}}$. Consequently, one obtains

$$F(m) = -2m\left\langle f(\mathbf{s})\, A_{+}(\mathbf{s})\right\rangle_m - 2\left\langle f(\mathbf{s})\, A_{-}(\mathbf{s})\right\rangle_m . \tag{4.41}$$

A number of interesting general facts may be proved by simple algebra from this result.

The linear system

Consider a one-dimensional system with spatially symmetric NN attractive ($K > 0$) interactions. A rather general realization for the rate is

$$f(\mathbf{s}) = 1 + \sigma s_{x+1}s_{x-1} + \sigma_2(s_{x+1} + s_{x-1}). \tag{4.42}$$

Several familiar cases are subsumed in this expression: $\sigma \neq 0$ and $\sigma_2 = 0$ corresponds to a rate considered by de Masi *et al.*, and to the Metropolis algorithm, while $\sigma = \sigma_2 = 0$ corresponds to the rates introduced by Glauber (1963) and Kawasaki (1972). One has for (4.42) that

$$F(m) = \sigma_2\alpha + m\left[\alpha(\sigma+1) - 1\right] + m^2\sigma_2(\alpha - 2) - m^3\sigma, \tag{4.43}$$

where $|\sigma| \leqslant 1$ and $|\sigma_2| \leqslant \frac{1}{2}(1+\sigma)$ is implied by the positivity of $f(\mathbf{s})$, and the condition of stability in (4.35) requires that

$$\alpha(\sigma+1) - 1 + 2\sigma_2(\alpha-2)\,m^* - 3\sigma\left(m^*\right)^2 < 0, \tag{4.44}$$

where m^* follows from $F(m^*) = 0$.

For $\sigma_2 \equiv 0$ in (4.42), the system exhibits a second-order phase transition if and only if $\sigma = \sigma(K) > 0$. The critical temperature, $T_C = J/k_B K_C$, satisfies

$$\frac{1}{\tanh(2K_C)} - 1 = \sigma(K_C) \equiv \sigma_C. \tag{4.45}$$

The case $T_C = 0$ is not excluded; it occurs, for example, for the Metropolis algorithm (see below). The magnetization is

$$m^* = \begin{cases} 0 & \text{for} \quad \sigma < \sigma_C \\ \pm\sqrt{\alpha + (\alpha - 1)/\sigma} & \text{for} \quad \sigma \geqslant \sigma_C \end{cases} . \tag{4.46}$$

For *very fast* ($\epsilon \to 0$) and completely random ($T' \to \infty$) diffusion, the phase transition is of the classical (Landau) type.

The same system, i.e., (4.42), with $\sigma \equiv 0$ and $|\sigma_2| \leqslant \frac{1}{2}$ behaves differently. No phase transition occurs for any value of σ_2, but the magnetization is given by

$$m^* = \frac{\alpha - 1 + \sqrt{(1-\alpha)^2 + 4\sigma_2^2 \alpha (2-\alpha)}}{2\sigma_2 (2-\alpha)}. \tag{4.47}$$

That is, m^* is always nonzero, and has the sign of σ_2.

Two-dimensional systems

Consider a two-dimensional system with $K > 0$, and the general rate

$$f(\mathbf{s}) = \sum_{k=1}^{5} \sigma_k \alpha_k (\mathbf{s}) . \tag{4.48}$$

Here,

$$\begin{aligned} &\alpha_0 = \sigma_0 = 1, \quad \alpha_1 = S_\mathbf{i} S_\mathbf{j}, \qquad \alpha_2 = P_\mathbf{i} P_\mathbf{j}, \\ &\alpha_3 = S_\mathbf{i} + S_\mathbf{j}, \quad \alpha_4 = P_\mathbf{i} + P_\mathbf{j}, \quad \alpha_5 = S_\mathbf{i} P_\mathbf{j} + S_\mathbf{j} P_\mathbf{i}, \end{aligned} \tag{4.49}$$

with

$$S_\mathbf{n} \equiv \frac{1}{2}(s_{\mathbf{x}+\mathbf{n}} + s_{\mathbf{x}-\mathbf{n}}), \quad P_\mathbf{n} \equiv s_{\mathbf{x}+\mathbf{n}} s_{\mathbf{x}-\mathbf{n}}, \quad \mathbf{n} = \mathbf{i}, \mathbf{j}. \tag{4.50}$$

It follows that

$$F(m) = m f_1(m^2), \quad f_1(X) = A + BX + CX^2, \tag{4.51}$$

is the most general function F having all solutions of the form $(m^*)^2 = \sigma \geqslant 0$, and (4.51) occurs for $\sigma_3 = \sigma_5 = 0$ in (4.48). If the latter holds, one obtains

$$\left.\begin{aligned} &A = 2(2\alpha - 1) + \alpha\left(2 - \tfrac{1}{2}\alpha\right)\sigma_1 + 4\alpha\sigma_4, \\ &B = -2\alpha^2 - \left(\alpha^2 - 2\alpha + 2\right)\sigma_1 + 2\alpha(2-\alpha)\sigma_2 - 4\left(\alpha^2 - \alpha + 1\right)\sigma_4, \\ &C = -2\sigma_2 - \tfrac{1}{2}\sigma_1\alpha^2, \end{aligned}\right\} \tag{4.52}$$

relating K and the coefficients in (4.48) and (4.51). Stability implies that

$$\left.\begin{aligned} &A m^* + B\left(m^*\right)^3 + C\left(m^*\right)^5 = 0, \\ &A + 3B\left(m^*\right)^2 + 5C\left(m^*\right)^4 \leqslant 0. \end{aligned}\right\} \tag{4.53}$$

It is convenient to define two temperatures, $T_C^{(1)}$ and $T_C^{(2)}$, at which the solutions $m^* = 0$ and $m^* \neq 0$, respectively, become unstable, as follows:

$$\left.\begin{aligned} & A\left(T_C^{(1)}\right) = 0, \\ & A\left(T_C^{(2)}\right) + B\left(T_C^{(2)}\right)(m^*)^2 + C\left(T_C^{(2)}\right)(m^*)^4 = 0, \\ & A\left(T_C^{(2)}\right) + 3B\left(T_C^{(2)}\right)(m^*)^2 + 5C\left(T_C^{(2)}\right)(m^*)^4 = 0. \end{aligned}\right\} \tag{4.54}$$

For $T_C^{(1)} = T_C^{(2)}$, either $m^* = 0$ (the transition is of second order) or else $m^* \neq 0$ if $B(T_C^{(1)}) + C(T_C^{(1)}) = 0$ (the transition is of first order), while m^* is always nonzero (corresponding to a first-order transition) for $T_C^{(1)} \neq T_C^{(2)}$.

Influence of the reaction rate

The case of repulsive interactions, $K < 0$, is not very interesting for $\epsilon \to 0$. It may be shown that $m^* = 0$ for any dimension d and any reaction rate that depends only on the q neighbors of $s_{\mathbf{x}}$ and has the symmetry $c(\mathbf{s};\mathbf{x}) = c(-\mathbf{s};\mathbf{x})$, where $-\mathbf{s} = \{s_{\mathbf{x}} \to -s_{\mathbf{x}}, \forall \mathbf{x}\}$; a proof of this is in Garrido *et al.* (1989).

Consider the case of arbitrary d and $K > 0$ for some specific realizations of the reaction rate:

(i) The Glauber (1963) rate corresponds to $f(\mathbf{s})$ =const.> 0 in (4.38). One obtains from (4.41) that

$$F(m) = (1+\alpha m)^d (1-m) - (1-\alpha m)^d (1+m) \equiv m f_2\left(m^2\right). \tag{4.55}$$

Therefore, $m = 0$ for $\alpha < \alpha_C$, where

$$\alpha_C \equiv \alpha\left(T_C^{(1)}\right) \equiv \tanh\left(2J/k_B T_C^{(1)}\right) = \frac{1}{d}. \tag{4.56}$$

That is, $T_C = T_C^{(1)} = T_C^{(2)} = 0$ for $d = 1$, and there is a second-order phase transition at $T_C > 0$ for any $d > 1$. One obtains $T_C^{(1)} \to \infty$ as $d \to \infty$, and the critical behavior is always classical.

(ii) As indicated above, one may *force* the one-dimensional system to present a phase transition by using a rate function that favors low energy states. That is, consider

$$c(\mathbf{s};\mathbf{x}) = 1 - \upsilon s_x (s_{x-1} + s_{x+1}) + \upsilon^2 s_{x-1} s_{x+1}, \tag{4.57}$$

$\upsilon \equiv \tanh K$ as in (4.42) with $\sigma_2 \equiv 0$. Then

$$\frac{\partial m(r;t)}{\partial t} = \frac{\partial^2 m}{\partial r^2} - \frac{\mathrm{d}}{\mathrm{d}m}\left[\frac{1}{2}\upsilon^2 m^4 - (2\upsilon - 1)m^2\right], \tag{4.58}$$

where r is the position along the line. Equation (4.58) has a unique solution for any initial magnetization, $m(r,0)$. This implies that $m^* = 0$ is a time-independent solution that is stable for $0 \leqslant \upsilon < \upsilon_C = \frac{1}{2}$. It becomes unstable at $\upsilon = \upsilon_C$, and two homogeneous solutions arise for $\upsilon > \upsilon_C$ with magnetizations $m^* = \pm \upsilon^{-1}\sqrt{1-\upsilon/\upsilon_C}$.

The two-site correlation function corresponding to the fluctuating field described in §4.3 is

$$C\left(r,r';t\right) = \left[1 - m^2(r,t)\right]\delta\left(r-r'\right) + 8\upsilon \int_0^t d\tau \sqrt{8\pi\tau}\exp\left[-\frac{\left(r-r'\right)^2}{8\tau} - 4\left(1-\frac{\upsilon}{\upsilon_C}\right)\tau\right] \quad (4.59)$$

in the one-phase region. For $\upsilon < \upsilon_C$, this transforms as $t \to \infty$ into $\upsilon\sqrt{\upsilon_C - \upsilon} \times \exp\left[-2\left(\upsilon_C - \upsilon\right)|r-r'|\right]$. Consequently, the fluctuations grow exponentially with time for $\upsilon > \upsilon_C$ while they grow like $t^{1/2}$ for $\upsilon = \upsilon_C$. The details of the long-time behavior are rather involved, requiring one to go beyond linear Gaussian fluctuations. The long-time evolution of the microscopic state, before taking the hydrodynamic limit, departs from the prediction of (4.58), i.e., $m(r;t) = 0$. We refer the reader to de Masi, Presutti, & Vares (1986) for details.

The rate (4.57) may be generalized to arbitrary dimension as

$$f(\mathbf{s}) = \prod_{i=1}^{d}\left(1 + \upsilon^2 s_{\mathbf{x}+\mathbf{i}} s_{\mathbf{x}-\mathbf{i}}\right). \quad (4.60)$$

One has for this case that

$$F(m) = (1+\upsilon m)^{2d}(1-m) - (1-\upsilon m)^{2d}(1+m). \quad (4.61)$$

This agrees with (4.55) if one substitutes $d \to 2d$ and $J \to \frac{1}{2}J$ in this relation; the critical value is now $\upsilon_C = \tanh(2K_C) = (2d)^{-1}$.

(iii) In order to illustrate further the effect of the reaction rate on the steady state, one may consider $f(\mathbf{s})$ independent of $s_\mathbf{x}$ and its NNs. One finds that $F(m) = \langle f(\mathbf{s})\rangle_m F^{(i)}(m)$, where $F^{(i)}(m)$ is the polynomial obtained for case (i), equation (4.55). In addition to the solutions for the latter, there follow now the solutions of $\langle f(\mathbf{s})\rangle_m = 0$ that may be as varied as one wishes by choosing $f(\mathbf{s})$ appropriately.

(iv) The rate introduced by Kawasaki (1972) may be defined as

$$c(\mathbf{s};\mathbf{x}) = \left\{1 + \exp[H(\mathbf{s}^\mathbf{x}) - H(\mathbf{s})]\right\}^{-1}, \quad (4.62)$$

i.e., $f(\mathbf{s}) = A_+^{-1}$. Consequently, $F(m) = -2m - 2\left\langle A_-(\mathbf{s})\,A_+^{-1}(\mathbf{s})\right\rangle_m$, and one obtains $F(m) = 2m(\alpha - 1)$ for $d = 1$, and

$$F(m) = -m(2 - 8B_1^+ - 8B_1^- m^2), \quad B_1^\pm = \frac{1}{8}[\tanh(4K) \pm 2\alpha] \quad (4.63)$$

for $d = 2$. Therefore, $m^* = 0$ for $d = 1$, while one has $m^* = 0$ if $B_1 \leqslant \frac{1}{4}$, and $m^* = \pm\sqrt{(\alpha^3 - \alpha^2 + 2\alpha - 1)\,\alpha^{-3}}$ if $B_1 \geqslant \frac{1}{4}$ for $d = 2$. The latter implies a transition of second order with T_C the solution of $\alpha_C^3 - \alpha_C^2 + 2\alpha_C = 1$, i.e., $K_C = 0.3236$. For $d = 3$,

$$F(m) = -2m[1 - 6B_1^+ - 20B_2m^2 - 6B_1^- m^4] \tag{4.64}$$

with $B_1^\pm = \frac{1}{32}[\tanh(6K) \pm 4\tanh(4K) + 5\alpha]$ and $B_2 = \frac{1}{32}[\tanh(6K) - 3\alpha]$: a second-order transition occurs at T_C such that $3\alpha_C^5 - 3\alpha_C^4 + 9\alpha_C^3 - 4\alpha_C^2 + 3\alpha_C = 1$, i.e., $K_C = 0.197$. (The explicit computation of the relevant averages for this and for the next case may be found in the appendix of Garrido *et al.* (1989).)

(v) The Metropolis rate, $c(\mathbf{s};\mathbf{x}) = \min\{1, \exp[-H(\mathbf{s}^\mathbf{x}) + H(\mathbf{s})]\}$, corresponds to

$$f(\mathbf{s}) = \exp\left[-K\left|\sum_{i=1}^{d}(s_{\mathbf{x}+\mathbf{i}}s_{\mathbf{x}-\mathbf{i}})\right|\right]\prod_{i=1}^{d}\cosh\left[K(s_{\mathbf{x}+\mathbf{i}}s_{\mathbf{x}-\mathbf{i}})\right]. \tag{4.65}$$

One gets from (4.41) that

$$F(m) = 2m\left[2B_1 - B_2^+ - B_2^- m^2\right], \tag{4.66}$$

where

$$\left.\begin{aligned} B_1 &= \tfrac{1}{2}\kappa^{1/2}\sinh(2K),\\ B_2^\pm &= \tfrac{1}{2}[\kappa^{1/2}\cosh(2K) \pm 1], \end{aligned}\right\} \tag{4.67}$$

and $\kappa \equiv \exp(-4K)$, for $d = 1$, i.e., the only stable solution is $m^* = 0$, and no phase transition occurs. For $d = 2$,

$$F(m) = 2m[4B_1 - B_2 + m^2(4B_3 - 6B_4) - m^4B_5], \tag{4.68}$$

where $16B_1 = 3 - 2\kappa - \kappa^2$, $16B_2 = \kappa^2 + 4\kappa + 11$, $16B_3 = -(1-\kappa)^2$, $16B_4 = \kappa^2 - 1$, and $16B_5 = \kappa^2 - 4\kappa + 3$. Thus $m^* = 0$ is stable for $\kappa > \kappa_C^{(1)}$, where $5(\kappa_C^{(1)})^2 + 12\kappa_C^{(1)} = 1$ (which yields $\kappa_C^{(1)} = 0.0806$) . In addition, there is a solution m_+ given by $m_+^2(3-\kappa) = 1 + 5\kappa + 2[(1 - 7\kappa + 3\kappa^2 - 5\kappa^3)(1-\kappa)^{-1}]$, which is stable for $\kappa < \kappa_C^{(2)}$, where $5(\kappa_C^{(2)})^3 - 3(\kappa_C^{(2)})^2 + 7\kappa_C^{(2)} = 1$ (i.e., $\kappa_C^{(2)} = 0.1501$). Therefore, the transition is discontinuous; the same is true for $d = 3$.

(vi) Consider an external magnetic field, so that the Hamiltonian is $H(\mathbf{s}) = -K\sum_{|\mathbf{x}-\mathbf{y}|=1} s_\mathbf{x}s_\mathbf{y} - h\sum_\mathbf{x} s_\mathbf{x}$, and

$$c(\mathbf{s};\mathbf{x}) = f(\mathbf{s})\left\{A_+\left[\cosh(h) - s_\mathbf{x}\sinh(h)\right] - A_-\left[\sinh(h) - s_\mathbf{x}\cosh(h)\right]\right\}. \tag{4.69}$$

Assume that $f(\mathbf{s})$ =const. as for (i), where the existence of a critical point at $\alpha_C = d^{-1}$ (for $h = 0$) was shown. It follows that $m \sim Ah^{1/\delta}$ as $h \to 0$, with $\delta = 3$, confirming the classical nature of the phase transition, and $A^3 = 3d^2/(d^2 - 1)$. Incidentally, the latter reveals that $d \to 1$ is again singular, and that $A \to 0$ as $d \to \infty$. On the other hand, one obtains

$$m(1 - 2\kappa + m^2\kappa^2) = \tanh(h)[1 - 2m^2\kappa + m^2\kappa^2] \tag{4.70}$$

for $f(\mathbf{s})$ as in (ii). For small fields, this reduces to $m(1 - 2\kappa) \approx h - m^3\kappa^2$; thus, $m = (4h)^{1/3} + \mathcal{O}(h)$ near κ_C, while the prediction far from κ_C is that $m \approx h(1 - 2\kappa)$ when m is small enough and $\kappa < \frac{1}{2}$.

Related lattice gases

We now describe two lattice gases that are closely related to the reaction–diffusion system in §4.3. One consists of particles on a line; in addition to jumping to a NN empty site at a rate of ϵ^{-2}, a particle may be created on such a site at rate $\frac{1}{2}$. This model is characterized by

$$c(\mathbf{s}; x) = \frac{1}{8}(1 - s_x)(2 + s_{x-1} + s_{x+1}); \tag{4.71}$$

see Ben Jacob *et al.* (1985), Lebowitz (1986), Bramson *et al.* (1986), Kerstein (1986). The macroscopic equation for this model is

$$\frac{\partial m(r;t)}{\partial t} = \frac{\partial^2 m}{\partial r^2} + \frac{1}{2}(1 - m^2). \tag{4.72}$$

Equations with this structure have been studied as a simple case admitting travelling front solutions for all speeds $|c| \geq c^*$ ($c^* = 2$ in the present case) of the form

$$u_c(r - ct) = \frac{1}{2}[1 + m(r,t)]; \tag{4.73}$$

here, $u_c(r)$ satisfies

$$u_c''(r) + cu_c'(r) + u_c(r)[1 - u_c(r)] = 0, \tag{4.74}$$

with $u_c(r) \to 1$ as $r \to -\infty$ and $u_c(r) \to 0$ as $r \to \infty$, for $c > 0$. This may model a one-dimensional cross-section of a flame front with convection if one lets $s_x = \pm 1$ represent burned and unburned cells, respectively. Moreover, (4.72) may be seen to contain a selection principle effect that may be relevant to understanding pattern selection, for example, in describing dendritic growth of a solid front moving into the melt (Ben Jacob *et al.* 1985).

In the second case of interest, the dynamics proceeds by exchanges only. If the rate is spatially symmetric, the associated macroscopic equation is

(4.18) with $F = 0$, i.e., a linear diffusion equation. If the rate is asymmetric and $d = 1$, the system follows Burger's equation for a simple asymmetric exclusion process,[10] i.e.,

$$\frac{\partial m(r;t)}{\partial t} = \frac{1}{2}(1-2p)\frac{\partial}{\partial r}(1-m^2), \tag{4.75}$$

which is the analog of the Euler equations for fluids. This result requires that the probability of jumps to the right is $q > \frac{1}{2}$, and that both time and space are scaled by ϵ^{-1}. In order to obtain a viscous dissipation term $\sim \partial^2 m/\partial r^2$ on the r.h.s. of (4.75) one would need to make the asymmetry $q - \frac{1}{2}$ of order ϵ, and to scale time like ϵ^{-2}. It is remarkable that such a difference in scaling needed for the nondissipative and dissipative cases also occurs in other representations of fluid behavior, as discussed by Lebowitz (1986).

Finite-temperature diffusion

We consider in the remainder of this section a diffusion process in the limit $\epsilon \to 0$ at a *finite* temperature T'. One has that $\partial m(t)/\partial t = F(m) = -2\langle s_{\mathbf{x}} c(\mathbf{s};\mathbf{x})\rangle$, with $m = \langle s_{\mathbf{x}}\rangle$, where the averages are to be computed with probability (4.26). Consider first the reaction rate $c(\mathbf{s};\mathbf{x}) = 1 - \frac{1}{2}\alpha s_x(s_{x+1} + s_{x-1})$ corresponding to rate (i) with $d = 1$. It follows that $F(m) = 2m(\alpha - 1)$, so that the only solution is $m^* = 0$. For $d = 1$, which is the only exactly solvable case, rate (ii) is more interesting. After some algebra, this leads to

$$F(m) = -2\left[m(1-2\upsilon) + \upsilon^2\langle s_{x-1}s_x s_{x+1}\rangle\right]. \tag{4.76}$$

The three-site correlation, defined with respect to the equilibrium probability distribution in (4.10), may be computed by the transfer matrix method. One obtains

$$\langle s_{x-1}s_x s_{x+1}\rangle = \frac{m}{\varpi^2}\left[2\varpi\cosh\eta - 1 - \frac{3}{e^{4K'}}\right], \tag{4.77}$$

where $K' \equiv J/k_B T'$, $\varpi \equiv \cosh\eta + m^{-1}\sinh\eta$, and η is related to the magnetization by $m^{-2} - 1 = (e^{2K'}\sinh\eta)^{-2}$. Therefore, $m^* = 0$ is a stable solution as long as $\upsilon \leqslant \alpha_C \equiv (1 + e^{2K'})/(3 + e^{2K'})$ for any T'. Otherwise, i.e., for $2K > \ln(2 + e^{2K'})$, there is a stable nonzero solution given by

$$m^* = \lambda\left[\lambda^2 + (1-\lambda^2)\,e^{-4K'}\right], \tag{4.78}$$

[10] See chapters 2, 3, and 6, Lebowitz (1986), and de Masi, Presutti, & Scacciatelli (1986).

where

$$\lambda^2 \equiv 1 - \left[\frac{2\upsilon(1-\upsilon)}{(1-2\upsilon-\upsilon^2)\sinh(2K') + 2\upsilon^2\cosh(2K')}\right]^2, \qquad (4.79)$$

which corresponds to a second-order transition. As expected, $T' \to 0$ and $T' \to \infty$ lead, respectively, to the results for equilibrium and for random diffusion, namely, $\alpha \leqslant 1$ as obtained by Glauber (1963), and $\alpha \leqslant \frac{1}{2}$ as obtained by de Masi *et al.* (1986).

It is also interesting that there is, for any K', a *critical reaction temperature*

$$K_C = \frac{1}{2}\ln\left(e^{2K'}+2\right), \qquad (4.80)$$

while K must exceed $\frac{1}{2}\ln 3$ in order to have a *critical diffusion temperature*

$$K'_C = \frac{1}{2}\ln(e^{2K}-2). \qquad (4.81)$$

That is, the system exhibits a line of critical points where the critical exponents are classical. Thus classical behavior, which previously appeared to be associated with the infinite-T' limit, here is a consequence rather of the *fast rate limit* $\epsilon \to 0$. This is also supported by simulation results for finite ϵ presented in the next section.

In the limit $T' \to \infty$ ($K' = 0$), the system is characterized by the absence of correlations between sites, as discussed in §4.3. As one might have expected, this is no longer true for finite K'. This is illustrated by computing the two-site correlation for any given K', for $d = 1$ and rate (ii). One finds in this case that

$$\langle s_0 s_n\rangle = \frac{1}{\omega_+ + \omega_-}\left[\left(\omega_+^* + \omega_-^*\right)^2 - 4\left(\frac{\omega_-}{\omega_+}\right)^n \omega_+^*\omega_-^*\right], \qquad (4.82)$$

where

$$\omega_\pm = \frac{e^{K'} \pm \sqrt{\lambda^2 e^{2K'} + (1-\lambda^2)\,e^{-2K'}}}{\sqrt{1-\lambda^2}}, \qquad \omega_\pm^* = \omega_\pm - e^{K+\eta}. \qquad (4.83)$$

In the one-phase region, one has $\eta = 0$, $\lambda = 0$, and it follows that

$$\langle s_0 s_n\rangle = \exp\left(-\frac{n}{\xi_1}\right), \qquad \xi_1 = \frac{-1}{\ln(\tanh K')}, \qquad (4.84)$$

i.e., spatial correlations at high temperatures decay exponentially, unlike in the DLG (note that magnetization is not conserved here). In the two-phase region, an expansion around $m = 0$, corresponding to $\lambda = 0$, yields

$$\langle s_0 s_n\rangle = \left(\lambda e^{2K'}\right)^2 + \left[1 - \left(\lambda e^{2K'}\right)^2\right]\exp\left[-n\left(\frac{1}{\xi_1} + \frac{1}{\xi_2}\right)\right], \qquad (4.85)$$

with

$$\xi_2 = \frac{1}{\lambda^2 \, e^{3K'}} \, . \tag{4.86}$$

Consequently, for any given T', the length ξ_1 remains finite, while ξ_2 diverges as $m \to 0$, as expected. Moreover, one finds

$$\langle s_1 s_2 s_3 \rangle = \frac{2\, e^K \sinh \eta}{\omega_+^2 \,(\omega_- + \omega_+)} \left[\omega_+^2 - 4\omega_+ e^K \cosh \eta + 4 \cosh (2K)\right] \tag{4.87}$$

for the three-site correlation function.

This section has illustrated a few general properties of nonequilibrium macroscopic states for lattice systems, and some interesting information about several specific hydrodynamic situations has emerged. The reader is invited to work out further cases with a similar approach.

4.5 Simulation results

The preceding section describes the hydrodynamic limit ($\epsilon \to 0$) of the reaction–diffusion lattice gas. It is complemented here with the principal results from a series of MC simulations for finite values of ϵ. The interest here lies in varying values for the relative speed of diffusion and not only in very fast diffusion. This provides a comprehensive picture that may be compared with the approximate theories in §4.6; some questions arise that remain to be fully understood.

The relevant algorithm is a simple extension of the usual one in MC simulations. One chooses an arbitrary random initial configuration on a given lattice with periodic boundary conditions, and iterates the following steps. First, a site $\mathbf{x}$ is selected at random, and a random number ζ is generated in [0,1]. If $\zeta < p$, $s_{\mathbf{x}}$ is exchanged with a randomly chosen NN, in accord with the (diffusion) rate $c(\mathbf{s}; \mathbf{x}, \mathbf{y})$, which involves the temperature T'; if $\zeta \geqslant p$, the *flip* $s_{\mathbf{x}} \to -s_{\mathbf{x}}$ is performed instead, using the (reaction) rate $c(\mathbf{s}; \mathbf{x})$, as if the spin were in contact with a heat bath at temperature T. This corresponds to the master equation (4.7). (The case considered in §4.3 to obtain the hydrodynamic limit corresponds to $\epsilon^{-2} = p/d\,(1-p)$, except for a rescaling of time units.) Cluster algorithms, widely used to accelerate equilibrium simulations, have not been used here, to avoid any uncontrolled influence of the dynamical process on the steady state. Of course, one needs to specify the reaction and diffusion rates, which influence the steady state as demonstrated above. Due to considerations of efficiency, most of the MC data were obtained with both $c(\mathbf{s}; \mathbf{x}, \mathbf{y})$ and $c(\mathbf{s}; \mathbf{x})$ of the Metropolis form. It is clear, however, that the study of further choices for the rates is desirable. Only attractive ($J > 0$) interactions and one-dimensional and square lattices have been analyzed so far to our knowledge.

Linear systems

González-Miranda *et al.* (1987) studied the reaction–diffusion lattice gas in $d = 1$, with $T' \to \infty$, and Metropolis reaction rate. Their findings are consistent with exact results for $p \to 1$; see case (v) in §4.4. A system with $p = 0.95$ and $N = 2500$ (so that finite-size effects are expected to be small) exhibits no phase transition. In fact, the magnetization curve $m(T)$ found for $p = 0.95$ is identical to the equilibrium one; (simulations seem to indicate, however, that slow diffusion induces some systematic differences in the behavior of the energy — for example, density of particle–hole bonds). If interactions extend up to next NNs with $p = 0.95$, both the energy and m tend to reveal more structure than for NNs, but the behavior is still monotonic; neither the specific heat nor the magnetic susceptibility suggests the existence of a true phase transition.

More intriguing are the results of Garrido *et al.* (1989) for the (reaction) rate (4.57). These authors considered NN interactions on a lattice of 10^4 sites, $p = 0.1$, 0.5, 0.75, and 0.95, and $T' \to \infty$. At the two largest p values, long-range order is observed, even at finite T; the onset of order (as revealed by m or by the temperature derivative of the energy, for instance) is rather sharp; figure 4.1 presents some of the evidence. This is consistent with the fact that the same system with $p \to 1$ has a (classical) phase transition of second order for $T' \leqslant \infty$ (see §§4.4 and 4.6). The situation for $p = \frac{1}{2}$ is not so clear, and it is quite ambiguous for $p = 0.1$. At all of the p values investigated, the simulations require very long times due to their slow evolution; the relaxation time increases with decreasing p. This reflects how difficult reaching an ordered state via diffusion in a chain is. In summary, present evidence favors the existence of a second-order phase transition *at least* for $p > \frac{1}{2}$; see figure 4.2. Further investigation is required before one can reach a more definite conclusion.

In one-dimensional, equilibrium systems with short-range interactions, phase transitions are forbidden by a general theorem (van Hove 1950; Dyson 1968; Ruelle 1969). One may argue, in the light of the theory in §4.3, that fast diffusion produces a local condition that induces mean-field behavior which, in a sense (Lebowitz & Penrose 1966), is associated with long-range interactions. The situation may not be so simple, however. In fact, there is some evidence that order might occur for a certain rate at finite values of p, while the system remains disordered at finite temperatures for most reaction rates (see equations (4.42) and (4.43)) in the fast-rate limit for diffusion. Therefore, it is interesting to mention the case of a two-dimensional system in which (completely random) exchanges are not restricted to NNs but extend to arbitrary range. This modification appears to generate mean-field-type effective interactions and classical

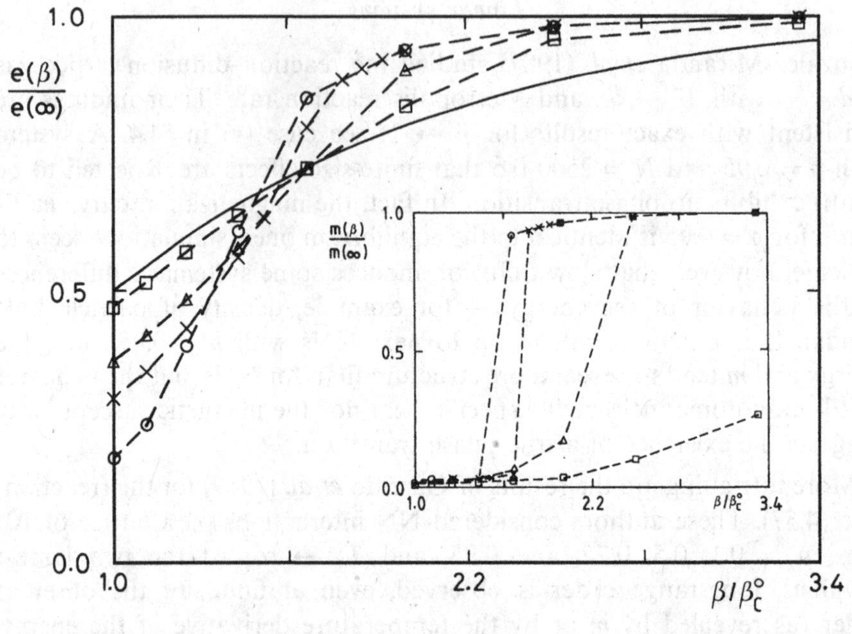

Fig. 4.1. The energy (main graph) and magnetization m (inset), as a function of $\beta \equiv (k_B T)^{-1}$ for $d = 1$, NN interactions, $T' \to \infty$, and reaction rate (4.57) in simulations. Symbols: $p = 0.1$ (□), 0.5 (△), 0.75 (×), and 0.95 (o); the solid line is the equilibrium result for $p = 0$; the dashed lines are a guide to the eye. β is in units of $(1/2J)\log 3$, the corresponding critical value for $p \to 1$ and $T' \to \infty$. (From Garrido *et al.* (1989).)

critical behavior for any $p > 0$.[11] There is, furthermore, a proof that no long-range order exists for $d = 1$ if the reaction rate is of Glauber type, while a classical phase transition has been observed numerically at finite temperature for other rates even if $p < 1$ (Droz, Rácz, & Tartaglia 1990, Xu, Bergersen, & Rácz 1993). It is interesting that a relation may exist between the nature of the random exchange process and the range of effective interactions (Bergersen & Rácz 1991).

Square lattice

The study by González-Miranda *et al.* (1987) for two dimensions, NN interactions, $T' \to \infty$, (reaction) Metropolis rate, and varying p indicates that a naive extrapolation of the exact results in §4.4 is unsafe. That is, the case $p \to 1$, for which a first-order transition has been found, is

[11] Droz, Rácz, & Tartaglia (1991); see also Droz, Rácz, & Schmidt (1989) for some mean-field effects in a one-dimensional system that exhibits no long-range order.

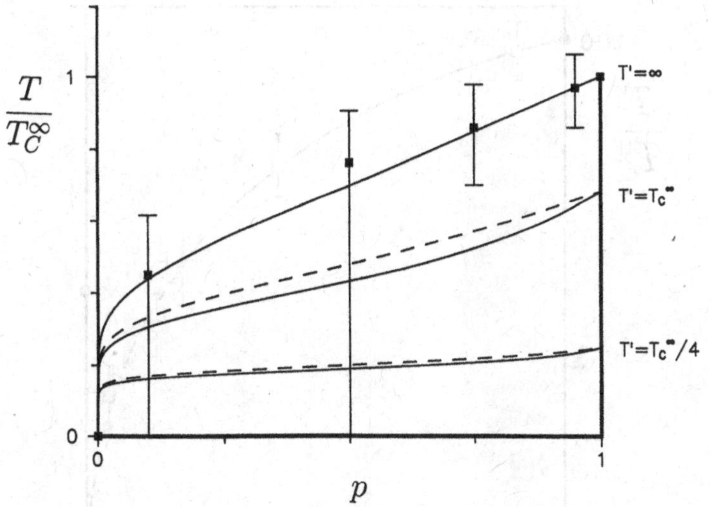

Fig. 4.2. The phase diagram for the one-dimensional system of figure 4.1 as one varies the diffusion temperature T'. The p-axis coincides with the equilibrium zero-T critical point, for which $p < 1$ is a trivial parameter. The right-most vertical line is a line of critical points for $p \to 1$ and any $T' > T$. The solid and dashed curves correspond, respectively, to diffusion rates of the Metropolis and Kawasaki types for different values of T', as indicated, within the mean-field approximation of §4.6. The symbols represent MC data whose error bars are consistent with having a phase transition for $p > \frac{1}{2}$ only. $T_C^\infty \equiv 2J\,(k_B \log 3)^{-1}$ is the exact critical temperature for $p \to 1$ and $T' \to \infty$. (From Alonso *et al.* (1993).)

rather singular. MC data reveal that the transition is of first order for $p > p_t \approx 0.83$ only: $p_t(T' \to \infty)$ is a nonequilibrium *tricritical point;* the transition is second-order for $p < p_t$; see figure 4.3.

The transition for $p < p_t$ appears to be of the Ising type in MC simulations; a MC renormalization group study by Wang & Lebowitz (1988) for $p = \frac{1}{4}$ and $\frac{1}{2}$ reveals Ising behavior as well. Thus p_t may be interpreted as marking a sudden change from ordinary Ising critical behavior to the first-order mean-field-type transition found in the bifurcations in (4.18) about the $m = 0$ solution. For the same system with the Glauber reaction rate, the bifurcations in (4.18) give a second-order transition, as for $p = 0$. Therefore, there should be no abrupt change in the nature of the phase transition as one varies p in this case. In fact, the renormalization group analysis of Wang & Lebowitz gives consistency with Ising behavior for $p = \frac{1}{4}, \frac{1}{2}$, and 0.95; the convergence is very poor for $p = 0.95$, however, perhaps reflecting a crossover from Ising to classical behavior, which is known to occur for $p \to 1$. It would be interesting to study this crossover in detail.

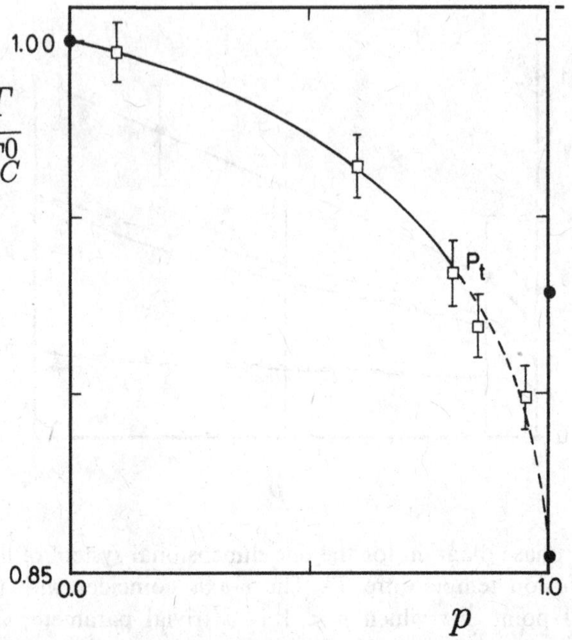

Fig. 4.3. MC (□) and exact (•) results for the phase diagram of the two-dimensional system with NN interactions, $T' \to \infty$, and a reaction rate of the Metropolis type. The solid line (and • for $p = 0$) corresponds to transitions of second order; the dashed line (and • for $p = 1$) indicates first-order transitions. Temperature is in units of the Onsager critical temperature.

Systems with reaction rates of the Kawasaki or de Masi types, for which a classical second-order transition occurs for $p \to 1$, may be analyzed in light of the dynamic universality argument advanced by Grinstein, Jayaprakash, & He (1985) (for further details see chapter 8). In any case, one may conclude again both that the reaction rate is important in determining the steady state and that the limit $p \to 1$ (rather than $T' \to \infty$) induces classical behavior, thus extending some exact results reported in §4.4. Nevertheless, it is difficult to draw further conclusions about the relations between the system parameters p and $c(\mathbf{s}, \mathbf{x})$ and the nature of the resulting phase transition, for $p < 1$.

Alonso, Marro, & González-Miranda (1993) have discovered an even more varied situation for the same system with finite T', in accordance with the exact results at the end of §4.4. When the reaction and diffusion rates are both of the Metropolis form, at temperatures T and T', respectively, three distinct types of behavior appear:

(*A*) A continuous phase transition at $T_C = T_C(p, T')$ (see table 4.1), with apparently Ising critical exponents, occurs for any T' if p is not very

Table 4.1. MC estimates of T_C (the critical temperature for reactions) as a function of T' and p for $d = 2$, and the type of behavior at each phase point as defined in the main text. Temperatures are in units of T_C^0, the Onsager critical temperature. Error bars are smaller than ± 0.01; the results for $T' = \infty$ are exact.

T'	$p = 0.60$	$p = 0.80$	$p = 0.85$	$p = 0.95$	$p = 1.00$
0.1	1.008(A)	1.000(A)	0.992 (A)	0.915(B')	—
0.5	—	—	—	0.942(B')	—
1	1.005(A)	—	1.005(A)	1.000(A)	—
2.5	—	—	—	1.030(A)	—
3	—	—	1.006(A)	—	—
4	—	—	—	0.993(A)	—
5	0.992(A)	—	—	0.975(A)	—
7	—	—	0.985(B)	0.955(B)	—
10	0.981(A)	0.970(A)	0.965(B)	0.935(B)	—
∞	0.96(A)	0.94(A)	0.92(B)	0.90(B)	0.855(B)

large, say $p \leqslant 0.8$, for $p = 0.85$ if $T' < 7$ (temperatures are here in units of T_C^0, the Onsager temperature), and for $p = 0.95$ excluding the ranges $T' < 1$ and $T' > 5.2$. We have described this type of behavior above for $p < p_t$ when $T' \to \infty$. It seems to be the most general case; see figure 4.4.

(B) The phase transition is of first order, with well-defined discontinuities, metastable states and, apparently, coexisting phases. This occurs for $p \geq 0.85$ and $T' > 7$, for example; see figure 4.4. It has been seen for $p > p_t$ when $T' \to \infty$.

(B') The situation for larger p, for example, $p = 0.95$ and $T' < 1$, is less clear-cut. It seems that the transition is discontinuous, but some qualitative differences with type (B) are observed. On the practical side, MC simulations are hampered by very slow evolution and long-lived metastable states. Moreover, while m is discontinuous as in case (B), the discontinuities, if any, in the energy and specific heat are much weaker. This is similar to the one-dimensional case shown in figure 4.1, where a transition of second order may occur. The structure of the magnetic susceptibility near the transition temperature for type (B') is much more pronounced than for type (B), and a larger degree of local order is detected by other means. It may be mentioned also that the regions of the phase diagram in which MC simulations depict type (B') behavior exhibit second-order transitions within the mean-field approximation of §4.6. Two-site correlation functions are observed to decay exponentially at large distances at high enough temperature, in agreement with the

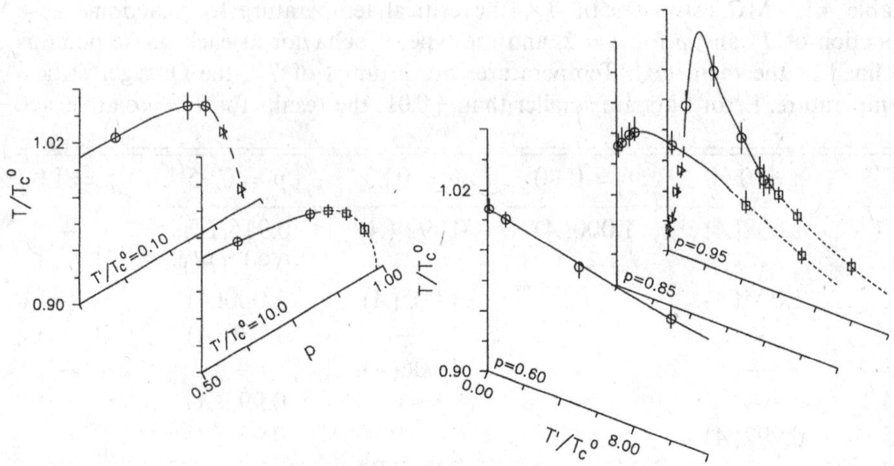

Fig. 4.4. Sections, as indicated, of the MC phase diagram for the two-dimensional system with both reaction and diffusion rates of Metropolis type, for temperatures T and T', respectively. The nature of the corresponding phase transition is represented by different symbols, o, □, and ▷, respectively, for types (A), (B), and (B'). The solid lines correspond to critical points; the dashed lines correspond to discontinuous transitions. (From Alonso *et al.* (1993).)

analytical computations in §4.4 for $p \to 1$, as expected for a nonconserved order parameter (§2.5).

4.6 Kinetic cluster theory

The following sections describe a cluster mean-field theory of the reaction–diffusion system.[12] The method generalizes the one described in chapter 3 to reaction–diffusion competing dynamics.

The minimum size of D (§3.1) that allows for competition between reaction and diffusion consists of the *interior* I with only two NN sites surrounded by the *surface* S with $2(2d-1)$ sites outside I. The chief interest is in a first-order approximation in which only pair (NN) correlations within D are considered; i.e., one only needs to monitor the particle density, $p(\bullet) \equiv N^{\bullet}/N = n$, and the density of particle–particle pairs, $p(\bullet,\bullet) \equiv N^{\bullet\bullet}/dN = u$ (using an obvious notation). In practice, however, it is convenient to define $p(\circ) \equiv N^{\circ}/N = 1 - n$ and $p(\circ,\circ) \equiv N^{\circ\circ}/N \equiv$

[12] The theory in the present section was developped by Dickman (1987) for $T' \to \infty$, and by Alonso *et al.* (1993) for finite T'. Bellon & Martin (1989) have studied competition between thermal and random diffusion within a mean-field approximation to evaluate the influence on the transition of *cascade effects* consisting of several atoms shifted at once.

$z = 1 - 2n + u$; then the magnetization is $m = 2n - 1$, and either u or $e \equiv p(\bullet, \circ) = n - u$ may serve as a measure of the energy. The configurations of D may be characterized by spin variables at the two sites of I, s_j with $j = 1, 2$, and by the number of (surface) sites that are NN of j and are in state $+1$, N_j. The associated probabilities are (as in equation (3.7))

$$Q_D(s_1, s_2; N_1, N_2) = p(s_1, s_2) \prod_{j=1}^{2} \binom{2d-1}{N_j} p(\bullet \mid s_j)^{N_j} \; p(\circ \mid s_j)^{2d-1-N_j}, \tag{4.88}$$

where $p(s \mid s') = p(s, s')\, p(s')^{-1}$. Equations for n and u follow after recognizing that the dynamics consists here of a competition between two mechanisms:

- exchanges (E) at temperature T' with probability $p\psi(\delta H)$, and
- reactions (R) at temperature T with probability $(1-p)\,\phi(\delta H)$.

The rates $\psi(\delta H)$ and $\phi(\delta H)$ correspond to exchanges and reactions, respectively; δH is the associated change in energy.

Consider separately the actions of mechanisms E and R. The variation with time of α (which represents either n or u) induced by R is

$$\frac{d\alpha}{dt} = \sum_{s_1, s_2 = \pm 1} \sum_{N_1, N_2 = 0}^{2d-1} \delta\alpha\, \phi(\delta H)\, Q_D(s_1, s_2; N_1, N_2), \tag{4.89}$$

where $\delta\alpha$ is the change in α due to the inversion of s_i. (Here we have used translation–invariance.) One of the summations may be done, giving

$$\frac{d\alpha}{dt} = \sum_{s=\pm 1} p(s) \sum_{v=0}^{2d} \delta\alpha\, \phi(\delta H) \binom{2d}{v} p(\bullet \mid s)^{v} \; p(\circ \mid s)^{2d-1-v}. \tag{4.90}$$

This is precisely what one would have written by considering directly the minimum domain for spin flips, i.e., only one interior site (and its NN surface sites). One obtains for $\alpha \equiv n$ (with $\delta\alpha \equiv \delta n = -s$) that

$$\frac{dn}{dt} = \mathscr{F}(n, u; T) \tag{4.91}$$

with

$$\mathscr{F}(n, u; T) \equiv \sum_{v=0}^{2d} \binom{2d}{v} \tilde{\phi}(v - d)\, (1-n)^{v} \left[\frac{z^{2d-v}}{(1-n)^{2d-1}} - \frac{u^{2d-v}}{n^{2d-1}} \right], \tag{4.92}$$

where $\tilde{\phi}\,[s(\nu-d)] \equiv \phi\left[4Js(\nu-d)/k_BT\right]$. For $\alpha \equiv u$ (with $\delta u = -\nu s(4d-1)^{-1}$) one finds

$$(4d-1)\frac{\mathrm{d}u}{\mathrm{d}t} = \mathcal{G}(n,u;T) \tag{4.93}$$

with

$$\mathcal{G}(n,u;T) \equiv \sum_{\nu=0}^{2d}\binom{2d}{\nu}\tilde{\phi}(\nu-d)\,(1-n)^{\nu}\left[\frac{\nu z^{2d-\nu}}{(1-n)^{2d-1}} - \frac{(2d-\nu)\,u^{2d-\nu}}{n^{2d-1}}\right]. \tag{4.94}$$

Next, consider the action of E on the domain D, the interchange $s_1 \rightleftarrows s_2$ of the spin variables at the two interior sites. This leaves n constant but (for $s_1 = -s_2$) changes u by $s_1(N_2-N_1)/(4d-1)$ and H by $4Js_1(N_1-N_2)/k_BT'$. Consequently, one obtains (see equation (3.8))

$$(4d-1)\frac{\mathrm{d}u}{\mathrm{d}t} = \mathcal{H}\left(n,u;T'\right) \tag{4.95}$$

with

$$\begin{aligned}\mathcal{H}\left(n,u;T'\right) \equiv 2\sum_{N_1,N_2=0}^{2d-1}&\binom{2d-1}{N_1}\binom{2d-1}{N_2}(N_2-N_1)\\ &\times u^{N_1}z^{2d-1-N_2}\frac{(1-n)^{2d-N_1+N_2}}{[n(1-n)]^{2d-1}}\psi(N_1-N_2),\end{aligned} \tag{4.96}$$

where a sum over s_1 has been performed. Finally, one obtains for the competition between the processes E and R that

$$\frac{\mathrm{d}n}{\mathrm{d}t} = (1-p)\,\mathcal{F}(n,u;T), \tag{4.97}$$

$$(4d-1)\frac{\mathrm{d}u}{\mathrm{d}t} = (1-p)\,\mathcal{G}(n,u;T) + p\mathcal{H}\left(n,u;T'\right), \tag{4.98}$$

where $\mathcal{F}(T)$, $\mathcal{G}(T)$, and $\mathcal{H}(T)$ are defined in (4.92)–(4.96). The properties of the steady state follow from $\mathrm{d}n/\mathrm{d}t = 0$ and $\mathrm{d}u/\mathrm{d}t = 0$ which, for the realizations of the rate that are of interest, reduce to two polynomials in $\eta \equiv \exp\left(-2J/k_BT\right)$.

Suppose the rate functions $\phi(X)$ and $\psi(X)$ and the values (p, T') are given. The system should be fully disordered for high enough T, corresponding to $m = 0$ ($n = \frac{1}{2}$). Consequently, $z = u$, and $\mathcal{F}\left(\frac{1}{2}, u; T\right) = 0$, and the only condition remaining at high T is

$$(1-p)\,\mathcal{G}(n,u;T) + p\mathcal{H}\left(n,u;T'\right) = 0. \tag{4.99}$$

This determines u for a given T. Stability requires that $\left(\partial\mathcal{F}/\partial n\right)_{n=\frac{1}{2},\,u} < 0$. The breakdown of this last condition marks a phase transition. More precisely, one may define $T_1(p, T')$ as the solution of $\left(\partial\mathcal{F}/\partial n\right)_{n=\frac{1}{2},\,u} = 0$, where u follows from (4.99), and $T_2(p, T')$ as the solution of $\left(\partial\mathcal{F}/\partial n\right)_{n\neq\frac{1}{2},\,u} = 0$, where (n, u) is the stationary solution of (4.97)–(4.98). The transition is of second order if $T_1 = T_2 \equiv T_C$. Otherwise, it is of first order. $m = 0$ is the only stable solution for $T > T_2$, a unique $m \neq 0$ solution exists for $T < T_1$, and there are two locally stable solutions for $T_1 < T < T_2$. (One or the other of the two solutions occurs depending on the initial condition.) This method gives no information on which one of them is globally stable but locates a region of *metastable behavior* between T_1 and T_2.

Some general results

First, we note that some information on steady states may be obtained analytically for arbitrary d. In particular, the known equilibrium (mean-field) results follow for $p = 0, 1$. For $p = 0$, (4.99) reduces to $\mathcal{G}\left(\frac{1}{2}, u; T\right) = 0$. This leads to $(2u)^{-1} - 1 = \eta$ if detailed balance, $\phi(X) = \eta^{2X}\phi(-X)$, is used. After combining with $\left(\partial\mathcal{F}/\partial n\right)_{n=\frac{1}{2},\,u} = 0$, one gets $u = d\,[2\,(2d-1)]^{-1}$; consequently, $T_C = 2J/k_B \ln[d\,(d-1)]$ which is the Bethe critical temperature. Furthermore, (4.97)–(4.98) reproduce the ordered states of the Bethe theory for $T < T_C$.

On the other hand, $p = 1$ implies that $\mathcal{H}\left(n, u; T'\right) = 0$ and $n =$const. We note that

$$\mathcal{H}\left(n, u; T'\right) = \sum_{N_2 > N_1} (N_2 - N_1)\ \psi\,(N_1 - N_2)\ \tilde{F}_{N_1,N_2}, \tag{4.100}$$

where $\tilde{F}_{N_1,N_2}$ has the structure

$$\begin{aligned}\tilde{F}_{N_1,N_2} &\equiv Q_D\,(\bullet,\circ; N_1, N_2) - Q_D\,(\bullet,\circ; N_2, N_1)\,\zeta^{2(N_2-N_1)} \\ &= \tilde{K}_{N_1,N_2}\left\{(1-n)^{2(N_2-N_1)} - \left[\zeta^2 u\,(1-2n+u)\right]^{N_2-N_1}\right\}.\end{aligned} \tag{4.101}$$

Here, $\zeta \equiv \exp\left(-2J/k_B T'\right)$ and the coefficient $\tilde{K}$ is positive. Consequently, the condition $\mathcal{H}\left(n, u; T'\right) = 0$ leads to $(1-n)^2 = \zeta^2 u\,(1-2n+u)$. That is, the energy is

$$e = \frac{\zeta\,(2m^2-1) + \sqrt{1-m^2\,(1-\zeta^2)}}{\zeta + \sqrt{1-m^2\,(1-\zeta^2)}} \tag{4.102}$$

which corresponds to the (equilibrium) Bethe–Peierls solution for given m and T'. This is independent of transition rates, as expected for $p = 1$, and of the spatial dimension, which is a feature of the pair approximation. Setting $T = T'$ (i.e., $\zeta = \eta$) reproduces the equilibrium solution above for $p = 0$, namely, $\zeta^2 u(1-2n+u) = (1-n)^2$, which causes the diffusion term $\mathscr{H}(n,u;T)$ (with T' replaced by T) to vanish.

More intriguing is the (nonequilibrium) limit $p \to 1$. According to (4.97)–(4.98), the time scale for the variations of m or, equivalently, n induced by the reaction process is of order $(1-p)^{-1}$ which diverges for $p \to 1$. In contrast to such slow relaxation of m, the energy changes very rapidly in this limit due to the (fast) diffusion process, whose time scale is of order unity. More explicitly, the slow variation of m allows the diffusion process to equilibrate the system at temperature T' before any significant change of m occurs. This produces at each time a condition of *local equilibrium* at temperature T' with e given by (4.102). This is the condition described in §4.3, except that local equilibrium is characterized by the condition $u(1-2n+u) = \left[(1-n)/\zeta\right]^2$ within the present approximation. The steady states follow from the conditions $\mathscr{F}(n,u;T) = 0$ and $\mathscr{H}\left(n,u;T'\right) = 0$. That is, they are also local equilibrium states (4.102) with no dependence on the diffusion rate $\psi(X)$. $T_1 = T_1(p \to 1, T')$ follows from

$$\sum_{v=0}^{2d} \binom{2d}{v} \tilde{\phi}\, \zeta^{v} \left[v - 1 - (2d - v)\,\zeta\right] = 0, \tag{4.103}$$

which has a strong dependence on both $\phi(X)$ and d, as discussed below. One obtains $u = [2(1+\zeta)]^{-1}$ for $m = 0$.

Next, we apply the method to the reaction–diffusion lattice gas for $d = 1$ and 2. These two cases behave differently from each other within the present mean-field approximation as a consequence of differences within the respective domain D.

One-dimensional systems

Equation (4.96) leads to

$$\mathscr{H}\left(n,u;T'\right) = \psi(-1)(1-n)\,\frac{(1-n)^2 - \zeta^2 u(1-2n+u)}{n^3(1-n)^3} \tag{4.104}$$

for $d = 1$. This implies that the existence of a phase transition does not depend on the diffusion rate. Different realizations of $\psi(X)$ modify only the relative speed of the diffusion process, i.e., produce a different *effective* value for p, according to the last term in equation (4.98). On the other hand, one concludes from (4.99) and $\left(\partial\mathscr{F}/\partial n\right)_{n=\frac{1}{2},\,u} = 0$ that:

- a critical point exists at $T_C = 0$ for any p and T' if $\phi(X)$ is of the Kawasaki type;
- the zero-T critical point disappears for $\phi(X)$ of the Metropolis type, unless either $T' = 0$ or else $p = 0$; otherwise, only $m = \pm 1$ solutions occur for $T = 0$;
- for either $\phi(X) = e^{-\frac{1}{2}X}\left[\cosh\left(J/k_B T\right)\right]^2$ or $\phi(X) = e^{-\frac{1}{2}X}$, a surface of critical points occurs at $T_C = T_C(p, T')$ which is the solution of

$$\frac{4\rho\gamma^2(1-\gamma)^2}{(2\gamma\zeta+\gamma-\zeta)(2\gamma\zeta-\gamma-\zeta)} = \psi(-1)\frac{p}{1-p}. \tag{4.105}$$

Here,

$$\rho = \begin{cases} \gamma^{-1} & \text{for} \quad \phi(X) = e^{-\frac{1}{2}X} \\ 4(1+\gamma)^{-2} & \text{for} \quad \phi(X) = e^{-\frac{1}{2}X}\cosh\left(J/k_B T\right) \end{cases} \tag{4.106}$$

and $\gamma \equiv \exp\left(-2J/k_B T_C\right)$.

Two important facts follow from this. On the one hand, the form of ψ is irrelevant for the limiting cases $T' = 0, \infty$ and $p = 0, 1$, and it induces only minor quantitative changes otherwise. On the other hand, the present mean-field approximation for $T' \to \infty$ leads to the same transition points as in simulations. The prediction of (4.105) is that the phase transition is of second order, in agreement with the exact result for $p \to 1$, which is consistent with the MC data reported in §4.5. Thus it seems that the mean-field solution is an accurate description, at least for $p \geq \frac{1}{2}$.[13] In fact, there is qualitative agreement between mean-field and MC results for the order parameter and for the specific heat, at least for $p \geq \frac{1}{2}$, as illustrated in figure 4.5. Two further observations support this conclusion: (1) the present theory reproduces the fact revealed by both exact and simulation results, that the one-dimensional system has no phase transition at finite T for, for example, Metropolis and Kawasaki reaction rates; and (2) (4.105) reduces for $p \to 1$ to $\gamma = \zeta(1+2\zeta)^{-1}$ which is the exact solution for arbitrary T' (§4.4).

The agreement between the mean-field and exact theories indicates that the former is independent of the size of domain D for $d = 1$; therefore, the only approximation involved is the restriction to pair correlations, which are in fact the only ones when local equilibrium sets in for $p \to 1$. The proof is as follows. Assuming homogeneity (which is implicit in the

[13] Very slow evolution precluded a definite conclusion about the possible existence of long-range order for smaller p (§4.5).

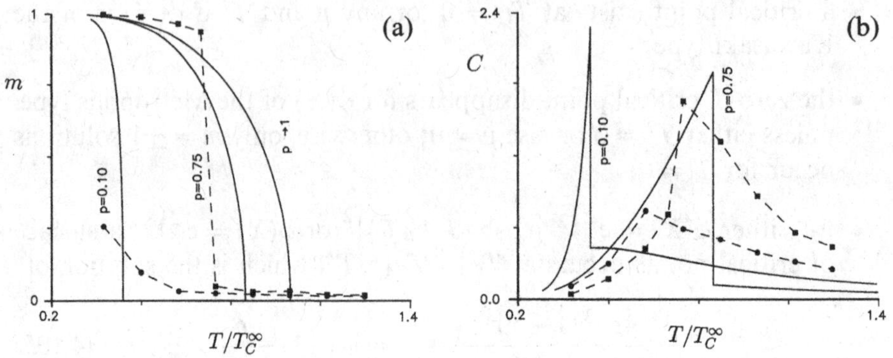

Fig. 4.5. (a) The order parameter m and (b) the *specific heat* C obtained from the energy fluctuation both for $d = 1$, $T' \to \infty$, and the reaction rate $\phi(X) = e^{-\frac{1}{2}X}$. The solid lines indicate mean-field predictions for various p. The symbols represent MC data (full square for $p = 0.75$, full circle for $p = 0.10$); the dashed lines are a guide to the eye. (From Alonso *et al.* (1993).)

method), the exact equation for the magnetization for $d = 1$ and $p \to 1$ follows as

$$\frac{d\langle s_x \rangle}{d\tau} = -2 \sum_{\mathbf{s}_D} s_x Q_D(\mathbf{s}_D)\, \phi(\delta H) \tag{4.107}$$

(§4.3) after summing over $\mathbf{s} - \mathbf{s}_D$; here, $\mathbf{s}_D$ represents any configuration of D, and

$$Q_D(\mathbf{s}_D) = \sum_{\mathbf{s}-\mathbf{s}_D} \mu_{eq}^{T'}(\mathbf{s}; m) = p(s_x)\, p(s_{x-1} \mid s_x)\, p(s_{x+1} \mid s_x)\,. \tag{4.108}$$

One may write this as in equation (4.90), i.e.,

$$\frac{d\langle s_x \rangle}{dt} = -2 \sum_{s_x=\pm 1} \sum_{v=0}^{2} s_x Q_D(s_x, v)\, \tilde{\phi}(s_x(v-1)), \tag{4.109}$$

where the definition of $Q_D(s_x, v)$ follows by comparing this to (4.90). For $n = \frac{1}{2}(1 + \langle s_x \rangle)$, this transforms into (4.91) which (together with the result $\zeta^2 u(1 - 2n + u) = (1 - n)^2$) is the solution of the present method, i.e., the exact solution for $p \to 1$ is indeed reproduced. This argument also suggests that one might expect the same to be valid, at least approximately, for a range of p values smaller than unity. It is clear, however, that, lacking more general exact results, further simulations are needed to reach a definite conclusion regarding this issue.

The square lattice

The relevant equations are now

$$\phi(-2)\left(\gamma^4-\omega^4\right)+2\phi(-1)\,\omega\left(\gamma^2-\omega^2\right)$$
$$=6\frac{p}{1-p}u\omega(\omega-\zeta)\left[1-2(1-\zeta)u\right]\alpha, \tag{4.110}$$

where $\omega \equiv -1+1/(2u)$, and

$$\phi(-2)\left[(3u-2)\gamma^4+3\omega^4u\right]-2\phi(-1)\,\omega^2\left[6u\gamma^2+\omega(1-6u)\right]$$
$$-6\phi(0)\,\omega^2(1-3u)=0, \tag{4.111}$$

where $\gamma \equiv \exp\left(-2J/k_BT_1\right)$, and

$$\alpha \equiv \psi(-3)\left(\omega^4+\zeta^2\omega^2+\zeta^4\right)+\psi(-2)\,4\omega u\left(\omega^2+\zeta^2\right)+\psi(-1)\,5\omega^2u \tag{4.112}$$

contains the dependence on the diffusion rate. These equations transform into two polynomials of γ for the realizations of ϕ of interest. They are polynomials of second order in γ for the Metropolis reaction rate.[14] For $p=0$, (4.111) leads to $\gamma=\omega$ which yields the Bethe critical temperature $\gamma=\frac{1}{2}$ when used in equation (4.110), independent of the rates, as expected. For $p\to 1$, (4.111) leads to $\zeta=\omega$ which produces in (4.110):

$$\phi(-2)\left[3\zeta^4-(1+4\zeta)\gamma^4\right]-\phi(-1)\,4\left[3\zeta^2\gamma^2+\zeta^3(\zeta-2)\right]$$
$$+\phi(0)\,6\zeta^2(1-2\zeta)=0. \tag{4.113}$$

This, which corresponds to (4.103) for $d=2$, gives $T_1=T_1(T')$ for different ϕs. One finds

$$\gamma^2=\zeta\frac{\sqrt{6+20\zeta+19\zeta^2-4\zeta^3}-6\zeta^2}{1+4\zeta} \tag{4.114}$$

for Metropolis rates; this reproduces the known solution $\gamma^2=0.081$ as $T'\to\infty$.

The present mean-field solution has an interesting property found for $d=1$ above, i.e., it reproduces the exact solution as long as both $T'\to\infty$ and $p\to 1$ hold. The reason is that $T'\to\infty$ induces local equilibrium, where most correlations are suppressed, which makes the pair approximation exact and the method independent of D. Therefore, the only actual restriction is the implicit assumption about homogeneity. Some specific results for this case are as follows.

---

[14] One may check that (4.110)–(4.111) reduce to the equations of Dickman (1987) as $T'\to\infty$.

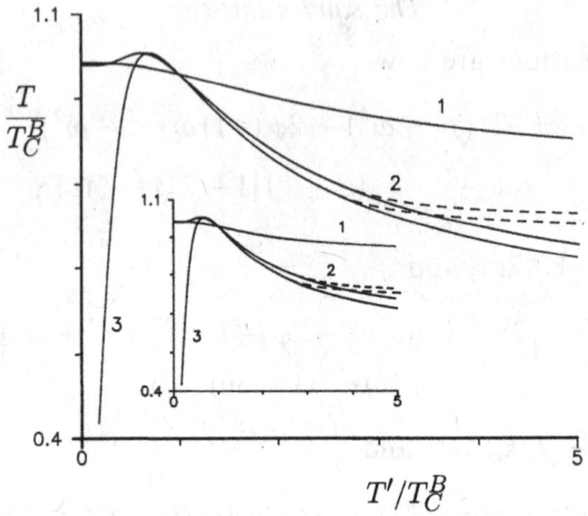

Fig. 4.6. The mean-field phase diagram for the two-dimensional system. The solid and dashed lines represent $T_1(p, T')$ and $T_2(p, T')$, respectively. The transition is of first order if $T_1 \neq T_2$, and of second order otherwise. $T_C^B \equiv 2J/k_B \log 2$ is the equilibrium Bethe critical temperature. The main graph corresponds to the case in which the system evolves with Metropolis reaction and diffusion rates for different values of p; curves 1, 2, and 3 are for $p = \frac{1}{2}$, $p = 0.95$, and $p \to 1$, respectively. The inset depicts the same behavior for a diffusion rate of the Kawasaki type.

The system exhibits a transition of second order if the reaction process is implemented by either Kawasaki or van Beijeren–Schulmann rates; the critical point is located at $J/k_B T_C = 0.3236$ and 0.2554, respectively. The transition is of first order for the Metropolis reaction rate, however. The main equation for this case is

$$\frac{1}{m}\frac{\mathrm{d}m}{\mathrm{d}t} = \left(5 + 10m^2 + m^4\right)\gamma^4 + 4\left(3 - 2m^2 - 4m^4\right)\gamma^2 - \left(1 + 2m^2 - 3m^4\right). \tag{4.115}$$

One obtains T_1 and T_2 as the solutions, respectively, of $\gamma^2 = 0.081$ and $5\gamma_2^6 - 3\gamma_2^4 + 7\gamma_2^2 = 1$, where $\gamma_2 \equiv \exp\left(-2J/k_B T_2\right)$.

The situation under more general conditions is very varied, for example, both first- and second-order transitions may occur for a given rate. A systematic study proceeds by computing $m = m(T)$ and $e = e(T)$ for selected choices of p, T', $\phi(X)$, and $\psi(X)$. Figures 4.6–4.8 summarize the picture obtained if both rates are of the Metropolis form. The following facts may be noted:

(1) $T_1(p, T')$ exists for any p and T'; $m \neq 0$ solutions occur for $T < T_1$.

(2) The critical value of u is always larger than $\frac{1}{3}$, for any p, for $T < T_C^B$, the Bethe critical temperature, as shown in figure 4.8. This suggests

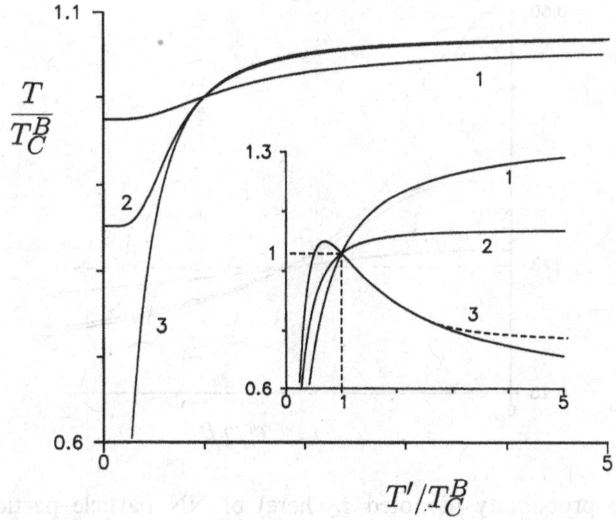

Fig. 4.7. The main graph corresponds to the same situation as in the main graph of figure 4.6 but for a reaction rate of the Kawasaki type. (The results for a Kawasaki diffusion rate are indistinguishable from those shown here.) The inset corresponds to $p \to 1$ for $\psi(X) = \min\{1, e^{-X}\}$ and different reaction rates: curve 1 is for $\phi(X) = e^{-\frac{1}{2}X}$, curve 2 is for $\phi(X) = (1 + e^X)^{-1}$, and curve 3 is for $\phi(X) = \min\{1, e^{-X}\}$.

that inhomogeneous solutions with $m = 0$ then exist, due to a tendency of the diffusion mechanism to induce phase segregation.

(3) One may distinguish up to three qualitatively different types of behavior for $p \to 1$ (see curves 3 in the main graph of figure 4.6 and in the inset of figure 4.7:

(A) transitions of second order for $T_C^B < T' < 2.574 T_C^B$;

(B) transitions of first order for $T' > 2.574 T_C^B$; and

(C) transitions of second order for $T' < T_C^B$; the corresponding solutions are inhomogeneous (as described above) near $T_C(T')$, and one observes for very small values of T', say $T' < 0.1 T_C^B$, that (i) sharp variations of both m and e with T occur near $T_C(T')$; (ii) jumps for e are an order of magnitude smaller than for m; (iii) numerical integration reveals very slow relaxation; this may reflect full phase segregation, which makes the system rather insensible to attempted exchanges at low T'.

In each of these cases, u and $\sigma = \frac{1}{4}uz\,(1-n)^{-2}$ depend on T' but are independent of T given that any short-range order is dominated by the

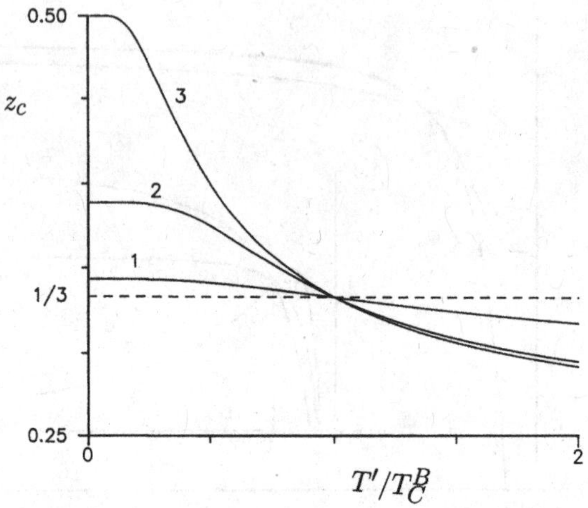

Fig. 4.8. The probability (denoted z_C here) of NN particle–particle pairs at critical temperature $T_1(T')$ for different values of p (curves 1, 2, and 3 are for $p = \frac{1}{2}$, 0.95, and 1, respectively), for Metropolis reaction and diffusion rates. This illustrates the asymmetry around $T' = T_C^B$, discussed in the text.

diffusion process for $p \to 1$. The similarity between mean-field cases (A), (B), and (C) and the corresponding MC studies discussed in §4.4 is also noteworthy. Case (C), in which $T_C(T') \to 0$ as $T' \to 0$, corresponds to the behavior denoted as type (B') above (the weak discontinuities reported would then be associated with 'experimental' difficulties due to very slow evolution).

(4) The case $p = 0.95$ (see curve 2 in the main graph of figure 4.6) is qualitatively similar to the one for $p \to 1$, except that:

(i) The *tricritical point* separating (A) and (B) is now at $T' = 2.622T_C^B$;

(ii) There is no indication of behavior (C) (which is still evident in the equations for $p = 0.999$ and $T' = 0.05T_C^B$), and some reduction in $T_C(T')$ occurs, but not a tendency for it to approach zero, as $T' \to 0$. Both u and σ depend on T; $\sigma(T)$ decreases monotonically with T, and exhibits a discontinuity for case (B).

(5) As p is further decreased, the region (A) broadens, so that, for example, the tricritical point for $p = \frac{1}{2}$ is at $8.152T_C^B$, and there is no first-order transition for $p < 0.364$. Moreover, $T_1(T')$ decreases monotonically with T'; see curve 1 in the main graph of figure 4.6.

In general, the steady state depends qualitatively on $\phi(X)$. This is evident from the inset of figure 4.7, where the curves correspond to different realizations of $\phi(X)$; the main graph of figure 4.7 illustrates the phase diagram for varying values of p, for a Kawasaki reaction rate. The main results for this case may be summarized as follows:

(i) The transitions are always of second order, with $T_C = T_1(T')$ increasing monotonically with T'.

(ii) The situation for $T' \to \infty$ is qualitatively similar to that in the simulation of Wang & Lebowitz (1988) for a Glauber reaction rate. In particular, T_C increases monotonically with p.

(iii) As $p \to 1$, short-range order is determined by the diffusion process at temperature T', and one may distinguish two cases: The behavior is of type (A) for any $T' \geq 0.1T_C^B$, approximately, while it is of type (C) (as for the Metropolis rate above) when $T' \geq 0.1T_C^B$.

(iv) The behavior is always of type (A) for any $p < 1$.

The consideration of different realizations for $\psi(X)$ is also interesting. The inset in figure 4.6 illustrates the case of Metropolis reaction and Kawasaki diffusion rates (which is to be compared with the main graph). One obtains the same qualitative behavior as in the main graph of figure 4.7 when both rates are of the Kawasaki type. We conclude that independent of p, different choices for the diffusion rate seem to induce only minor quantitative changes, unlike the case for the reaction rate.

4.7 Summary of static properties

The reaction–diffusion lattice gas presented in §4.2 is one of the most intriguing microscopic models at hand exhibiting nonequilibrium phenomena. It merits interest as an idealized representation of natural systems in which both dynamical and static aspects are relevant to macroscopic behavior and, more specifically, of systems in which an interplay of reaction and diffusion processes induces time variations and spatial gradients. A considerable amount of information has already been obtained for this model, and hydrodynamic-like equations have been rigorously derived for it. Furthermore, a large variety of ordering phenomena emerge as one varies the parameters. The combined picture from exact results, MC simulations, and mean-field approximations is a bit complex; obtaining a comprehensive description thus requires some effort. In this section, such an attempt is made by collecting some of the results for steady states and phase transitions; some specific questions are discussed, and possible lines of study are sketched.

One of the main conclusions is that the effects of the reaction and diffusion rates on stationary properties are quite distinct. Assuming that reaction and diffusion are governed, respectively, by rates $(1-p)\,\phi\left(\delta H/k_B T\right)$ and $p\psi\left(\delta H/k_B T'\right)$, we find that different realizations of $\psi(X)$ induce only minor changes. For example, the *effective* value of p may be modified; it has been proved for some familiar cases that $\psi(X)$ is irrelevant for the limiting cases $T' \to 0, \infty$.[15] The choice of $\phi(X)$, by contrast, has a very decisive influence on the nature of the steady state. For example, the one-dimensional system exhibits long-range order below $T_C(p, T') > 0$ for a family of reaction rates that includes $\phi = 1-\alpha s_x[s_{x-1}+s_{x+1}]+\alpha^2 s_{x-1}s_{x+1}$, where $\alpha \equiv \tanh\left(J/k_B T\right)$. This family of rates has the effect of strongly favoring low-energy configurations. The Metropolis algorithm and other choices induce a zero-temperature critical point, or none at all. It also follows that correlations for $d = 1$ decay exponentially with distance in agreement with expectations.

Mean-field theory suggests that the phase transition exhibited by the one-dimensional system at $T_C(p, T') > 0$ is of second order, in agreement with exact results for fast, completely random diffusion ($T' \to \infty$ and $p \to 1$). In addition, the latter predicts that $m = \left(1/\alpha\right)\sqrt{2\alpha - 1} > 0$ for $\alpha > \alpha_C = \frac{1}{2}$. This is similar to the result obtained for $d = 2$ after using a certain generalization of the Glauber rate. It is also suggested by mean-field theory that long-range order occurs for $d = 1$ at finite T' for any $p > 0$, while MC simulations for $T' \to \infty$ indicate that the transition occurs for $p > \frac{1}{2}$ only, and apparent discontinuities have been observed. Simulations of this one-dimensional case are relatively difficult, however, and this observation may be an artefact due to very slow relaxation. Thus, it would be interesting to perform further numerical studies of this and related cases to compare the prediction in §4.5 with MC data for $d = 1$, to clarify the nature of the phase transition as one varies p, its relation to the classical one occurring for $p \to 1$, and the possible existence of discontinuities and/or a tricritical point.

As expected, the mean-field equations provide an accurate description of many properties of the reaction–diffusion lattice gas. The following mean-field predictions are in accord with exact results. The reaction process induces a slow relaxation of m, on a time scale of order $(1-p)^{-1}$, while diffusion produces very rapid changes of the energy, on a time scale of order p^{-1}. Consequently, a condition of *local equilibrium* (in which most correlations are suppressed) at temperature T' sets in. The mean-field description is naturally independent of the size of the involved

15 These results may not hold, however, for realizations of $\psi(X)$ not considered here, for example, the cases in which H involves further interactions or detailed balance is not satisfied.

domain D as long as $d = 1$, so that the only approximation invoked is a restriction to pair correlations, which is exact at local equilibrium. Thus one obtains perfect agreement with the exact solution, for arbitrary T', in one dimension. The same occurs for $d = 2$ when $T' \to \infty$, which again induces local equilibrium.

For $d = 2$, analytical and MC studies describe a phase transition whose nature depends on both $\phi(X)$ and p, for example, it may vary from second- to first-order as T' is increased. The following types of behavior are observed in simulations when both reaction and diffusion mechanisms are of the Metropolis form: for $p = 0.95$, the transition is of second order so long as $1 < T' \lesssim 5$ (in units of the Onsager critical temperature), the transition is of first order for $T' \gtrsim 6$, and discontinuities are relatively weak, while long-lasting metastable states and very slow evolutions occur, for $T' < 0.8$. That is, the temperature of the (relatively fast) diffusion process is crucial. The speed of the diffusion process is also critical: the two-phase coexistence region disappears as p is decreased, and only continuous transitions occur for $p < 0.83$. The mean-field description (which agrees with the exact solution for $p \to 1$ and $T' \to \infty$ only) reveals the same qualitative picture; nevertheless, the third-mentioned class of behavior shows up as a second-order transition in this instance, and the states are inhomogeneous as $p \to 1$ (but not for $p < 1$). Further study of coexisting phases is motivated by both simulation and mean-field results in this chapter. The Kawasaki diffusion rate does not modify the steady state qualitatively (within the mean-field description). The use of the same rate for the reaction mechanism induces an important change, however. Namely, only second-order transitions occur, with T_C an increasing function of both p and T'.

A finite-size scaling analysis of the simulation data for Metropolis rates reveals that critical behavior is indistinguishable from that of the two-dimensional Ising model. This is consistent with exact results for $T' \to \infty$ and $p \to 1$ indicating a phase transition of first order when $\phi(X)$ is of the Metropolis type, and of second order otherwise. The latter is classical given that $T' \to \infty$, and $p \to 1$ induce mean-field behavior. (Mean-field behavior has been shown to be a consequence of infinite-range diffusion for any $p \leqslant 1$ as well.) The limit $p \to 1$ is singular, however: a relatively slower process with $T' \to \infty$ induces for Metropolis rates a *tricritical point* at $p > 0.83$, so that the phase transition is of first order for $p > 0.83$ only. No similar systematic study has been done for other choices of $\phi(X)$, except for a MC renormalization group analysis which suggests that no first-order transition occurs for Glauber rates when $p < 1$, in agreement with the exact result for $p \to 1$. It would be interesting to resolve the issue of possible changes of the second-order transition as one decreases p from the fast diffusion limit; one might expect a changeover from classical to

Ising-like critical behavior. It would also be interesting to derive transport coefficients for finite T'. The onset of disorder due to diffusion in an antiferromagnetic-like system might also be an interesting phenomenon to study given the related result in §4.4.

Finally, it is worth mentioning some work related to the above. Mendes & Lage (1993) studied the one-dimensional Blume–Emery–Griffiths model subject to reaction–diffusion dynamics. These authors report on the phase diagram and hydrodynamics for this three-state model. Nonequilibrium systems with more than two states have been studied also by Kinzel (1985), Ziff *et al.* (1986) (see chapter 5), Cornell *et al.* (1991), and Marqués (1993), for instance. The last work considers, in particular, a two-dimensional Ising-like system with an average number of mobile vacancies. Glotzer, Stauffer, & Jan (1994) have studied a binary A–B mixture model that simultaneously undergoes spinodal decomposition and reaction $A \rightleftharpoons B$. Systems undergoing reaction are investigated further in subsequent chapters.

5
Catalysis models

The model introduced by Ziff, Gulari, & Barshad (1986) (ZGB) for the oxidation of carbon monoxide (CO) on a catalytic surface has provided a source of continual fascination for students of nonequilibrium phase transitions. This manifestly irreversible system exhibits transitions from an active steady state into absorbing or 'poisoned' states, in which the surface is saturated by oxygen (O) or by CO. The transitions attracted wide interest, spurring development of numerical and analytical methods useful for many nonequilibrium models, and uncovering connections between the ZGB model and such processes as epidemics, transport in random media, and autocatalytic chemical reactions.

The literature on surface reactions continues to expand as variants of the ZGB scheme are explored. In this chapter we do not attempt to give even a partial survey; we define the model, examine its phase diagram, and describe mean-field and simulation methods used to study it.[1]

5.1 The Ziff–Gulari–Barshad model

To begin, let us describe some facts about the oxidation of CO, a catalytic process of great technological importance; see Engel & Ertl (1979). (An immediate poison is converted into a global one!) The reaction, which is catalyzed by various platinum-group metals, proceeds *via* the Langmuir–Hinshelwood mechanism: to react, both species must be chemisorbed. CO molecules adsorb end-on, and require a relatively small area. O_2 molecules require a larger cluster of sites (both atoms bind to the surface), but dissociate upon adsorption — the two atoms are free to react independently with nearby CO molecules. CO_2 molecules desorb upon formation, leaving behind newly-vacated adsorption sites.

[1] For reviews see Evans (1991), and Zhdanov & Kasemo (1994).

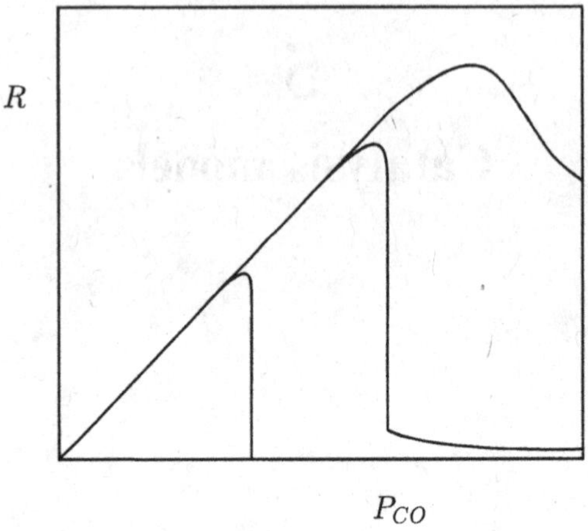

Fig. 5.1. Typical dependence of the rate of CO_2 production on a Pt surface as a function of the partial pressure of CO. The curves represent different temperatures, increasing from left to right.

Experiments on the catalyzed oxidation of CO (Ehsasi *et al.* 1989, Ertl & Koch 1972, Christmann & Ertl 1972, Golchet & White 1978, Matsushima, Hashimoto, & Toyoshima 1979) yield steady-state coverages, and the CO_2 production rate, R, as a function of the partial pressure P_{CO} of CO in the chamber. At sufficiently low temperatures, R grows with increasing P_{CO}, but then suffers a sudden drop when P_{CO} exceeds a certain value, signalling a nonequilibrium phase transition to a CO-saturated surface (see figure 5.1). The CO–O_2–Pt system also exhibits oscillations and waves under certain conditions (Field & Burger 1985, Ertl, Norton & Rüstig 1982, Cox, Ertl & Imbihl 1985).

The transition between nonequilibrium steady states in the CO–O_2–Pt system arises because CO blocks oxygen adsorption. The ZGB model incorporates this feature in a simple set of kinetic rules, defining an interacting particle system as well as a recipe for simulations. The catalytic surface is represented by a two-dimensional lattice (the square lattice is considered here). The transition rules (shown schematically in figure 5.2) involve a single parameter Y, which represents the probability that the next molecule arriving at the surface is CO. Since CO adsorbs end-on it needs only a single vacant site; O_2 requires a NN pair of vacant sites. After each adsorption event, the surface is cleared of any CO–O NN pairs, so that the reaction rate for CO_2 formation is, in effect, infinite. (A more precise definition of the model is given in the following section.) The ZGB rules do not allow CO_2 readsorption and dissociation, or nonreactive desorption, and so are manifestly irreversible, representing a system maintained far from equilibrium. (This is a typical mode of

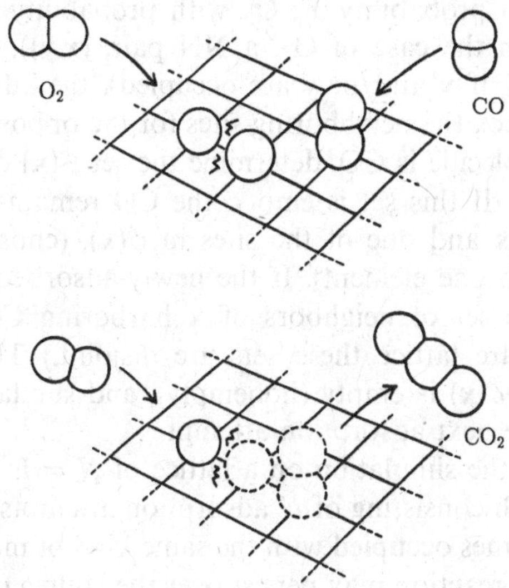

Fig. 5.2. Schematic of adsorption and reaction processes in the ZGB model.

operation in catalytic processes: steady supply of reactants and removal of products prevent the system attaining equilibrium; see §4.1.)

The ZGB model doesn't provide a detailed, realistic description of the reaction, just the essential kinetics. In statistical mechanics a great deal of insight has been gained using simplified models (§2.2), and in keeping with this approach, the ZGB model includes a minimum of detail. However, in one respect, it is *too* symmetric: just as CO blocks adsorption by O, so a surface covered with O atoms cannot adsorb CO. The latter phenomenon is not observed in CO oxidation on platinum-group metals: an oxygen-saturated surface can still take up CO. (This is not to say that oxygen poisoning is ruled out in principle, in all catalytic reactions.) One might easily eliminate oxygen poisoning from the model, for example by forbidding O atoms to occupy neighboring sites, and requiring instead that O_2 adsorb at *second* neighbor pairs, thereby leaving at least half of the surface available for CO. But this, as we shall see, would deprive the model of one of its most intriguing features. Since we are primarily concerned with nonequilibrium phase transitions, rather than with the details of catalytic kinetics, we shall concentrate on the ZGB model despite its variance with experiment as regards oxygen poisoning.

5.2 The phase diagram

To define the ZGB model more precisely, we specify the adsorption and reaction events comprising the dynamics. First choose the adsorbing species — CO with probability Y, O_2 with probability $1 - Y$ — and a lattice site **x** (or, in the case of O_2, a NN pair, (**x**,**y**)), at random. If **x** is occupied (for O_2, if **x** and/or **y** are occupied), the adsorption attempt fails. Otherwise, check the neighboring sites for the opposite species. If the newly-adsorbed molecule is CO, determine the set $\mathcal{O}(\mathbf{x})$ of NNs of **x** that harbor an O atom. If this set is empty, the CO remains at **x**; otherwise it reacts, vacating **x** and one of the sites in $\mathcal{O}(\mathbf{x})$, (chosen at random if $\mathcal{O}(\mathbf{x})$ has more than one element). If the newly-adsorbed molecule is O_2, construct $\mathcal{C}(\mathbf{x})$, the set of neighbors of **x** harboring CO, and similarly $\mathcal{C}(\mathbf{y})$. (On the square lattice these sets are disjoint.) The O atom at **x** remains (reacts) if $\mathcal{C}(\mathbf{x})$ is empty (nonempty), and similarly for the atom at **y**. Proceed to the next adsorption attempt.

Suppose we run the simulation on a lattice of $N = L^2$ sites, for a total of S MC steps (each consisting of N adsorption attempts). It may happen that every site becomes occupied with the same kind of molecule (O or CO 'poisoning'); or the reaction may persist over the course of the simulation, meaning there is always at least one vacant site. Simulations (Ziff, Gulari, & Barshad 1986; Meakin & Scalapino 1987; Jensen, Fogedby, & Dickman 1990) reveal that for $Y < y_1 \approx 0.391$ the lattice eventually poisons with O, while for $Y > y_2 \approx 0.5256$ CO poisoning occurs. A reactive steady state is possible for $y_1 < Y < y_2$. Configurations typifying various Y values are shown in figure 5.3. As the plot of steady-state concentrations (figure 5.4) shows, the transition from the O-poisoned to the reactive state is continuous, while the CO transition is strongly discontinuous. As Y is increased beyond y_2 the CO coverage jumps from just a few percent to unity, and the CO_2 production rate, which grows steadily with Y in the active phase, falls suddenly to zero. Strictly speaking, no *finite* system has a reactive steady state, as it will surely fluctuate into one of the absorbing states, given enough time. For $y_1 < Y < y_2$, however, such fluctuations become exponentially rare (with increasing lattice size), so that in practice, for $L \geq 40$ or so, one is dealing with a very long-lived, reproducible state.

Some aspects of the phase diagram can be understood on the basis of general arguments (Evans 1991). To begin, it is helpful to spend a moment thinking about the simpler monomer–monomer (AB) process, i.e., the ZGB model with *each* species requiring only a single site (Ziff, Gulari, & Barshad 1986; Fichthorn & Ziff 1986). Let p_A denote the probability that a molecule arriving at the surface is of type A, and let θ_v be the fraction of vacant sites. In table 5.1 we list the possible outcomes of an adsorption/reaction event, and their effect on $X \equiv N_A - N_B$. Here $Q\,(\mathrm{A}|\mathrm{v})$ denotes the conditional probability that at least one neighbor of site **x** is occupied by A, given that **x** is vacant, etc.

Collecting terms, we see that X evolves via a one-dimensional random walk with absorbing boundaries at $X = \pm N$ (N is the number of sites).

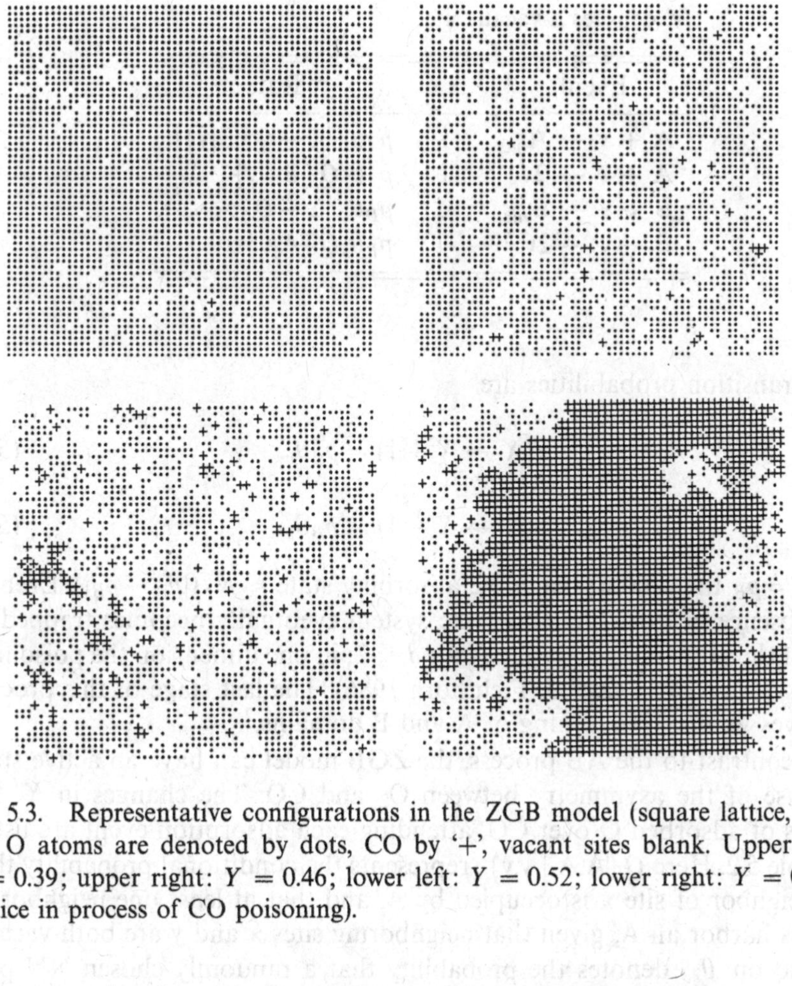

Fig. 5.3. Representative configurations in the ZGB model (square lattice, $L = 60$). O atoms are denoted by dots, CO by '+', vacant sites blank. Upper left: $Y = 0.39$; upper right: $Y = 0.46$; lower left: $Y = 0.52$; lower right: $Y = 0.527$ (lattice in process of CO poisoning).

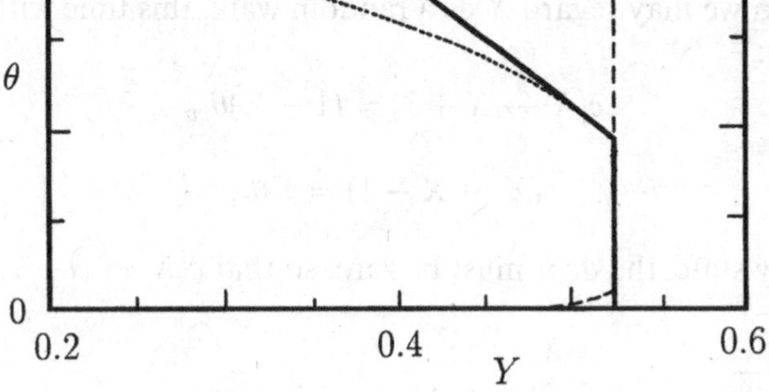

Fig. 5.4. Coverage in the ZGB model. Solid line: from simulations; dashed line: from simulations; dotted line: as predicted by pair mean-field theory.

Table 5.1. The AB monomer–monomer reaction. 'v' denotes a vacant site.

Event	ΔX	Rate
$A + v \to A_{ads}$	1	$p_A\theta_v\left[1 - Q\left(B\vert v\right)\right]$
$A + v \to 2v$	1	$p_A\theta_v Q\left(B \mid v\right)$
$B + v \to B_{ads}$	-1	$p_B\theta_v\left[1 - Q\left(A\vert v\right)\right]$
$B + v \to 2v$	-1	$p_B\theta_v Q\left(A\vert v\right)$

The transition probabilities are

$$c(X \to X + 1) = p_A\theta_v, \tag{5.1}$$

$$c(X \to X - 1) = p_B\theta_v. \tag{5.2}$$

If $p_A \neq p_B$ the time to reach the absorbing state $\approx N/(|p_A - p_B|)$. Even in the absence of drift ($p_A = p_B$), the system eventually becomes trapped at one of the absorbing boundaries for $d \leqslant 2$; in one dimension the poisoning time grows $\propto N^{2.1}$ (Ziff & Fichthorn 1986). The last stage of this process involves gradual coarsening of A and B domains.[2]

In contrast to the AB process, the ZGB model can have an active state because of the asymmetry between O_2 and CO. The changes in X, the excess of adsorbed O over CO, attending each adsorption event are listed in table 5.2. Here $Q\left(\tilde{A},A \mid v,v\right)$, represents the conditional probability that no neighbor of site **x** is occupied by A, and that at least one neighbor of **y** does harbor an A, given that neighboring sites **x** and **y** are both vacant, and so on. θ_{vv} denotes the probability that a randomly chosen NN pair has both sites vacant.

Once again we may regard X as a random walk, this time with transition probabilities:

$$c(X \to X + 2) = (1 - Y)\theta_{vv} \tag{5.3}$$

$$c(X \to X - 1) = Y\theta_v, \tag{5.4}$$

In a steady state, the drift must be zero, so that $c(X \to X - 1) = 2c(X \to$

[2] The effect of clustering in this model, due to an attractive interaction amongst particles of one species, was explored by Silverberg, Ben-Shaul, & Rebentrost (1985) and Silverberg & Ben-Shaul (1987). The structure and dynamics of an A/B interface have been studied by Kang & Weinberg (1993,1994).

Table 5.2. The ZGB model. 'A' represents CO, 'B' oxygen.

Event	ΔX	Rate
$B_2 + 2v \to 2B_{ads}$	2	$(1-Y)\theta_{vv}Q(\tilde{A},\tilde{A}\|v,v)$
$B_2 + 2v \to 2v$	2	$2(1-Y)\theta_{vv}Q(\tilde{A},A\|v,v)$
$B_2 + 2v \to 4v$	2	$(1-Y)\theta_{vv}Q(A,A\|v,v)$
$A + v \to A_{ads}$	-1	$Y\theta_v[1-Q(B\|v)]$
$A + v \to 2v$	-1	$Y\theta_v Q(B\|v)$

$X + 2$), or

$$\frac{Y}{1-Y} = 2P(v \mid v), \tag{5.5}$$

where $P(v \mid v) = \theta_{vv}/\theta_v$ is the conditional probability that a site is vacant, given that one of its neighbors is vacant. Since $P(v \mid v) \leqslant 1$, there can be no steady state for $Y > 2/3$. Thus the lattice must poison with CO for $Y \geq y_2$, where $y_2 \leqslant 2/3$. The coverage will drift in the other direction, towards O poisoning, if $Y/(1-Y) < 2P(v \mid v)$. Should we expect O poisoning at some finite Y, or only in the limit $Y \to 0$? For the latter to occur, we must have $P(v \mid v) \to 0$ as $\theta_v \to 0$. However, new vacancies can only appear at sites adjacent to existing ones, so it is more plausible that $P(v \mid v)$ approaches a nonzero limit. One therefore expects the lattice to poison with O for $Y \leqslant y_1$, for some $y_1 > 0$.

CO poisoning is a first-order transition. That is, states with some range of CO coverages are unstable. To see why, suppose most sites are occupied by CO. Then O_2 can adsorb only at rare vacancy pairs, and O atoms will not remain long before they react with adsorbed CO. Every vacancy, by contrast, is vulnerable to occupation by CO. Hence configurations with most sites occupied by CO are unstable to complete CO saturation. On the other hand, a state with high O coverage is not necessarily unstable, because isolated vacancies are not susceptible to filling by O; only pairs are. At high O coverages, CO adsorption (which can happen at *any* vacancy, isolated or not) usually leads to a reaction; this provides a growth mechanism for vacancy clusters. So states in which the oxygen fraction is close to unity can be stable, and the O poisoning transition is continuous.

The O-poisoning transition is an example of a nonequilibrium critical point: spatial and temporal correlations become long-range in its vicinity. As noted in §2.5, we may introduce various critical exponents to characterize the system in the critical region. For example, the steady-state

Table 5.3. Site mean-field theory for the ZGB model. $\alpha \equiv (1-\theta_{CO})$.

Event	Rate	ΔN_O	ΔN_{CO}
$O_2 + 2v \rightarrow 2O_{ads}$	$(1-Y)\theta_v^2\alpha^2$	2	0
$O_2 + 2v \rightarrow O_{ads} + CO_2$	$2(1-Y)\theta_v^2\alpha(1-\alpha)$	1	-1
$O_2 + 2v \rightarrow 2CO_2$	$(1-Y)\theta_v^2(1-\alpha)^2$	0	-2
$CO + v \rightarrow 2O_{ads}$	$Y\theta_v(1-\theta_O)^4$	0	1
$CO + v \rightarrow CO_2$	$Y\theta_v\left[1-(1-\theta_O)^4\right]$	-1	0

fraction of vacant sites, $\overline{\rho}_v$, obeys a power law: $\overline{\rho}_v \propto (Y-y_1)^\beta$. Critical behavior is discussed further in §5.5.

5.3 Kinetic mean-field theory

As discussed in earlier chapters (see especially chapter 3), mean-field theory is often useful in getting a picture of the phase diagram. We therefore begin with the simplest mean-field description of the ZGB model (Dickman 1986, Fischer & Titulaer 1989, Dumont *et al.* 1990). At this level, the dynamical variables are the CO and O coverages, θ_{CO} and θ_O, respectively. (By *coverage* we mean the fraction of sites occupied by CO or by O.) Their rates of change are determined by considering the effects of the events listed in table 5.2. In general, the rate of a process is the product of an arrival rate per site (Y or $1-Y$), and the probability for finding a given site and its neighbors in the configuration required for that process. We approximate the probability of an n-site cluster in terms of probabilities for clusters of size $n' < n$; in the site approximation $n' = 1$, so there is no information about correlations. For example, the adsorption of CO without reaction requires a vacant site with four neighbors free of oxygen. In the site approximation the rate for this process is $Y\theta_v(1-\theta_O)^4$, where $\theta_v = 1-\theta_O-\theta_{CO}$ is the fraction of vacant sites. The contribution of a given process to $d\theta_i/dt$ is its rate times the change in the number of adsorbed molecules of type i. Enumerating all of the processes and their associated rates (see table 5.3), we obtain the equations governing the coverages:

$$\frac{d\theta_O}{dt} = 2(1-Y)\theta_v^2(1-\theta_{CO})^3 - Y\theta_v\left[1-(1-\theta_O)^4\right] \tag{5.6}$$

and

$$\frac{d\theta_{CO}}{dt} = Y\theta_v(1-\theta_O)^4 - 2(1-Y)\theta_v^2\left[1-(1-\theta_{CO})^3\right]. \tag{5.7}$$

These equations exhibit two kinds of steady states: absorbing states with $\theta_v = 0$, and active steady states.

To find the active steady-state solution, notice that

$$\frac{\mathrm{d}}{\mathrm{d}t}(\theta_O - \theta_{CO}) = 2(1-Y)\theta_v^2 - Y\theta_v, \tag{5.8}$$

which implies

$$\overline{\theta}_v = \frac{Y}{2(1-Y)} \equiv Z, \tag{5.9}$$

if $\theta_v \neq 0$. When substituted in equation (5.6) this yields

$$(Z + \overline{\theta}_{CO})^4 + (1 - \overline{\theta}_{CO})^3 - 1 = 0. \tag{5.10}$$

For $Z < Z_s = 0.638986$, corresponding to $Y = 0.561013 \equiv Y_s$, this equation has two solutions, θ_- and θ_+ in the interval [0, 1]. The former solution is locally stable, the latter unstable. The solutions coalesce at Y_s; above this value a CO-poisoned lattice is the only stationary solution. For $Y \leqslant Y_s$, starting from an empty lattice, equations (5.6) and (5.7) approach an active state with CO coverage θ_-. For $Y > Y_s$ the mean-field equations attain the CO-poisoned state. Fischer & Titulaer (1989) showed that in the one-site approximation, the CO-poisoned state is locally stable for $Y < Y_s$, and that its domain of attraction expands to include all possible initial values as $Y \rightarrow Y_s$.

Y_s marks a spinodal point, the limit of stability of the active phase (see chapter 3). In equilibrium, phase boundaries are reckoned on the basis of minimum free energy. Out of equilibrium there is usually no free energy function, and phase boundaries can only be defined through a dynamical criterion. For example, Ziff, Gulari, & Barshad (1986) simulated the model starting with half the lattice vacant, and half covered with CO; they found an active steady state for $y_1 < Y < y_2 = 0.525$. It is not possible to represent such a segregated initial state faithfully in a mean-field theory; the best one can do is to take initial coverages $\theta_{CO}(0) = 1/2$, $\theta_O(0) = 0$. One finds a first-order transition to the CO-poisoned phase at $Y = 0.4787$, about 9% less than the value found in simulations. A more reliable approach is an *inhomogeneous* mean-field theory, in which local coverages $\theta_{x,O}$ and $\theta_{x,CO}$ at sites x of a one-dimensional lattice evolve *via* the analog of equations (5.6) and (5.7) (Fischer & Titulaer 1989). (Coverages at site x are coupled to those at $x \pm 1$ and $x \pm 2$.) The idea is to study a system (assumed uniform in one direction) with an interface between reactive and CO-poisoned regions, and to determine the value of Y that permits a stationary interface. This analysis yields a coexistence value of $Y_c = 0.5197$, quite close to the simulation result.

Thus the simplest mean-field description gives a plausible account of CO poisoning, and yields quantitative results for the phase boundary if one studies an interface. However, it fails to predict O poisoning. Another flaw is that for $Y > y_2$ the solution converges to a CO-poisoned state with a small but *nonzero* O coverage (typically a few tenths of a percent). This unphysical situation results from ignoring the strict prohibition against CO–O NN pairs.

Site mean-field theory is readily extended to include nonreactive desorption of CO by including a term $-k\theta_{CO}$ on the r.h.s. of equation (5.7). (Desorption of CO occurs at a significant rate at higher temperatures; the rate for oxygen is much smaller.) With CO desorption there is no longer a CO-poisoned state, but for small k there is a discontinuous transition between low and high CO coverages. Above a critical desorption rate the coverages vary smoothly with Y (Brosilow & Ziff 1992a, Evans 1992, Tomé & Dickman 1993). The same phenomenon is observed in experiments: at high temperatures the CO desorption rate is large enough to remove the transition (see figure 5.1).

The site approximation is but the first in a sequence of *cluster* mean-field theories in which one follows the populations of $2, 3, \ldots, n$-site clusters. As discussed in chapter 3 for the driven lattice gas, the results generally improve with n, but there is no guarantee (ben-Avraham & Köhler 1992a, Garrido *et al.* 1996). For the ZGB model, the pair approximation offers a considerable advantage over site mean-field theory, as CO–O NN pairs can be excluded, and correlations which affect reactions and O_2 adsorption are partly taken into account. At this level the system is described in terms of NN pair fractions; there are four independent variables in this rather more complicated formulation.[3] The pair approximation predicts a continuous O-poisoning transition at $y_1 = 0.2497$ (versus 0.391 from simulation). For $Y > y_1$ the vacancy density grows linearly with Y, so that the mean-field value of β is 1. Higher-order approximations are expected to yield the same value, because the mean-field prediction for the order parameter $\theta_v(Y)$ crosses zero with a finite slope. In pair approximation the transition to CO poisoning occurs (for an initially half poisoned lattice) at $Y = 0.5241$, in very close agreement with simulation. The spinodal value Y_s is the same as in the site approximation.

The pair approximation is the simplest mean-field theory permitting study of diffusion (Jensen & Fogedby 1991). Evans (1993a) has determined the phase diagram in the high-diffusion-rate limit. Mean-field theory has also been used to determine the phase diagram in the case of a finite reaction rate (Considine, Takayasu, & Redner 1990; Köhler & ben-Avraham 1992). An important conclusion from both mean-field theory and simulations is that the phase diagram for a finite reaction rate is qualitatively the same as in the ZGB model. In particular, the reactive window persists

[3] For details see Dickman (1986). Illustrations of pair and higher-order mean-field theory are found in chapters 3 and 6.

down to very low reaction rates (Köhler & ben-Avraham 1992). Another interesting application is an analysis of poisoning times in finite systems *via* a mean-field approximation to the master equation (ben-Avraham *et al.* 1990b). This work supports the idea that for Y in the reactive range, the lifetime grows exponentially with lattice size.

5.4 Simulation methods

In this section we examine simulation methods used for studying phase transitions in the ZGB and related models. As discussed in chapter 2, simulations near a critical point are hampered by finite-size effects and slow relaxation. Long simulations in the vicinity of a poisoning transition are especially problematic since the system may abruptly fluctuate into the absorbing state. Before considering specific strategies, we mention a simple modification that can improve efficiency by orders of magnitude. If adsorption sites are chosen at random, then many trials must be rejected, and most are wasted if $\theta_v << 1$. Choosing from a list of vacant sites eliminates much of this waste: CO adsorption attempts are sure to be successful, and the success rate of O_2 adsorption is also improved. When using a vacancy list, the time per event is no longer constant: selection from a list of N_v vacant sites corresponds to a time increment of N_v^{-1} in the original formulation, since on average N/N_v 'blind' trials are required to hit a vacant site.

In steady-state simulations, the goal is to determine the stationary coverages $\overline{\theta}_O$ and $\overline{\theta}_{CO}$ as functions of Y and L (on a lattice of $L \times L$ sites). Obviously, a preliminary requirement is that the data represent the steady state. Suppose one computes mean coverages over blocks of D lattice updates. As noted in chapters 2 and 4, the portion of the data that must be discarded as representing an initial transient can be estimated by plotting the block averages and estimating the onset of steady conditions, or by computing autocorrelation functions for the coverages. Uncertainty estimates can be obtained from the standard deviation of the mean, computed over a set of blocks, provided their duration is $\geq \tau$, the relaxation time (Binder 1979). τ may again be estimated by computing the autocorrelation function. A simple alternative is to evaluate standard deviations for a sequence of increasing block sizes (Flyvbjerg & Petersen 1989).

Finite-size effects are a familiar problem in simulations of critical phenomena. The usual problems — strong dependence of coverages upon L, shifting of the (apparent) critical point from its limiting ($L \to \infty$) value, and the rounding of critical singularities — affect the ZGB model near the O-poisoning transition, when $L < \xi$, the (infinite system) correlation length. A conservative procedure is to use a sequence of lattice sizes, say

$L = 2^n l$, and to discard data that are not reproduced when L is doubled. Instead of simply rejecting data for small lattices, one may use results for various L in a finite-size scaling analysis, as described in chapter 6.

In the simulations described thus far, the CO probability Y is fixed whilst the coverages fluctuate. Ziff & Brosilow (1992) devised an alternative approach, in which the CO coverage is fixed and Y fluctuates. (One may regard the usual fixed-Y simulations as analogous to an equilibrium grand canonical ensemble simulation, in which the density fluctuates subject to a fixed chemical potential, while constant coverage simulations are more like canonical ensemble simulations. Note, however, that away from equilibrium we do not have a theorem establishing the strict equivalence of 'ensembles.') Constant coverage simulations are particularly useful near the first-order transition, and also allow one to study the metastable extension of the reactive state. The simulation procedure is modified in only one respect: the choice of the adsorbing molecule. Let $\tilde{\theta}_{CO}$ be the desired or 'target' value of the CO coverage θ_{CO}. Whenever $\theta_{CO} < \tilde{\theta}_{CO}$ an attempt is made to adsorb CO; if $\theta_{CO} > \tilde{\theta}_{CO}$ O_2 adsorption is attempted. Once a steady state is attained, θ_{CO} is effectively pinned at $\tilde{\theta}_{CO}$. The Y value corresponding to this coverage is estimated from $\overline{Y}$, the fraction of CO adsorption attempts (successful or not) over the run. While the rule for choosing the next adsorbing molecule may seem arbitrary, it should yield results identical to the usual method for large systems, since the identity of the next adsorbing molecule in any small region of the lattice is random. Using the constant coverage method, Ziff & Brosilow determined the location of the first-order transition to high precision: $y_2 = 0.52560(1)$,[4] and studied metastable states, interfaces between active and CO-poisoned regions, and the effects of CO desorption (Brosilow & Ziff 1992a). This method should prove useful in simulations of other first-order nonequilibrium transitions.

As mentioned above, the ZGB model with CO desorption at rate k exhibits, for small k, a discontinuous transition between high and low CO coverage phases, while for larger k the transition disappears. An efficient method for determining the critical value $k_c(L)$ marking the boundary between these regimes is to study coverage histograms (Marro *et al.* 1985; Landau 1990). The range of CO coverages is divided into bins of width δ, and the frequencies $N(i)$ for each bin are recorded. For $k < k_c$, there is a range of Y (around the 'coexistence' value) for which the histogram $N(i)$ has a bimodal structure, indicating distinct phases. As Y is varied the relative heights of the peaks should shift, while their positions remain constant. For $k > k_c$ there is only a single peak, while just at the critical

[4] Here and in what follows, numbers in parentheses denote the uncertainty in the last figure.

point $N(i)$ exhibits a plateau, reflecting large fluctuations in coverage. This approach, which has been used for some time in equilibrium studies, was applied to the ZGB model with CO desorption, revealing an *Ising-like* critical point at $k_c = 0.04060(5)$ (Tomé & Dickman 1993).

One approach to characterizing the critical behavior of the ZGB model (at the oxygen-poisoning transition) is through the exponents β and $\nu_\perp$. However, a precise determination is difficult in steady-state simulations. The *time-dependent* simulation method (Grassberger & de la Torre 1979, Grassberger 1982) follows the evolution from an initial configuration very near the absorbing state, and yields exponents describing critical dynamics (see §6.2). In time-dependent simulations of the ZGB model (Jensen *et al.* 1990), one generates a large set of trials, all starting from the same configuration: a lattice covered with oxygen save for one NN vacancy pair; the evolution is followed up to time t_m. The scaling hypothesis (see chapter 6) forming the basis for this approach asserts that at the critical point, the survival probability, $P(t)$, the mean number of vacancies, $n(t)$, and the mean-square spread of vacancies, $R^2(t)$, follow asymptotic power laws (Grassberger & de la Torre 1979),

$$P(t) \propto t^{-\delta}, \tag{5.11}$$

$$n(t) \propto t^{\eta}, \tag{5.12}$$

and

$$R^2(t) \propto t^{z}. \tag{5.13}$$

Off-critical evolutions show clear deviations from these power laws, permitting the critical point y_1 to be fixed within narrow limits. For equations (5.11)–(5.13) to hold over the course of the simulation, it is necessary that the lattice be sufficiently large that vacancies never appear at the boundaries for $t \leqslant t_m$. Since the fraction of vacant sites remains quite small, use of a vacant site list is now imperative. The log–log plots of figure 5.5 show the power laws characteristic of the critical evolution. The exponents δ, η, and z may be estimated to good accuracy by studying the *local slopes*

$$\delta(t) = \frac{\ln[P(rt)/P(t)]}{\ln r}, \tag{5.14}$$

etc., as shown in figure 5.6 (in this study, $r = 5$). This analysis serves to minimize the effects of finite-time corrections to the power laws.

Time-dependent simulations have also been used to study spreading of the active phase from a lattice nearly poisoned with CO, for $Y \leqslant y_2$. This 'epidemic analysis' reveals that the survival probability and the number of vacant sites are governed by power laws at the first-order transition at $Y = y_2$ (Evans & Miesch 1991). However, the exponents are nonuniversal, and exhibit a strong dependence on the reaction rate.

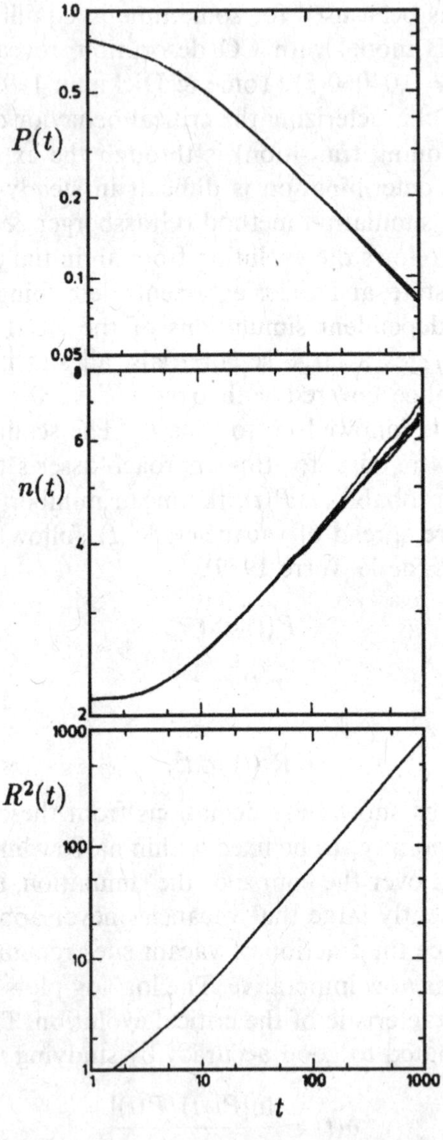

Fig. 5.5. Evolution of the survival probability, $P(t)$, mean number of vacant sites, $n(t)$, and mean square distance of spreading of vacant sites, $R^2(t)$, in time-dependent simulations of the ZGB model. Each graph contains four curves — hardly distinguishable from each other — with (from bottom to top) $Y = 0.3905$, 0.3906, 0.3907, and 0.3908 (Jensen *et al.* 1990).

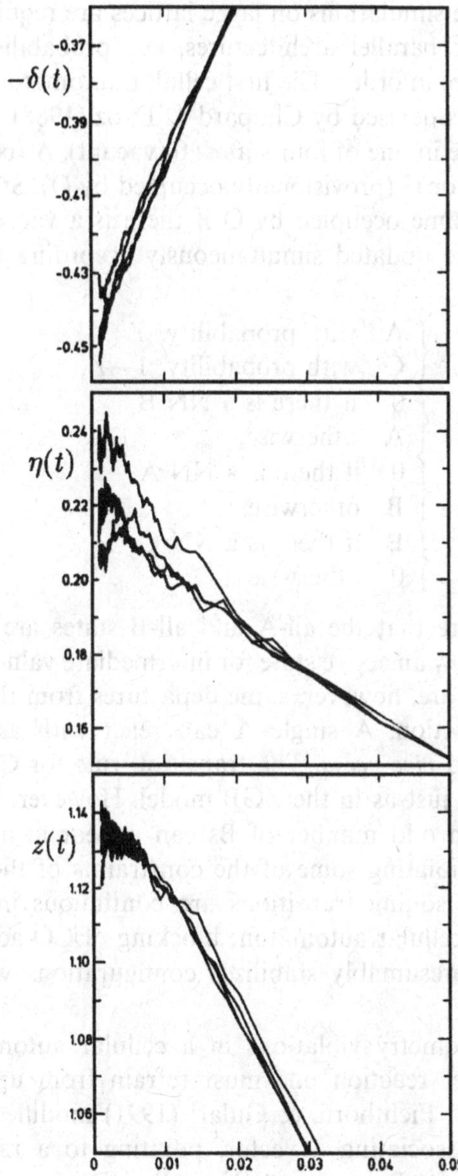

Fig. 5.6. Local slopes $-\delta(t), \eta(t)$, and $z(t)$ associated with the data of figure 5.5 (Jensen *et al.* 1990).

In order to derive reliable steady-state properties near the O-poisoning transition, extensive simulations on large lattices are required. Algorithms suited to vector or parallel architectures, i.e., probabilistic cellular automata, are therefore in order. The first cellular automaton for the CO–O_2 surface reaction was devised by Chopard & Droz (1988). Each site in the square lattice can be in one of four states: 0 (vacant), A (occupied by CO), B (occupied by O), or C (provisionally occupied by O). State C represents a site that will become occupied by O if there is a vacant NN for O to adsorb. All sites are updated simultaneously according to the following transition rules:

$$
\begin{aligned}
0 &\rightarrow \begin{cases} \text{A} & \text{with probability } Y \\ \text{C} & \text{with probability } 1-Y, \end{cases} \\
\text{A} &\rightarrow \begin{cases} 0 & \text{if there is a NN B} \\ \text{A} & \text{otherwise,} \end{cases} \\
\text{B} &\rightarrow \begin{cases} 0 & \text{if there is a NN A} \\ \text{B} & \text{otherwise,} \end{cases} \\
\text{C} &\rightarrow \begin{cases} \text{B} & \text{if there is a NN C} \\ 0 & \text{otherwise.} \end{cases}
\end{aligned}
$$

These rules ensure that the all-A and all-B states are absorbing, and simulations also show an active state for intermediate values of Y, as in the ZGB model. There are, however, some departures from the stoichiometry of the original reaction. A single A can react with as many as four neighboring Bs, and *vice versa.* The transition rule for C prevents single Bs from adsorbing, just as in the ZGB model. However, in contrast with the ZGB model, an odd number of Bs can appear at a single step. As a consequence of violating some of the constraints of the reaction, both the CO- and O-poisoning transitions are continuous in the Chopard–Droz probabilistic cellular automaton; blocking of CO adsorption by the transient C state presumably stabilizes configurations with a high CO coverage.

To avoid stoichiometry violations in a cellular automaton model of the monomer–dimer reaction one must refrain from updating all sites simultaneously. Ziff, Fichthorn, & Gulari (1991) modified the Chopard–Droz scheme by associating a vector, pointing to a randomly chosen neighbor, with each A, B, or C site. (New directions are assigned at each time step.) The reactions $\text{A}+\text{B} \rightarrow 2\text{v}$ and $2\text{C} \rightarrow 2\text{B}$ are only allowed between NN sites with vectors directed toward each other. (That is, on average only $1/16$ of the pairs are able to react at any step.) Simulations confirm that by eliminating the possibility of a molecule participating in more than one reaction, the transitions are of the same order as in the original ZGB model.

Another cellular automaton that preserves the stoichiometry of the ZGB

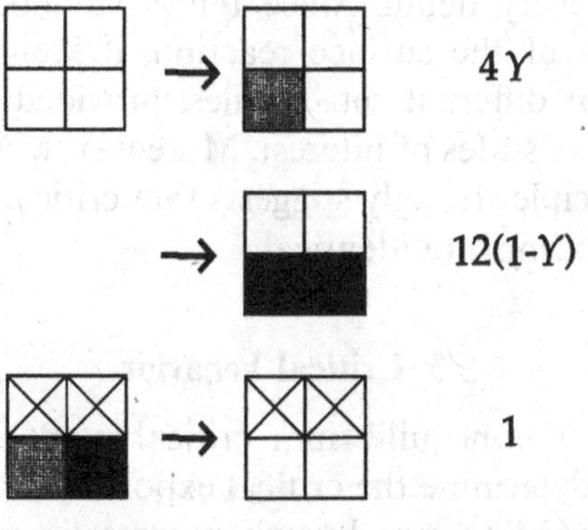

Fig. 5.7. A few of the transitions in the Mai–von Niessen cellular automaton for the CO + O_2 reaction. Vacant sites are represented by white squares, CO by gray, and O by black. '×' denotes a site in an unspecified state. The expressions represent (unnormalized) transition probabilities.

model was devised by Mai & von Niessen (1991). The lattice is partitioned into 2 × 2 *blocks* of sites; multiple reactions are avoided by updating blocks rather than individual sites. The complete list of block transition probabilities is rather long; here we only display some representative examples (see figure 5.7). The probability that a CO molecule adsorbs at one of the sites in a vacant block is taken as $4Y/A$, where $A(Y)$ is a normalization factor. Similarly, the probability of two COs adsorbing in such a block is assigned the value $12Y^2/A$. O_2 adsorbs only at empty NN pairs within a block; the probability of O_2 adsorbing in an empty block is taken (somewhat arbitrarily) to be $12(1-Y)/A$. CO–O NN pairs within a block are removed with probability 1. And if all four sites are occupied by the same species, no change occurs within that block. On a given iteration, all blocks whose lowest, leftmost site is of the form $\mathbf{x} = (2i, 2j)$ are updated. On subsequent iterations the grid defining the blocks is shifted, so that, for example, the lowest, leftmost site is of the form $\mathbf{x} = (2i+1, 2j)$, and so on. Thus any CO–O pair will react after at most four iterations.

Compared with the scheme devised by Ziff, Fichthorn, & Gulari, this probabilistic cellular automaton offers some advantages of computational efficiency at the cost of transition probabilities somewhat further removed from those of the original model. The method has also been extended to include surface diffusion (Chopard, Droz, & Kolb 1989, Mai & von Niessen 1993). In each of the probabilistic cellular automata, transition probabilities are chosen to parallel those of the ZGB model, but are not precisely equivalent. Indeed there appears to be no way for probabilistic cellular automaton rules to reproduce the dynamics of a sequentially updated system in every detail. Since the ZGB model itself represents a vast simplification of the surface reaction, it seems perfectly sensible to study a somewhat different set of rules, provided they reproduce the absorbing and reactive states of interest. Moreover, while there is no proof, the universality principle strongly suggests that critical exponents for PCA and sequential algorithms are identical.

5.5 Critical behavior

The appearance of a nonequilibrium critical point in the ZGB model prompted efforts to determine the critical exponents, and to learn whether this transition belonged to any known universality class. Early on, the mean-field value ($\beta = 1$) was ruled out by simulations, which placed β in the range 0.6–0.7 (Meakin & Scalapino 1987).[5] After studies of a simpler system (Dickman & Burschka 1988) suggested a connection between oxygen poisoning and *directed percolation* (chapter 6), a strong case for this identification was advanced by Grinstein, Lai, & Browne (1989). Time-dependent simulations provided further numerical support (Jensen *et al.* 1990).

The *A model* studied by Dickman & Burschka (1988) is a single-component system that retains the essential features of the ZGB model near the O-poisoning transition. Each site of the lattice is either vacant or occupied. Particles continually rain down on the lattice, so that vacant sites become occupied at rate λ, independent of the neighboring sites. Desorption, by contrast, is cooperative: the desorption rate is unity for particles with at least one vacant neighbor, zero otherwise. The particles are analogous to O atoms in the ZGB model, which cannot leave the lattice while surrounded by other atoms; the absorbing state is devoid of vacant sites.

Both simulations and mean-field theory predict a continuous 'poisoning' transition in the A model. The system is simple enough to permit precise analysis using series expansions (Dickman 1989a, Dickman & Jensen 1991), as well as simulations. These studies show that the critical exponents of the A model are the same as for directed percolation. In fact, the A model is a variant of the contact process (see chapter 6), which belongs to the directed percolation universality class. There are, however, two differences between the ZGB system and the A model that might spoil the correspondence: pairwise destruction of vacancies (in O_2 adsorption), and the presence of two components. Grinstein *et al.* (1989) argued that these

[5] In fact one may identify *two* exponents, β_v and β_{CO}, which describe the growth of $\overline{\theta}_v$ and $\overline{\theta}_{CO}$, respectively, for $Y > y_1$. While these exponents may be determined independently in simulations, we expect them to be identical, on the basis of universality.

distinctions are irrelevant to critical behavior. Their reasoning is based upon an important principle proposed by Janssen (1981) and Grassberger (1982): *continuous transitions to an absorbing state belong generically to the directed percolation universality class.* We refer to this as the *DP conjecture*; its basis is a coarse-grained formulation to be discussed in §6.6. Grinstein *et al.* proposed a continuum description of the ZGB model (in the vicinity of O-poisoning):

$$\frac{d\mathbf{\Theta}}{dt} = A\mathbf{\Theta} - \mathbf{\Theta}B\mathbf{\Theta} + \cdots + D\nabla^2\mathbf{\Theta} + \mathbf{X}, \tag{5.15}$$

where $\mathbf{\Theta} \equiv (\theta_v, \theta_{CO})$, and the elements of A and B are functions of Y. $\mathbf{X}$ is a Gaussian noise term whose covariance matrix is proportional to the local density of vacancies, and so vanishes in the absorbing state. The critical point y_1 is marked by neutral stability, i.e., the larger eigenvalue of matrix A vanishes, while (for stability) B remains positive. Barring an extraordinary coincidence, only one of the modes 'goes critical' at y_1 (the other eigenvalue of A remains negative) so that in the vicinity of the critical point, an effective single-component description holds, and the critical behavior is the same as in the contact process. This prediction is supported by numerical exponent estimates: time-dependent simulations (Jensen *et al.* 1990) yield $\delta = 0.452(8), \eta = 0.224(10)$, and $z = 1.139(5)$, in very good agreement with the values for directed percolation (see table 6.1). Steady-state simulations of probabilistic cellular automata yielded $\beta \approx 0.55$ (Chopard & Droz 1988) and, in a more extensive study, $\beta = 0.58(2)$ (Mai & von Niessen 1991), again in close agreement with the directed percolation exponent. As expected on the basis of the DP conjecture, the critical behavior of the ZGB model is insensitive to diffusion (Jensen & Fogedby 1991), and to anisotropy in the adsorption rates (Yu, Browne, & Kleban 1991).

The phase diagram and critical behavior of the ZGB model are now rather well understood, but several puzzles remain. There are surprising changes when the ZGB model is placed on a fractal substrate. Motivated by the observation that catalytic surfaces may be fractal, Albano (1990) studied the model on percolation clusters, and found that *both* the CO- and O-poisoning transitions are continuous. (Casties, Mai, & von Niessen (1993) confirmed this, but found that CO poisoning resumes its discontinuous nature when CO diffusion is permitted.) Similar results are found on diffusion-limited aggregation (DLA) clusters (Mai, Casties, & von Niessen 1993), on randomly diluted lattices (Hovi *et al.* 1992), and on the Sierpiński gasket (Albano 1994a; Tretyakov & Takayasu 1991). An explanation based on the role of so-called 'red bonds' which provide the sole link between 'blobs' of multiply-connected sites was proposed by ben-Avraham (1993). Another modification that renders CO poisoning continuous is a repulsive interaction reducing the adsorption rate for CO at sites with one or more NNs occupied by CO (Satulovsky & Albano 1992). Other work has examined nucleation and the nature of interfaces in the ZGB model (Brosilow, Gulari, & Ziff 1993; Evans & Ray 1994). The interface dynamics belongs to the KPZ class (Kardar, Parisi, & Zhang 1986, Krug & Spohn 1992). Two particularly interesting systems are the dimer–trimer and dimer–dimer models (Köhler & ben-Avraham 1991, Maltz & Albano 1992, Albano 1992, 1994b), the latter representing surface-catalyzed formation of water from H_2 and O_2. These models have multiple absorbing states, a topic examined in §9.4.

6
The contact process

The preceding chapter showed how absorbing state transitions arise in catalytic kinetics. Having seen their relevance to nonequilibrium processes, we turn to the simplest example, the *contact process* (CP) proposed by T.E. Harris (1974) as a toy model of an epidemic (see also §8.2). While this model is not exactly soluble, some important properties have been established rigorously, and its critical parameters are known to high precision from numerical studies. Thus the CP is the 'Ising model' of absorbing state transitions, and serves as a natural starting point for developing new methods for nonequilibrium problems. In this chapter we examine the phase diagram and critical behavior of the CP, and use the model to illustrate mean-field and scaling approaches applicable to nonequilibrium phase transitions in general. Closely related models figure in several areas of theoretical physics, notable examples being Reggeon field theory in particle physics (Gribov 1968, Moshe 1978) and directed percolation (Kinzel 1985, Durrett 1988), discussed in §6.6. We close the chapter with an examination of the effect of quenched disorder on the CP.

6.1 The model

In the CP each site of a lattice (typically the d-dimensional cubic lattice, $\mathbf{Z}^d$) represents an organism that exists in one of two states, healthy or infected. Infected sites are often said to be 'occupied' by particles; healthy sites are then 'vacant.' The infection spreads through NN contact, an infected site passing the disease to its healthy neighbors at rate λ/q on a lattice with coordination number q. Infected sites recover at unit rate, and are immediately susceptible to reinfection. The rates for the one-dimensional CP are shown in figure 6.1. Since an organism must have a sick neighbor to become infected, the disease-free state is absorbing; persistence of the epidemic depends on the infection rate λ. We may

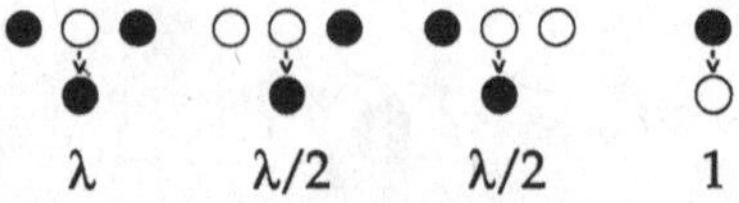

Fig. 6.1. Rates for the one-dimensional CP.

view the CP as a process in which some transient condition or excitation spreads by a short-range influence, the excited species becoming extinct unless it spreads sufficiently rapidly. The boundary between survival and extinction is marked by a critical point that typifies transitions into an absorbing state.

Although the CP is a continuous-time Markov process, a discrete-time formulation is often employed in simulations. Each step involves randomly choosing a process — creation with probability $v = \lambda/(1+\lambda)$, annihilation with probability $1-v$ — and a lattice site $\mathbf{x}$. In an annihilation event, the particle (if any) at $\mathbf{x}$ is removed. Creation proceeds only if $\mathbf{x}$ is occupied and a randomly chosen NN $\mathbf{y}$ is vacant; if so, a new particle is placed at $\mathbf{y}$. Time is incremented by Δt after each step, successful or not. (Normally one takes $\Delta t = 1/N$ on a lattice of N sites, so that a unit time interval, or MC step, corresponds, on average, to one attempted event per site.) To simulate the continuous-time process one replaces the fixed interval Δt with an exponentially-distributed waiting time. While they differ somewhat at short times, the discrete and continuous formulations share the same stationary properties and long-time dynamics. The simulation strategies for surface reaction and other nonequilibrium models described in previous chapters are equally applicable to the CP. In particular, efficiency is greatly improved through use of a particle list, analogous to the vacancy list advocated in simulations of the ZBG model. If $\mathbf{x}$ is chosen from a list of N_{occ} sites, $\Delta t = 1/N_{occ}$.

6.2 The phase diagram

Can the epidemic persist indefinitely, or must it become extinct? Under what conditions is the stationary density — the fraction of occupied sites — nonzero? A mean-field analysis gives a preliminary answer. Let $\sigma_{\mathbf{x}}(t) = 1(0)$ when site $\mathbf{x}$ is occupied (vacant) at time t. The equation of motion for $\rho(\mathbf{x}, t) \equiv \text{Prob}[\sigma_{\mathbf{x}}(t)=1]$ is

$$\frac{\mathrm{d}}{\mathrm{d}t}\rho(\mathbf{x}, t) = -\rho(\mathbf{x}, t) + \frac{\lambda}{q}\sum_{\mathbf{y}} \text{Prob}[\sigma_{\mathbf{x}}(t)=0, \sigma_{\mathbf{y}}(t)=1], \tag{6.1}$$

where the sum is over the NNs of $\mathbf{x}$. (Evidently the first term on the r.h.s. represents annihilation, while the second accounts for creation at a vacant site $\mathbf{x}$ by a particle at $\mathbf{y}$.) This relation is exact, but requires that we know the two-site probability appearing in the birth term. The equations for two-site probabilities bring in those for three, and so on. The simplest mean-field approximation truncates this infinite hierarchy at the level of $\rho(\mathbf{x}, t)$ by treating the occupancy of each site as statistically independent. If we also assume spatial homogeneity ($\rho(\mathbf{x}) \equiv \rho$), equation (6.1) becomes the Malthus–Verhulst equation:

$$\frac{d\rho}{dt} = (\lambda - 1)\rho - \lambda\rho^2. \tag{6.2}$$

For $\lambda \leqslant 1$ the only stationary solution is the vacuum, $\overline{\rho} = 0$. (The bar denotes a stationary value.) However, for $\lambda > 1$ there is also an *active* stationary state with $\overline{\rho} = 1 - \lambda^{-1}$. Notice that for $\lambda > \lambda_c = 1$ the active state is stable, the vacuum unstable. λ_c marks a critical point: the stationary density changes in a continuous but singular manner at λ_c. $\overline{\rho}$ is the stationary order parameter; it is nonzero only for $\lambda > \lambda_c$. If we let $\Delta \equiv \lambda - \lambda_c$, then for $\Delta > 0$, $\overline{\rho} = \Delta + \mathcal{O}(\Delta^2)$, according to equation (6.2). Near a critical point the order parameter follows a power law,

$$\overline{\rho} \propto \Delta^\beta \qquad (\Delta > 0). \tag{6.3}$$

Thus mean-field theory yields $\beta = 1$. It also predicts critical slowing-down —divergence of the relaxation time τ at the critical point. From equation (6.2) we see that for $\lambda \neq 1$,

$$\rho(t) - \overline{\rho} \propto \exp(-|1 - \lambda|t) \tag{6.4}$$

as $t \to \infty$, so that $\tau = |1 - \lambda|^{-1}$; for $\lambda = 1$ we have power-law decay: $\rho(t) \propto t^{-1}$. (Note the similarity with the mean-field critical behavior found for the branching process in chapter 1.)

Suppose we include a source of particles: spontaneous infection at vacant sites, at rate h, in addition to infection by contact. This of course eliminates the absorbing state. The mean-field equation for the density is now

$$\frac{d\rho}{dt} = (\lambda - 1)\rho - \lambda\rho^2 + h(1 - \rho), \tag{6.5}$$

and the stationary solution is

$$\overline{\rho} = \frac{1}{2\lambda}\left[\lambda - h - 1 + \sqrt{(\lambda - h - 1)^2 + 4\lambda h}\right]. \tag{6.6}$$

The source removes the phase transition, just as an external magnetic field does in the Ising model. At the critical point, $\lambda = 1$, we have $\overline{\rho} \propto \sqrt{h}$ as $h \to 0$. Defining the field exponent δ_h through $\overline{\rho}(\lambda_c, h) \propto h^{1/\delta_h}$, we see that mean-field theory gives $\delta_h = 2$.

This simple mean-field analysis provides some insight into the phase diagram, and turns out to be qualitatively correct, though (not surprisingly) quantitatively wrong regarding the exponents and λ_c. (More precisely, the mean-field exponents are incorrect for spatial dimensions $d < d_c = 4$ (Obukhov 1980). d_c denotes the *upper critical dimension*, above which fluctuations are too weak to change the critical exponents from their mean-field values.) The error lies in treating different sites as independent; in fact they are highly correlated. Alternatively, we may view the mean-field picture as a model in which rapid diffusion or stirring destroys any correlations generated by the reactions, as considered in detail in chapter 4 for the reaction diffusion model. The mean-field analysis given above is essentially Schlögl's (first) model of a well-stirred autocatalytic chemical system (Schlögl 1972).

There is an alternative formulation of mean-field theory that allows for spatial variation but neglects fluctuations. It is convenient to adopt a continuum description, in which $\phi(\mathbf{x}, t)$ represents the local particle density. This density field evolves via the same local dynamics as in equation (6.2), together with a diffusive coupling:

$$\frac{\partial \phi}{\partial t} = \Delta\phi - b\phi^2 + D\nabla^2\phi. \tag{6.7}$$

The precise values of b and D are not terribly important, so long as they are positive. With the addition of a suitable noise term, equation (6.7) is in fact the minimal-field theory of the CP (see §6.6). For the moment, we consider the noiseless version, a simple reaction–diffusion equation, which allows us to discuss spatial correlations. For $\Delta < 0$, the stationary value of ϕ is zero, so at long times ϕ is small and we may neglect the term $\propto \phi^2$. For $\Delta > 0$, we require the quadratic term for stability, but may consider deviations from the homogeneous steady state by introducing $\psi \equiv \phi - (\Delta/b)$, which obeys $\partial\psi/\partial t = -\Delta\psi - b\psi^2 + D\nabla^2\psi$. In either case then, neglect of the nonlinear term leads to an equation of the form

$$\frac{\partial \psi(\mathbf{x}, t)}{\partial t} = -|\Delta|\psi + D\nabla^2\psi, \tag{6.8}$$

which clearly describes a diffusing exponentially-decaying density. In one dimension, for instance, the Green's function for equation (6.8) is

$$G_0(\mathbf{x}, t) = (4\pi D t)^{-1/2} \exp\left(-|\Delta| t - \frac{x^2}{4Dt}\right). \tag{6.9}$$

Suppose there is a fluctuation at $\mathbf{x}$ at time zero. Then at time t points within a distance $\propto \sqrt{Dt}$ of $\mathbf{x}$ are correlated, as the effects of the excess density have diffused over this region. Since the lifetime of a fluctuation is $\propto |\Delta|^{-1}$, we should expect significant correlations between points with separations $\leqslant \xi \propto |\Delta|^{-1/2}$, in the steady state.

We can obtain a more explicit result by imagining that in the steady state (for $\Delta > 0$, of course), there is a noise source $\eta(\mathbf{x}, t)$, due to internal fluctuations, which has mean zero, and is 'delta-correlated':

$$\langle \eta(\mathbf{x}, t)\eta(\mathbf{x}', t')\rangle = \Gamma\overline{\phi}\delta(\mathbf{x}-\mathbf{x}')\delta(t-t'), \tag{6.10}$$

where the angle brackets denote a stationary average, and $\overline{\phi}$ is the stationary density (see §6.6). We can work out the stationary spatial correlation (in one dimension, for simplicity):

$$\begin{aligned} C(x_1, x_2) &\equiv \langle \psi(x_1, t)\psi(x_2, t)\rangle \\ &= \int_{-\infty}^{t} \mathrm{d}t_1 \int_{-\infty}^{t} \mathrm{d}t_2 \int \mathrm{d}y_1 \int \mathrm{d}y_2 G_0(x_1-y_1, t-t_1) \\ &\quad \times G_0(x_2-y_2, t-t_2)\langle \eta(x_1, t_1)\eta(x_2, t_2)\rangle \\ &= \frac{\Gamma\overline{\phi}}{4\pi D}\int_0^{\infty} \frac{\mathrm{d}\tau}{\tau}\int \mathrm{d}y \exp\left[-2\Delta\tau - \frac{(x_1-y)^2+(x_2-y)^2}{4D\tau}\right] \\ &= \frac{\Gamma\overline{\phi}}{2\pi\sqrt{2D}}\int_0^{\infty} \frac{\mathrm{d}\tau}{\sqrt{\tau}} \exp\left[-2\Delta\tau - \frac{(x_1-x_2)^2}{8D\tau}\right] \\ &= \frac{\Gamma\overline{\phi}}{8\sqrt{\pi\Delta D}} \exp\left[-\left(\frac{\Delta}{D}\right)^{1/2} |x_1-x_2|\right], \end{aligned} \tag{6.11}$$

which confirms that correlations decay on a scale $\propto |\Delta|^{-1/2}$.

The existence of a phase transition in the CP has been proven (Liggett 1985, Durrett 1988), as has the fact that the transition is continuous (Bezuidenhout & Grimmett 1990). These assertions are valid for spatial dimension $d \geq 1$, in the *infinite-size limit*. (Any *finite* system will eventually become trapped in the vacuum state!) It can also be shown that starting from a finite population, the survival probability and density die out exponentially for $\lambda < \lambda_c$, while for $\lambda > \lambda_c$ the radius of the populated region grows in a linear fashion. There are no exact results for λ_c (though there are bounds — see Liggett (1985)), but simulations (Grassberger & de la Torre 1979) and series analyses (Brower, Furman, & Moshe 1978, Essam *et al.* 1986, Dickman 1989a, Dickman & Jensen 1991, Jensen & Dickman 1993b) give $\lambda_c \simeq 3.2978$ for $d=1$, while β takes values near 0.277 and 0.58 for $d = 1$ and 2, respectively. The stationary density as a function of λ in the one-dimensional CP — from series derived by Dickman & Jensen (1991) — is shown in figure 6.2.

Consider the two-point (equal-time) correlation function

$$C(\mathbf{x}) \equiv \langle \sigma_{\mathbf{x}}\sigma_0\rangle - \overline{\rho}^2, \tag{6.12}$$

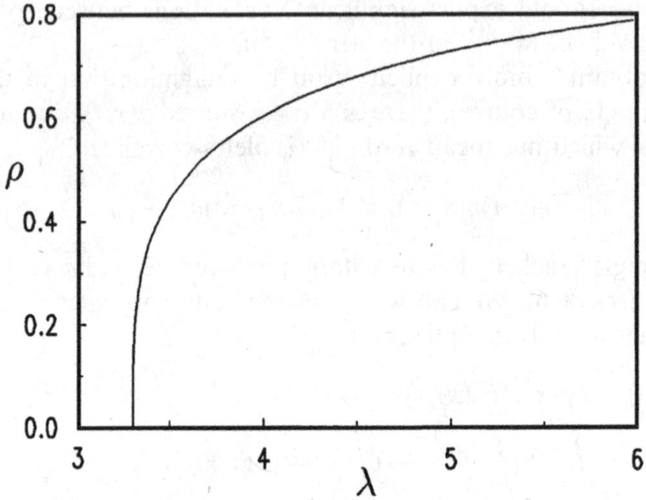

Fig. 6.2. Stationary density in the one-dimensional CP.

and the stationary time correlation function

$$C_s(t) \equiv \text{Prob}[\sigma_0(t_0+t)=1; \sigma_0(t)=1] - \overline{\rho}^2. \tag{6.13}$$

(Here we assume $\lambda > \lambda_c$, and $t_0 \to \infty$.) For large values of their arguments these correlations are expected to decay exponentially:

$$C(\mathbf{x}) \propto \mathrm{e}^{-|\mathbf{x}|/\xi}, \tag{6.14}$$

and

$$C_s(t) \propto \mathrm{e}^{-t/\tau}. \tag{6.15}$$

The correlation length ξ and relaxation time τ both diverge as $\lambda \to \lambda_c$; their critical behavior is characterized by exponents $\nu_\perp$ and $\nu_{||}$:

$$\xi \propto |\Delta|^{-\nu_\perp}, \tag{6.16}$$

and

$$\tau \propto |\Delta|^{-\nu_{||}}. \tag{6.17}$$

The notation comes from directed percolation (§6.6). Evidently $\nu_{||} = 1$ in the mean-field analysis of equation (6.2), while the discussion leading to equation (6.11) implies $\nu_\perp = 1/2$ in mean-field approximation.[1] Another static property with interesting critical behavior is the stationary variance of the density,

$$\chi \equiv L^d(\overline{\rho^2} - \overline{\rho}^2), \tag{6.18}$$

[1] For alternative mean-field approaches see Redner (1982) and De'Bell & Essam (1983).

where L is the linear extent of the system. The definition of χ includes a factor of L^d because ρ is the mean of $\sim (L/\xi)^d$ independent random variables. The variance of ρ is therefore $\propto L^{-d}$, and so the r.h.s. of equation (6.18) remains finite as $L \to \infty$. In the vicinity of the critical point fluctuations become very large: $\chi \propto |\Delta|^{-\gamma}$.[2]

6.3 Time-dependent behavior

The CP exhibits self-similar dynamics at the critical point. It is useful to focus on a particularly simple initial configuration, very near the absorbing state: a single particle at the origin of an otherwise empty lattice. A given trial may end after the first event, or it may survive for a considerable time. (In simulations trials stop at some maximum time, as discussed in chapter 5.) The basic features of the evolution are the survival probability, $P(t)$, and the mean number of particles

$$n(t) = \left\langle \sum_{\mathbf{r}} \sigma_{\mathbf{r}}(t) \right\rangle . \tag{6.19}$$

Here $\langle ... \rangle$ represents the mean over *all* trials starting with a single particle at time zero. (This includes trials that have already died at time t.) The spread of particles from the original seed is measured by

$$R^2(t) \equiv \frac{1}{n(t)} \left\langle \sum_{\mathbf{r}} r^2 \sigma_{\mathbf{r}} \right\rangle . \tag{6.20}$$

In the subcritical regime ($\lambda < \lambda_c$), both $P(t)$ and $n(t)$ decay exponentially. The descendants of the original seed spread diffusively *via* sequences in which a particle generates offspring at a neighboring site and subsequently disappears. Thus for $\lambda < \lambda_c$, $R^2 \propto t$. Let $\rho(\mathbf{r}, t; 0, 0)$ denote the conditional probability of finding a particle at $\mathbf{r}$ at time t, given a single particle at the origin, and all other sites vacant, at time zero. In the subcritical regime

$$\rho(r, t; 0, 0) \simeq e^{-r/\xi} e^{-t/\tau}, \tag{6.21}$$

for large r and t.

In the supercritical regime a trial may survive indefinitely: $P(t) \to P_\infty > 0$ as $t \to \infty$. In fact, the ultimate survival probability, P_∞, equals the stationary particle density $\overline{\rho}$ (Grassberger & de la Torre 1979).[3] For

---

[2] The notation is borrowed from equilibrium critical phenomena, where χ represents the *susceptibility* as well as the fluctuation in the order parameter density.

[3] The identity of the ultimate survival probability and the stationary density has been established rigorously for the CP, which enjoys a property known as *self-duality* (Durrett 1988). Numerical studies of a wide range of models with absorbing states show that P_∞ and $\overline{\rho}$ are proportional in the neighborhood of the critical point, i.e., both are governed by the exponent β.

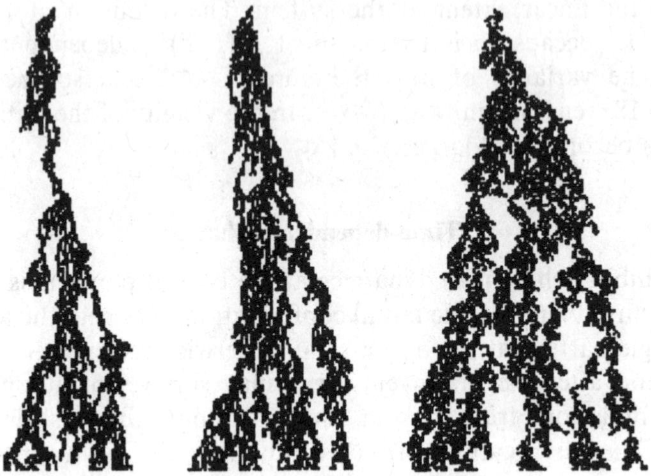

Fig. 6.3. CP trials starting from a single particle. Time increases downward. Left: $\lambda = 3.0$, $t_{max} = 200$; center: $\lambda = 3.2978$, $t_{max} = 400$; right: $\lambda = 3.5$, $t_{max} = 400$.

$\lambda > \lambda_c$, the boundary between the active region and the surrounding vacuum advances, on average, at a constant speed $v(\Delta)$; hence $n(t) \propto t^d$, and $R^2(t) \propto t^2$. Just at the critical point the process dies with probability one, but the mean lifetime diverges. In the absence of a characteristic time scale, the asymptotic evolution follows power laws, conventionally written

$$P(t) \propto t^{-\delta}, \tag{6.22}$$

$$n(t) \propto t^{\eta}, \tag{6.23}$$

and

$$R^2(t) \propto t^z. \tag{6.24}$$

Equations (6.22)–(6.24) apply at long times, and require that the system is large enough that particles do not encounter the boundary. A more complete description includes finite-time *corrections to scaling*, for example,

$$P(t) \propto t^{-\delta}(1 + at^{-\theta} + bt^{-\delta'} + \cdots). \tag{6.25}$$

There is good numerical evidence that $\theta = 1$ for the one-dimensional CP (Essam *et al.* 1986). Some typical trials are shown in figure 6.3, which contrasts a scrawny subcritical cluster with a robust supercritical one. The critical CP generates fractal clusters (Vicsek 1992). From equations (6.22) and (6.23) we see that the mean population in surviving trials

Table 6.1. Critical creation rate and exponents for the contact process in d dimensions. Numbers in parentheses denote uncertainties.

	$d=1$	$d=2$	$d=3$	$d=4$
λ_c	3.29785(2)[a]	1.6488(1)[d]	1.3169(1)[h]	
β	0.27649(4)[b]	0.583(4)	0.805(10)[h]	1
δ_h^{-1}	0.111(3)[c]	0.285(35)[e]	0.45(2)[e]	$\frac{1}{2}$
γ	0.54386(7)[b]	0.35(1)	0.19(1)	0
$\nu_{\parallel}$	1.73383(3)[b]	1.295(6)[f]	1.105(5)[h]	1
$\nu_{\perp}$	1.09684(6)[b]	0.733(4)	0.581(5)	$\frac{1}{2}$
δ	0.15947(3)[b]	0.4505(10)[g]	0.730(4)[h]	1
η	0.31368(4)[b]	0.2295(10)[g]	0.114(4)[h]	0
z	1.26523(3)[b]	1.1325(10)[g]	1.052(3)[h]	1

[a]Jensen & Dickman 1993b; [b]Jensen 1996; [c]Adler & Duarte 1987; [d]Grassberger 1989a; [e]Adler *et al.* 1988; [f]Grassberger & Zhang 1996; [g]Voigt & Ziff 1997; [h]Jensen 1992.

$n_s(t) \sim t^{\eta+\delta}$. Using $t \sim R^{2/z}$, we find $n_s \sim R^{d_f}$, with the fractal dimension $d_f = 2(\eta+\delta)/z$ (Grassberger 1989a).

Some elementary observations provide constraints on the exponents. Since the transition is continuous, the critical process must eventually die, so $\delta \geq 0$ (Bezuidenhout & Grimmett, 1990). (Since P_∞ and $\overline{\rho}$ are zero for $\lambda < \lambda_c$, $P_\infty(\lambda_c) > 0$ would imply a discontinuous transition.) In order to survive, the number of particles cannot be bounded as $t \to \infty$, hence $\eta + \delta \geq 0$. Since the population grows but the density tends to zero at long times, we expect z to fall between 1 and 2, the former limit corresponding to simple diffusion, the latter to spreading of a region with a nonzero density. We have introduced over half a dozen exponents, but shall see that they are not all independent. Exponent values for the CP are summarized in table 6.1.

6.4 Scaling theory

Scaling theory aims to provide a simplified description of a system in the vicinity of a critical point, as described for driven systems in §2.6. Here we follow the approach of Grassberger & de la Torre (1979), which describes the evolution from the initial state considered in the preceding section: a single particle at the origin.

In the *scaling regime*, i.e., sufficiently close to the critical point, at long times, and in large systems, the conditional probability of finding a particle at r, given that at time zero there was a particle at the origin, and that all other sites were vacant, obeys

$$\rho(r,t;0,0) \simeq t^{\eta - dz/2} F(r^2/t^z, \Delta t^{1/\nu_{\|}}). \tag{6.26}$$

The dependence upon position is demanded by symmetry and by the growth of the characteristic length scale $\propto t^{z/2}$. The time dependence of the scaling function F involves the ratio t/τ, since $\tau \propto \Delta^{-\nu_{\|}}$. Similarly the survival probability is expected to follow

$$P(t) \simeq t^{-\delta} \Phi(\Delta t^{1/\nu_{\|}}). \tag{6.27}$$

The reason for the prefactors becomes clear when we set Δ to zero: then equation (6.27) reduces to (6.22), while (6.26) yields (6.23) when integrated over space. Equation (6.24) results when we take the second moment (in space) of $\rho(r,t;0,0)$ with $\Delta = 0$:

$$R^2(t) \sim \frac{1}{n(t)} t^{\eta - dz/2} \int x^2 f(x^2/t^z) \mathrm{d}^d x \propto t^z . \tag{6.28}$$

For $\Delta < 0$ and large fixed t we expect $\rho(r,t;0,0) \simeq \mathrm{e}^{-r/\xi}$. Since $\xi \propto \Delta^{-\nu_\perp}$, this implies $F(u,v) \propto \exp(-C\sqrt{u}|v|^{\nu_\perp})$ for $v < 0$ (C is a positive constant). For ξ to be time-independent, we require the scaling relation

$$z = 2\nu_\perp/\nu_{\|}. \tag{6.29}$$

As noted above, $P_\infty \equiv \lim_{t\to\infty} P(t) = \overline{\rho}$, the stationary density. Now $\overline{\rho} \propto \Delta^\beta$, and in order for equation (6.27) to yield a finite limiting value for P, we must have

$$\delta = \beta/\nu_{\|} . \tag{6.30}$$

Consider next the local density at any fixed $\mathbf{r}$, with $t \to \infty$. For $\Delta > 0$ this must approach $P_\infty \overline{\rho}$. Since $\rho(r,t;0,0)$ in equation (6.26) represents the density averaged over *all* trials, we have that as $t \to \infty$,

$$\rho(r,t;0,0) \to P(t)\Delta^\beta \propto \Delta^{2\beta} , \tag{6.31}$$

which, together with (6.26) implies that for v large and positive, $F(0,v) \propto v^{2\beta}$; to remove the overall t-dependence we must have

$$4\delta + 2\eta = dz. \tag{6.32}$$

An expression such as this, connecting critical exponents with the dimensionality, is termed a *hyperscaling relation*. It is expected to hold for $d \leqslant d_c$. The scaling hypothesis yields three relations amongst the exponents; a relation for γ is derived in the following section. Finally, the relation $\delta_h + 1 = (\nu_{\|} + d\nu_\perp)/\beta$ (Kinzel 1983, 1985) allows us to conclude that of the eight exponents listed in table 6.1, only three are independent.

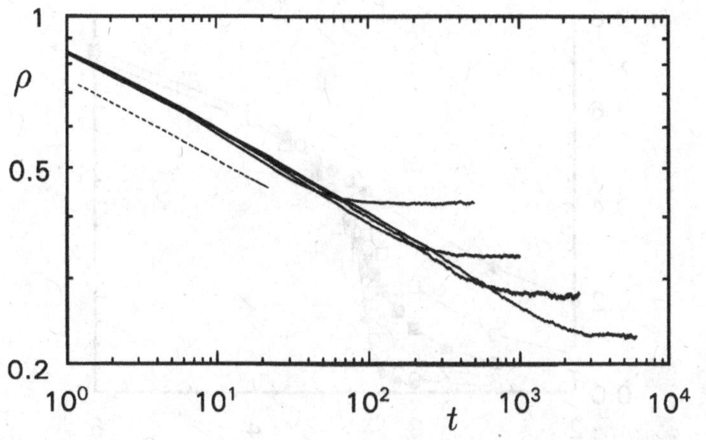

Fig. 6.4. Mean density in surviving trials versus t in the one-dimensional CP at $\lambda = \lambda_c$. From top to bottom, system sizes $L = 20, 50, 100, 200$. The slope of the straight line is -0.16.

6.5 Finite-size scaling

In the vicinity of the critical point, where ξ is large, intensive properties depend strongly on system size. As noted in §1.2, finite-size scaling analysis permits one to locate the critical point and estimate exponents using data on systems of several different sizes (Fisher 1971, Fisher & Barber 1972, Privman 1990; see §2.6). We again appeal to the notion that near the critical point, there is but a single pertinent length scale, $\xi \propto \Delta^{-\nu_\perp}$, so that the L-dependence of intensive properties is through the ratio L/ξ, conventionally represented by $\Delta L^{1/\nu_\perp}$.

In applying finite-size scaling theory to the CP and similar models a slight complication arises, namely that for a finite system the only true stationary state is the absorbing state. To learn about the active state from simulations of finite systems we study the *quasistationary state*, which describes the statistical properties of surviving trials following an initial transient. In practice, quasistationary properties are determined from averages over the surviving representatives of a large set of independent trials beginning with all sites occupied. After a transient whose duration depends on L and Δ, surviving-sample averages attain steady values; the latter converge to stationary values as $L \to \infty$. The data in figure 6.4 illustrate how the mean density $\rho_s(t)$ — the subscript denotes an average restricted to surviving trials — approaches its quasistationary value $\overline{\rho}(\Delta, L)$. In figure 6.5 we see the graph of $\overline{\rho}(\Delta, L)$ becoming sharper with increasing system size.

For large L and small Δ the quasistationary density may be written in

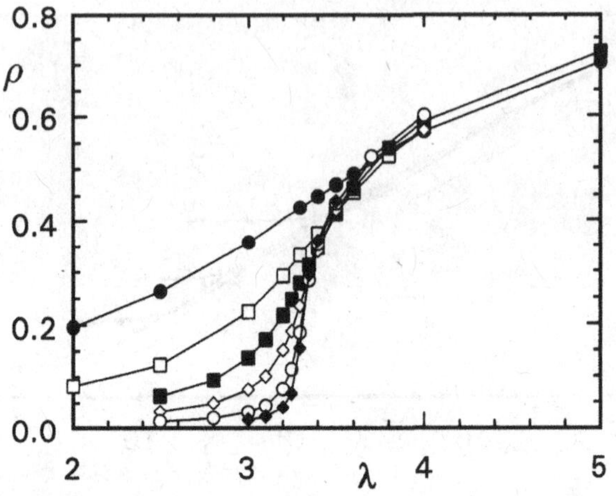

Fig. 6.5. Quasistationary density in the one-dimensional CP. From top to bottom (at left): $L = 20, 50, 100, 200, 500, 1000$.

the form

$$\overline{\rho}(\Delta, L) \propto L^{-\beta/\nu_\perp} f(\Delta L^{1/\nu_\perp}). \tag{6.33}$$

The scaling function $f(x) \propto x^\beta$ for large x (i.e., $L \gg \xi$), since $\overline{\rho} \propto \Delta^\beta$. The prefactor is required if $\overline{\rho}$ is to be independent of L in this limit. For $\Delta < 0$, the number of particles in a system that has yet to reach the absorbing state is $\mathcal{O}(1)$ (i.e., independent of L), so $\overline{\rho}(\Delta, L) \propto L^{-1}$, and therefore $f(x) \propto |x|^{-\nu_\perp+\beta}$ for x large and negative. Figure 6.6, a scaling plot of the data presented in figure 6.5, demonstrates these points: data for various system sizes collapse onto a single graph that follows the expected power laws. When the exponents and/or λ_c are not known, varying them to optimize the data collapse is a useful method for determining their values (Aukrust, Browne, & Webman 1990).

The critical point may also be found by examining the L-dependence of $\overline{\rho}(\lambda, L)$. When $\Delta = 0$ equation (6.33) yields

$$\overline{\rho}(0, L) \propto L^{-\beta/\nu_\perp}. \tag{6.34}$$

For $\Delta < 0$ (subcritical regime), $\overline{\rho}$ falls off as L^{-1}, while for $\Delta > 0$, $\overline{\rho}$ approaches a nonzero value as $L \to \infty$. Thus the power law, equation (6.34), obtains only at the critical point; plots of $\overline{\rho}$ versus L are therefore quite useful in locating λ_c.

Applied to the variance of ρ, finite-size scaling yields

$$\chi(\Delta, L) \propto L^{\gamma/\nu_\perp} g(\Delta L^{1/\nu_\perp}), \tag{6.35}$$

where now the scaling function $g(x) \propto x^{-\gamma}$ for large x, and $\chi(0, L) \propto L^{\gamma/\nu_\perp}$. Consider next the probability distribution $P(\rho, L)$ for the density in a

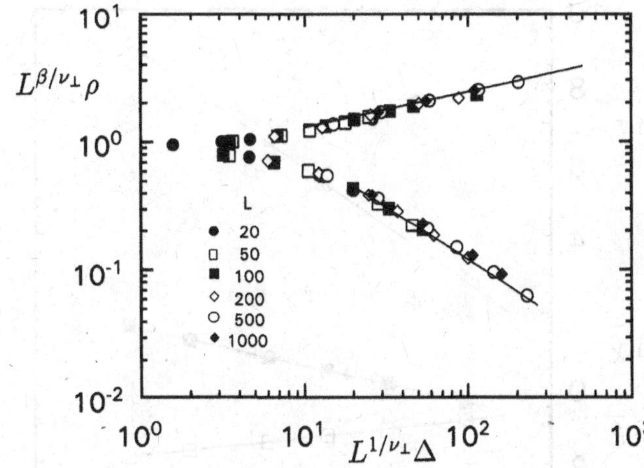

Fig. 6.6. Scaling plot of the quasistationary density in the one-dimensional CP. The slopes of the straight lines are 0.277 and −0.82.

critical system. It is reasonable to suppose that for $\rho \simeq \overline{\rho}$, P depends on ρ only through the ratio $\rho/\overline{\rho}$. Then a simple scaling form for the probability is

$$P(\rho, L) \propto L^{\beta/\nu_\perp} \mathscr{P}(\rho L^{\beta/\nu_\perp}), \tag{6.36}$$

where we used equation (6.34) for $\overline{\rho}$. The prefactor is demanded by normalization. (The maximum of P follows $\overline{\rho}$ toward zero, and becomes progressively sharper as $L \to \infty$.) The kth moment of the density is

$$\overline{\rho^k} = \int_0^1 \rho^k P(\rho, L) \mathrm{d}\rho \equiv I_k L^{-k\beta/\nu_\perp}, \tag{6.37}$$

where the weak dependence of the integral on L has been neglected in the large-L limit. Equation (6.37) implies, for $k = 2$, that $\chi(0, L) = L^d(\overline{\rho^2} - \overline{\rho}^2) \propto L^{(d\nu_\perp - 2\beta)/\nu_\perp}$, which, when combined with $\chi(0, L) \propto L^{\gamma/\nu_\perp}$, yields the hyperscaling relation:

$$\gamma = d\nu_\perp - 2\beta. \tag{6.38}$$

A further application of finite-size scaling concerns the mean lifetime of a system with a given initial density. Consider the evolution starting from a fully occupied lattice. The lifetime may be characterized in various ways, for example, by the time τ_H required for the survival probability to decay to one half. Since this is a relaxation time it scales (for $L \gg \xi$), as $\tau_H \propto |\Delta|^{-\nu_\parallel}$. Incorporating the L-dependence in the usual way yields

$$\tau_H(\Delta, L) \propto L^{\nu_\parallel/\nu_\perp} G(\Delta L^{1/\nu_\perp}) \tag{6.39}$$

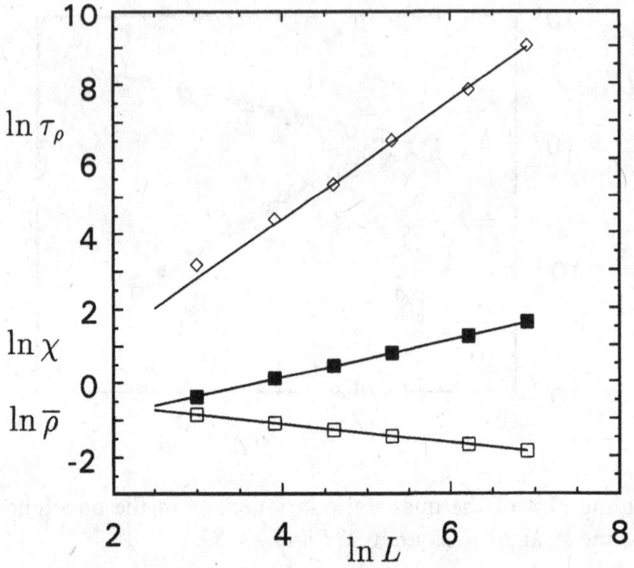

Fig. 6.7. Size dependence of (top to bottom) τ_ρ, χ, and ρ_s in the one-dimensional critical CP. The lines have slopes of 1.58, 0.497, and −0.252.

with $G(x) \propto x^{-\nu_\parallel}$ for large x. At the critical point, $\tau_H \propto L^{\nu_\parallel/\nu_\perp}$. Note also that if $\lambda < \lambda_c$ the half-life approaches a constant as $L \to \infty$, while for $\Delta > 0$ it grows exponentially with L. Studying τ_H as a function of L and λ is a valuable method for determining λ_c, and yields the exponent ratio $\nu_\parallel/\nu_\perp$.

Another characteristic relaxation time is τ_ρ, which governs the approach of $\rho_s(t)$ to its quasistationary value:

$$\rho_s(t) - \overline{\rho}(\Delta, L) \simeq C \exp(-t/\tau_\rho). \tag{6.40}$$

At the critical point, τ_ρ has the same size dependence as τ_H, but it is considerably smaller, permitting its determination in shorter runs. Another advantage associated with τ_ρ is that it peaks at λ_c, unlike τ_H, which grows monotonically with λ. In figure 6.7 simulation results for three critical properties are plotted versus L, confirming the expected power laws: $\overline{\rho}(0, L) \propto L^{-\beta/\nu_\perp}$, $\chi(0, L) \propto L^{\gamma/\nu_\perp}$, and $\tau_\rho(0, L) \propto L^{-\nu_\parallel/\nu_\perp}$.

Finally, consider the density $\rho(t, L)$, averaged over all trials at λ_c (including those that have reached the vacuum), starting from a fully occupied lattice. Assuming that the time dependence only involves the ratio $t/\tau \propto tL^{-\nu_\parallel/\nu_\perp}$, we write

$$\rho(\Delta{=}0, L, t) \propto L^{-\beta/\nu_\perp} g(tL^{-\nu_\parallel/\nu_\perp}). \tag{6.41}$$

(The prefactor is required for consistency with equation (6.34).) Now for times small compared to τ, the density should be *size-independent*, since the range of correlations has not yet grown comparable to L. Thus

$$\rho(0, L, t) \propto t^{-\delta} \qquad (t << L^{\nu_{||}/\nu_{\perp}}), \tag{6.42}$$

where we used $\delta = \beta/\nu_{||}$. This means that in a dense critical system, the initial decay of the density is governed by the same exponent as the asymptotic ($t \to \infty$) survival probability, starting near the vacuum (see figure 6.4).

6.6 Directed percolation

The CP describes a spatially distributed population of self-replicating mortal individuals. Many other models share these basic features, and have similar phase diagrams. Consider a discrete-time analog of the CP in which all sites are updated simultaneously. Then a d-dimensional system (whose sites are labeled $\mathbf{x}$), can be represented by a (d+1)-dimensional lattice ($\mathbf{x}$,t) whose layers record the configuration at times $t = 0, 1, 2, \ldots$. If site $\mathbf{x}$ is occupied in layer t, then $\mathbf{x}$ or one of its neighbors must be occupied in layer $t - 1$. (This removes the possibility of spontaneous creation.) We should permit $\mathbf{x}$ to be vacant in layer t even if it is occupied in layer $t - 1$ (annihilation). There are many ways in which this simultaneous updating might be implemented. The simplest, known as *directed percolation* (Broadbent & Hammersley 1957, Durrett 1988, Blease 1977, Deutscher, Zallen, & Adler 1983, Kinzel 1985), is illustrated in figure 6.8. The top row, representing the initial state, is connected by diagonal bonds to the row below. Each of these oriented bonds (and the ones connecting subsequent layers) is present with probability p, independent of the others. The rule that site ($\mathbf{x}$,t) is occupied if and only if it is connected to an occupied site in layer $t - 1$ leads to the transition probabilities shown in figure 6.9. These define *bond directed percolation* as a probabilistic cellular automaton (Wolfram 1983, Rujàn 1987).

DP may also be viewed as a static representation of directed flow in a random network, for example, percolation of water through porous rock in which a certain fraction of the channels is blocked, or electric current in a diluted diode network. Most of the questions we can ask about the CP have analogs in DP. Survival of a trial in the CP corresponds to *percolation*: the existence of a directed path extending arbitrarily far from the source. (Isotropic percolation has a vast literature; see Stauffer & Aharony (1992) for an introduction.) The percolation probability $P(p)$ is strictly zero for $p < p_c$, the percolation threshold, and above this threshold one finds $P \propto (p - p_c)^{\beta}$. For bond DP in 1+1 dimensions, $p_c \simeq 0.6446$

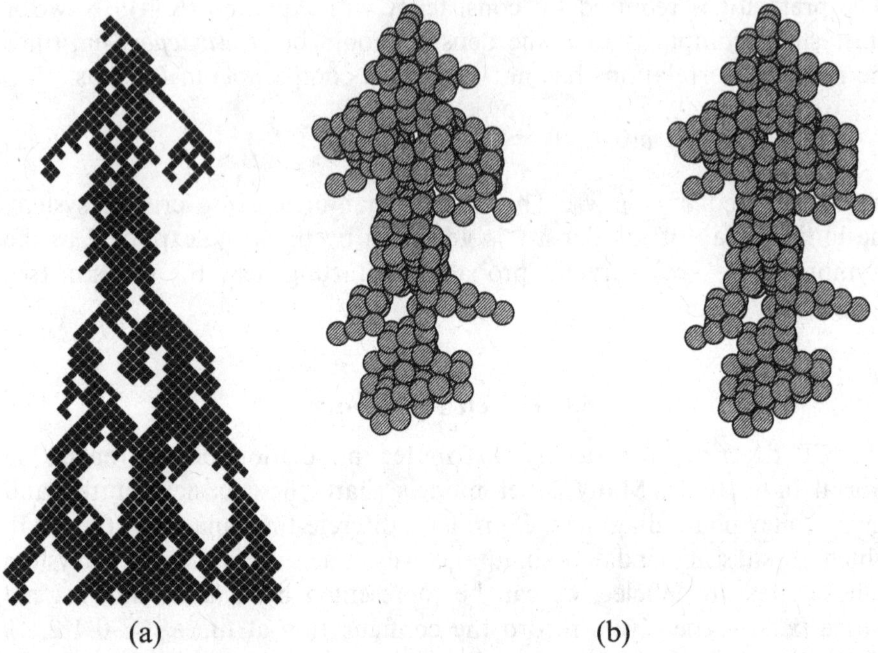

Fig. 6.8. Examples of bond DP, starting from a single site. (a) (1+1)-dimensional, $p = 0.65$, $t_m = 100$; (b) stereo-image pair showing a trial in 2+1 dimensions, $p = 0.26$, $t_m = 50$.

Fig. 6.9. Transition probabilities for bond DP with bond probability p.

(Jensen & Guttmann 1996). β and all the other critical exponents are the same for DP (in $d + 1$ dimensions) as for the d-dimensional contact process.

Although there is no way to map one onto the other, the CP and DP are equivalent as far as critical behavior is concerned. The two models belong to the universality class called 'directed percolation' or 'Reggeon field theory'.[4] The identity of CP, DP and Reggeon field theory critical exponents is strong evidence of universality: as long as the essential

[4] The latter is a particle physics model that shares the same formal properties; see Gribov (1968), Gribov & Migdal (1969), Abarbanel *et al.* (1975), Moshe (1978), Brower *et al.* (1978), Cardy & Sugar (1980).

features are the same, models with very different evolution rules have the same critical behavior. In this context the 'essential feature' is a continuous transition into an absorbing state. The *DP conjecture* asserts that such models belong generically to the DP universality class (Janssen 1981, 1985, Grassberger 1982). (We have overstated things a bit. One should demand that the evolution is *local*, and not subject to conservation laws — see chapters 2 and 9.)

As already noted in the context of DLG, the idea underlying universality is that many different models follow essentially the same evolution when viewed on the large length scales and long time scales that determine critical behavior. A continuum description of the CP and allied models (Janssen 1981, 1985), in the spirit of the time-dependent Landau–Ginzburg approach for critical dynamics (Hohenberg & Halperin 1977), adds noise to the spatially extended mean-field theory discussed in §6.2:

$$\frac{\partial \phi}{\partial t} = a\phi - b\phi^2 - c\phi^3 + \cdots + D\nabla^2\phi + \eta(\mathbf{x}, t). \qquad (6.43)$$

Here $\phi(\mathbf{x}, t)$ is the local order-parameter density, and η is a Gaussian noise that respects the absorbing state. This condition is imposed through the covariance: $\langle \eta(\mathbf{x}, t)\eta(\mathbf{x}', t')\rangle \propto \phi(\mathbf{x}, t)\delta(\mathbf{x} - \mathbf{x}')\delta(t - t')$.[5] The various coefficients in equation (6.43) are rates in a coarse-grained description, hence complicated functions of model parameters. (It is clear, of course, that a is an increasing function of λ.) The nature of solutions to equation (6.43) and its connection to the CP have been put on a firm basis (Mueller & Tribe 1994); numerical integration of a suitably discretized version yields DP critical exponents (Dickman 1994). Field-theoretic analysis (Janssen 1981) reveals that the cubic and higher-order terms are irrelevant to critical behavior so long as $b > 0$. The situation is analogous to that in equilibrium, where the Ising universality class is generic for models with a scalar order parameter and short-range interactions. The DP conjecture does not rule out higher-order transitions, but these require fine-tuning of the parameters, so that (for example) a and b vanish together (Ohtsuki & Keyes 1987b).

DP has been linked to a number of challenging dynamic phenomena. Of particular note is a conjecture relating the onset of turbulence — for example, in convection — to a DP-like infection process (Pomeau 1986; Chaté & Manneville 1986, 1990). The starting point is a deterministic partial differential equation (for example, the Swift–Hohenberg (1977)

[5] That the covariance is *linear* in $\phi(\mathbf{x}, t)$ can be understood by noting that the number of creation (or annihilation) events in a certain region, in a given time interval, is a Poissonian random variable, for which the variance is equal to the mean. As we coarse-grain over ever larger regions, these converge to Gaussian random variables.

equation) that describes convection, including a transition between laminar flow and spatiotemporal intermittency, in which turbulent regions are interspersed with laminar ones. The local dynamics drives a turbulent cell toward the laminar solution, but (diffusive) coupling of neighboring regions allows for the spread of turbulence prior to its dying out. This impressionistic description of spatiotemporal intermittency sounds very like the CP, hence Pomeau's suggestion that the onset of turbulence in spatially extended systems might correspond to a DP-like transition. (Note, in particular, that the fully laminar state is absorbing.) The connection between a deterministic continuum description and a discrete stochastic process is, however, far from transparent. Chaté & Manneville (1986, 1990) argue that the local dynamics are well represented by a *discrete-time map* of the form $X_{t+1} = f(X_t)$, where f possesses both a transient chaotic domain (for example, for $X \in [0, 1]$) and an absorbing one (such that once $X > 1$ it remains so for all future iterations). A spatially extended system is then modeled by a *coupled map lattice*, in which $X_{i,t+1}$, the state of cell i at time $t + 1$, is given by an average of $f(X_{i,t})$ and the corresponding values at neighboring sites. Simulations indicate that depending on the nature of the map f, the onset of spatiotemporal intermittency in the coupled map lattice may be either discontinuous or continuous, but that in the latter case the associated exponents are nonuniversal, so that the transition appears not to belong to the class of DP (or of any other known process). It is interesting to speculate on why this absorbing state transition remains outside the DP class; perhaps the deterministic nature of the process is responsible.

Other processes with a close connection to DP are the statistics of a pinned, driven interface (Tang & Leschhorn 1992, Horváth, Family, & Vicsek 1990, Buldyrev *et al.* 1992); self-organized criticality (Paczuski, Maslov, & Bak 1994); and damage spreading (Derrida & Weisbuch 1987, Grassberger 1995).

6.7 Generalized contact processes

In this section we consider two variants of the basic CP. One incorporates diffusion; the other relaxes the proportionality between birth rate and number of occupied neighbors. After defining the models, we use them to illustrate a cluster mean-field calculation. Then we discuss numerical results bearing on their critical behavior. For simplicity we confine the discussion to one-dimensional models.

In the *diffusive contact process* (DCP), particles execute a lattice random walk or *exclusion process* (Liggett 1985): they may jump to a randomly chosen, vacant NN. (We assume unbiased diffusion.) The jump rate is D; creation and annihilation proceed at their usual rates. One may interpret the DCP as an epidemic model for a nonsedentary population, or as an autocatalytic chemical reaction with diffusing molecules (see chapter 4).

The *generalized contact process* (GCP) (Durrett & Griffeath 1983; Katori & Konno 1993a,b), involves no new particle moves, but introduces a new parameter: the rate for creation at a (vacant) site with exactly one occupied neighbor is $\zeta\lambda$. Other processes retain their usual rates.[6] One naturally expects λ_c to be a decreasing function of ζ. Diffusion also enhances survival: for a given population, there are more particle–vacancy pairs, and so more particles may be created. On the other hand, there is no reason to expect a qualitative change in the phase diagram when we permit diffusion or set $\zeta \neq 1/2$: the vacuum is still absorbing, and there should be an active state for sufficiently large λ. (Except of course for $\zeta = 0$ in the GCP: there is no active stationary state for finite λ, since a vacancy cluster of two or more sites can never be reoccupied.) The infinite-D limit of the DCP corresponds to complete mixing: after each creation or annihilation event the system attains an uncorrelated distribution. In this limit the simple mean-field approximation discussed in the beginning of the chapter is correct.

We have seen that, in general, mean-field theory is not quantitatively accurate. (Indeed, the one-site approximation can say nothing about diffusion.) However, as discussed in previous chapters, one may define a series of mean-field approximations that retain the equations for the n-site joint probabilities, and truncate the hierarchy at order $n+1$. Numerical studies indicate that in many cases this sequence of *cluster mean-field theories* converges to the actual solution (ben-Avraham & Köhler 1992a). We shall work out the two-site approximation for the diffusive GCP.

In site mean-field theory we factorize the two-site joint probabilities into products of single-site probabilities; here we retain the two-site NN probabilities but factorize those for larger clusters. (We continue to assume spatial homogeneity.) So the three-site probability $p(\sigma_1, \sigma_2, \sigma_3)$ is approximated as the pair probability $p(\sigma_1, \sigma_2)$ times the conditional probability $p(\sigma_3|\sigma_2)$:

$$p(\sigma_1, \sigma_2, \sigma_3) \simeq \frac{p(\sigma_1, \sigma_2)p(\sigma_2, \sigma_3)}{p(\sigma_2)}. \tag{6.44}$$

The analogous approximation for the four-site joint probability is:

$$p(\sigma_1, \sigma_2, \sigma_3, \sigma_4) \simeq \frac{p(\sigma_1, \sigma_2)p(\sigma_2, \sigma_3)p(\sigma_3, \sigma_4)}{p(\sigma_2)p(\sigma_3)}. \tag{6.45}$$

Let • and ∘ stand, respectively, for occupied and vacant sites and, following the notation of chapter 3, let $u \equiv p(\bullet\bullet)$ denote the probability

---

[6] The basic CP corresponds to $\zeta = 1/2$; another well-studied case, called the A-model (Dickman & Burschka 1988) or the *threshold* CP (Liggett 1991) corresponds to $\zeta = 1$.

Table 6.2. Pair men-field theory for the diffusive GCP.

Process	Rate	$\Delta N_{\bullet}$	$\Delta N_{\bullet\bullet}$
●○● → ●●●	$\lambda w^2/(1-\rho)$	+1	+2
●○○ → ●●○	$2\lambda\zeta wz/(1-\rho)$	+1	+1
●●● → ●○●	u^2/ρ	−1	−2
●●○ → ●○○	$2uw/\rho$	−1	−1
○●○ → ○○○	w^2/ρ	−1	0
●●○○ → ●○●○	$Dzuw/\rho(1-\rho)$	0	−1
○●○○ → ○○●○	$Dw^3/\rho(1-\rho)$	0	+1

that both sites of a NN pair are occupied, and let $z \equiv p(\circ\circ)$. Note that $p(\circ\bullet) = p(\bullet\circ) = \rho - u \equiv w$ in a periodic system. The four probabilities must sum to unity, so that $z = 1 - 2\rho + u$, leaving only two independent quantities, say ρ and u, in pair mean-field theory. We begin the calculation by enumerating all processes in which ρ or u change (see table 6.2). Collecting results, one finds

$$\frac{d\rho}{dt} = \lambda\frac{\rho-u}{1-\rho}\,[2\zeta + (1-4\zeta)\rho - (1-2\zeta)u] \;-\; \rho, \tag{6.46}$$

and

$$\frac{du}{dt} = 2\lambda\frac{\rho-u}{1-\rho}[\zeta + (1-2\zeta)\rho - (1-\zeta)u] \;-\; 2u \;-\; D\frac{\rho-u}{\rho(1-\rho)}(u-\rho^2). \tag{6.47}$$

Consider first the stationary solutions for $\zeta = 1/2$. For $D = 0$ we have

$$\overline{\rho} = \frac{\lambda-2}{\lambda-1} \tag{6.48}$$

and

$$\overline{u} = \frac{\lambda-2}{\lambda}, \tag{6.49}$$

showing that $\lambda_c = 2$ for the basic CP in the pair approximation — an improvement over the one-site result, $\lambda_c = 1$, but still far from the correct value.[7] Notice also that the stationary pair density $\overline{u}$ is proportional to $\overline{\rho}$ not $\overline{\rho}^2$, indicating that the particles are strongly clustered. Clustering is suppressed when we allow diffusion: equation (6.47) shows that diffusion tends to reduce the difference between u and ρ^2. The ratio $\overline{u}/\overline{\rho}|_{\lambda=\lambda_c}$ decreases from one half, for $D = 0$, to zero, for infinite D.

[7] The reader is invited to derive the pair mean-field theory equations for the basic CP on a d-dimensional cubic lattice; they yield $\lambda_c = 2d/(2d-1)$.

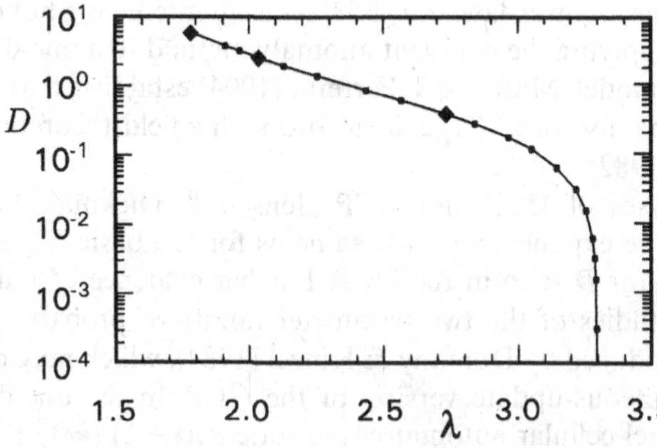

Fig. 6.10. Phase boundary of the DCP. Small circles: series expansion results; diamonds: MC simulations.

From the stationary solution for $\zeta = 1/2$, one finds the phase boundary of the DCP:

$$\lambda_c = 1 - \frac{D}{2} + \sqrt{1 + \left(\frac{D}{2}\right)^2}. \tag{6.50}$$

We recover the one-site mean-field result, $\lambda_c = 1$, for infinite D. The convergence to mean-field behavior has been studied by Katori (1994). The phase diagram of the DCP — obtained from series analysis and simulations (Jensen & Dickman 1993a) — is shown in figure 6.10.

The preceding example (and others scattered through this book) indicates that pair mean-field theory offers a considerable improvement over the one-site approximation, but the resulting estimates are still quite far from the correct value, for example, for λ_c. What can we learn from three-site and higher approximations? For the CP, the sequence of n-site approximations, $\lambda_c(n)$ converges to λ_c (Konno & Katori 1990; Ferreira & Mendiratta 1993); extrapolation of a series of cluster calculations therefore seems useful for absorbing-state transitions. However, while $\lambda_c(n)$ converges, the critical exponents do not change from their mean-field values. Pair and higher-order mean-field theories yield $\overline{\rho} \simeq A_n(\lambda - \lambda_c(n))$ near the transition, so that the mean-field value β_{MFT} is always unity. Although β_{MFT} doesn't approach the true β value for the CP, it is possible to extract useful exponent estimates from a sequence of cluster mean-field calculations by studying the growth of the critical amplitude A_n. This *coherent anomaly method* (Suzuki 1986) has been used quite effectively to estimate β in the CP and some variants (Konno & Katori 1990, Ferreira & Mendiratta 1993, Park *et al.* 1993). The effect of long-range interactions — specifically, creation not only at neighboring sites, but rather with a rate decaying as a power-law, $\propto r^{-(d+\alpha)}$ — is an interesting but challenging question. By applying the coherent anomaly method to a one-dimensional, long-range A-model, Marques & Ferreira (1994) established a crossover to mean-field behavior for $\alpha \leqslant 1/2$, in accord with a field-theoretic prediction (Grassberger 1982).

Series analyses of DCP and GCP (Jensen & Dickman 1993a, 1994) indicate that the exponents are the same as for the basic CP, for $\zeta > 0$ in the GCP, and for $D < \infty$ in the DCP. Further evidence of universality is provided by studies of the two-parameter family of probabilistic cellular automata introduced by Domany & Kinzel (1984), which may be regarded as the simultaneous-update version of the GCP. In the one-dimensional Domany–Kinzel cellular automaton the state $\sigma_x(t+1)$ (=0 or 1), of site x at time $t+1$ depends upon $T_x \equiv \sigma_x(t)+\sigma_{x-1}(t)$. If we let $\Pr[\sigma_x(t+1)=1] = h(T_x)$, then $h(0)=0$, $h(1)=p_1$, and $h(2)=p_2$. For each $p_2 < 1$ there is a critical value $p_{1,c}$, such that the vacuum (all sites in state 0) is the unique steady state for $p_1 < p_{1,c}$, whilst for $p_1 > p_{1,c}$ there is also an active stationary state with a nonzero density of sites in state 1. (Note that the line $p_2 = p_1$ corresponds to site DP, and $p_2 = 2p_1 - p_1^2$ to bond DP.) Numerical results (Bagnoli *et al.* 1992; Kohring & Schreckenberg 1992) indicate that the critical exponents do not vary along the critical line of the Domany–Kinzel model, for $p_2 < 1$.[8] These studies provide additional support for universality in models with a continuous transition to an absorbing state. It is natural to ask how far universality extends: what properties lead to an absorbing state transition *outside* the DP class? We return to this question in chapter 9.

6.8 Effects of disorder

Up to now, we have considered perfect, spatially homogeneous lattices. Since many-particle systems often incorporate disorder, it is of interest to study the effect of various kinds of randomness, for example, a random annihilation rate at each site, or a random creation rate associated with each bond. Similarly, disorder in the form of random fields or couplings has been studied extensively in equilibrium spin models (Ziman 1979, Stinchcombe 1983). A fundamental distinction is that between *quenched* disorder, in which the random variables describing disorder are fixed, and *kinetic* disorder, in which they evolve. (In equilibrium the latter case corresponds to *annealed* disorder.) Chapters 7 and 8 detail the effects of kinetic disorder in nonequilibrium spin models. In the present section we are concerned with quenched randomness.

[8] The critical endpoint ($p_1 = 1/2, p_2 = 1$), however, corresponds to *compact* DP (Essam 1989, Dickman & Tretyakov, 1995), discussed briefly in §1.1.

A particularly simple realization of disorder is *site dilution*. In the diluted CP sites are removed independently with probability x, i.e., there is a random variable $\eta(\mathbf{x})$ associated with each site of the lattice, taking values 0 and 1 with probabilities x and $1-x$, respectively. The diluted CP is the CP restricted to sites with $\eta = 1$; those with $\eta = 0$ are never occupied. There are several ways to implement the loss of sites. One (Marqués 1990) is to define a *site-dependent coordination number* $q(\mathbf{x}) = \sum_{\mathbf{y}} \eta(\mathbf{y})$ (the sum running over NNs of $\mathbf{x}$), and to take the rate of creation at $\mathbf{x}$, for $q(\mathbf{x}) > 0$, as $\lambda[\eta(\mathbf{x})/q(\mathbf{x})] \sum_{\mathbf{y}} \sigma_{\mathbf{y}}$. (This approach suggests that we are dealing with a random lattice in which the 'missing sites' were, in a sense, never there.) The implementation of Moreira & Dickman (1996), on the other hand, takes the rate of creation at $\mathbf{x}$ as $\lambda[\eta(\mathbf{x})/q] \sum_{\mathbf{y}} \sigma_{\mathbf{y}}$, i.e., using the coordination number of the undiluted lattice.

With site dilution there can be no active state for $x > 1 - p_c$, where p_c is the critical concentration for (ordinary) site percolation, because the CP is doomed to extinction on finite sets. Thus the diluted CP is mainly of interest for $d \geq 2$. In one dimension suitable implementations of disorder include a random creation rate (strictly > 0 at each site), or annihilation rate (Bramson, Durrett, & Schonmann 1991). Quenched site randomness may equally well be imposed upon DP, yielding a *disordered* DP model (Noest 1986).

The most obvious effect of dilution is to raise the critical creation rate λ_c. Mean-field analyses (Marqués 1990, Moreira 1996) and simulations (Moreira & Dickman 1996) indicate that $\lambda_c(x)/\lambda_c(0) \propto 1/(1-x)$ for small x. As $x \nearrow 1 - p_c$, $\lambda_c(x)$ increases to a *finite* limit $\leqslant \lambda_c$ for the one-dimensional CP. (For $x = 1 - p_c$ we have the CP on an incipient percolation cluster, which features a backbone at least as highly connected as a one-dimensional system.)

It is natural to ask whether quenched disorder changes the critical behavior. According to the Harris criterion (A.B. Harris 1974, Berker 1993), disorder is relevant if $d\nu \leqslant 2$, where ν is the correlation-length exponent of the pure model. A glance at the $\nu_\perp$ values in table 6.1 indicates that we should expect disorder to change the critical behavior of models belonging to the DP universality class, for $d < 4$. It turns out that disorder can change not only the exponents, but the very nature of criticality in these models. In a one-dimensional disordered CP, for example, Bramson *et al.* (1991) showed that there are two phase transitions! In their model the annihilation rate randomly assumes one of two values at each site. They showed that below a certain creation rate, say λ_1, there is no active steady state and that the creation rate must exceed a certain value $\lambda_2 > \lambda_1$ for there to be linear growth ($\propto t$) of the active region, starting from a single seed. (By 'active region' we mean the set $\{\mathbf{x}_l, \ldots, \mathbf{x}_r\}$, where $\mathbf{x}_l$ ($\mathbf{x}_r$) is the leftmost (rightmost) occupied site.) For $\lambda_1 < \lambda < \lambda_2$ survival from a single seed is possible (i.e., $P_\infty > 0$), but the growth of the active region is *sublinear*, taking an inordinately long time to expand through an 'unfavorable' zone, i.e., one whose local annihilation rate is higher than the average. There is evidence (Moreira & Dickman 1996) to support the conjecture of Bramson *et al.* that for $d \geq 2$ the disordered CP has a unique phase transition.

The first simulations of disordered DP (Noest 1986) yielded critical exponents quite different from those of the pure model, for example $\beta = 1.10(5)$, $\nu_{||} \simeq 2.0$, and $\nu_\perp \simeq 1.2$ in 2+1 dimensions. In addition to deriving numerical estimates, Noest also showed that disordered DP exhibits anomalously slow (i.e., power-law) relaxation for site probabilities in the range $p_c(0) < p < p_c(x)$, where $p_c(0)$ is the critical coupling for the pure model and $p_c(x)$ the critical point with dilution x. (The corresponding range for the diluted CP is $\lambda_c(0) < \lambda < \lambda_c(x)$.) In this *nonequilibrium Griffiths phase* the long-time dynamics are governed by atypical regions with unusually low concentrations of diluted sites, rendering the process locally supercritical (Griffiths 1969, Noest 1988). Briefly, the argument for power-law relaxation in the diluted CP is as follows. The probability of the seed landing in a favored region, of linear size L, in which the local density of diluted sites is such that $\lambda - \lambda_{c,eff} = \Delta$, is $\sim \exp(-AL^d)$. ($\lambda_{c,eff}$ is the critical creation rate for a system with the site density prevailing in this region.) The lifetime of the process in such a region $\sim \exp(BL^d)$. (Here the precise forms of A and B are unknown, but it is clear that they are positive, increasing functions of Δ for $\Delta > 0$.) It follows that at long times

$$P(t) \sim \max_{\Delta,L} \exp[-(AL^d + te^{-BL^d})] \sim \max_{\Delta} t^{-A/B} \sim t^{-\phi}, \tag{6.51}$$

where the last step defines a (nonuniversal) decay exponent ϕ.

Moreira & Dickman (1996) studied the critical dynamics of the diluted CP, and found that the survival probability $P(t)$, mean population $n(t)$, and mean-square spread $R^2(t)$ show a *logarithmic* dependence on t, rather than the power laws typical of critical dynamics in pure systems. Thus the critical exponents δ, η, and z are formally zero. This also means that the critical dynamics of the diluted CP is incompatible with the scaling hypothesis discussed in §6.4. Simulation data for several points on the critical line are shown in figure 6.11, which reveals that the onset of logarithmic time-dependence is earliest at the highest dilution. The simulations also confirm power-law relaxation in the Griffiths phase, and show a power-law (rather than exponential) relaxation of the survival probability to P_∞ for $\lambda > \lambda_c(x)$. Field-theoretic analyses of disordered DP may be found in Obukhov (1987) and Janssen (1997).

Further evidence that the effect of quenched disorder on absorbing-state

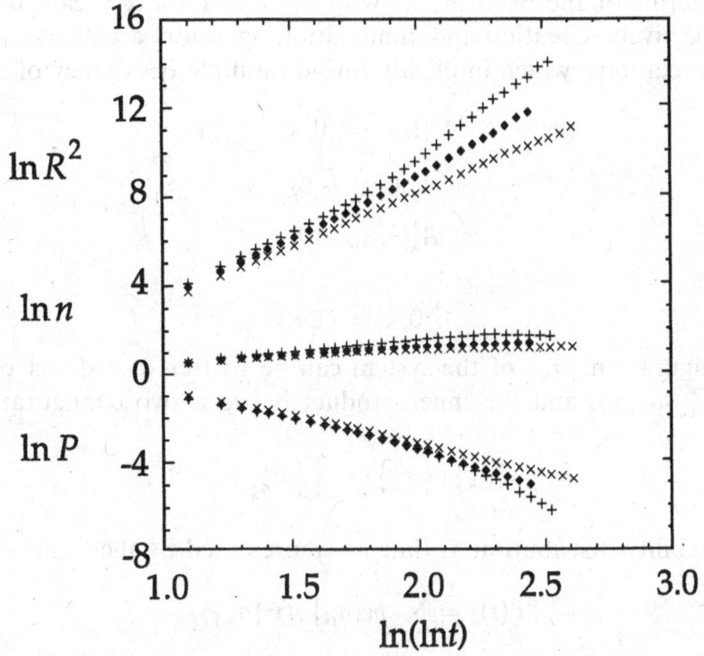

Fig. 6.11. Survival probability P, mean population n, and mean-square distance of particles from the origin R^2, in the critical diluted CP. $\times$: $x = 0.35$, $\lambda = 2.72$; $\diamond$: $x = 0.2$, $\lambda = 2.108$; $+$: $x = 0.05$, $\lambda = 1.7408$.

transitions promises to be a rich field of study comes from an analysis of the monomer–monomer model (see §5.2) incorporating random adsorption probabilities at each site (Frachebourg, Krapivsky, & Redner 1995). In this case disorder leads to a number of new phases, including reactive steady states where the pure model has none.

6.9 Operator methods

The CP and other models considered in this book are Markov processes with states corresponding to configurations of particles on a lattice. It is natural to represent the elementary events — creation, annihilation, diffusion — by means of operators. This permits us to represent the master equation governing each process in a compact form that lends itself to perturbative treatment.[9]

The configuration of the system is given by specifying the state of each

[9] In the present section we simply sketch the idea. For detailed expositions, see Peliti (1985), Dickman (1989a), and Jensen & Dickman (1993b).

site $\mathbf{x}$ in terms of the basis $|\sigma_{\mathbf{x}}, \mathbf{x}\rangle$ with $\sigma_{\mathbf{x}} = 0, 1$ for $\mathbf{x}$ vacant or occupied, respectively. Creation and annihilation operators are defined *via* the following relations, which implicitly forbid multiple occupancy of a site:

$$A_{\mathbf{x}}|1, \mathbf{x}\rangle = |0, \mathbf{x}\rangle, \tag{6.52}$$

$$A_{\mathbf{x}}|0, \mathbf{x}\rangle = 0, \tag{6.53}$$

$$A_{\mathbf{x}}^{\dagger}|1, \mathbf{x}\rangle = 0, \tag{6.54}$$

and

$$A_{\mathbf{x}}^{\dagger}|0, \mathbf{x}\rangle = |1, \mathbf{x}\rangle. \tag{6.55}$$

A configuration $\{\sigma_{\mathbf{x}}\}$ of the system can be written as a direct product $\{\sigma_{\mathbf{x}}\} = \prod_{\mathbf{x}} |\sigma_{\mathbf{x}}, \mathbf{x}\rangle$, and the inner product between two configurations is given by

$$\langle\{\sigma_{\mathbf{x}}\} \mid \{\sigma'_{\mathbf{x}}\}\rangle = \prod_{\mathbf{x}} \delta_{\sigma_{\mathbf{x}}, \sigma'_{\mathbf{x}}}. \tag{6.56}$$

The probability distribution at time t is represented by the 'state'

$$|\Psi(t)\rangle = \sum_{\{\sigma_{\mathbf{x}}\}} p(\{\sigma_{\mathbf{x}}\}, t) |\{\sigma_{\mathbf{x}}\}\rangle, \tag{6.57}$$

where the sum is over all configurations and $p(\{\sigma_{\mathbf{x}}\}, t)$ is the probability of configuration $\{\sigma_{\mathbf{x}}\}$ at time t. States must satisfy positivity and normalization to be physically admissible. Introducing the notation

$$\langle\ \ | \equiv \sum_{\{\sigma_{\mathbf{x}}\}} \langle\{\sigma_{\mathbf{x}}\}|, \tag{6.58}$$

(projection onto all possible states), normalization is expressed by

$$\langle\ \ |\Psi\rangle = 1, \tag{6.59}$$

while positivity demands

$$\langle\{\sigma_{\mathbf{x}}\}|\Psi\rangle \geq 0, \quad \forall\ \{\sigma_{\mathbf{x}}\}. \tag{6.60}$$

The probability distribution is governed by the master equation,

$$\frac{\mathrm{d}|\Psi(t)\rangle}{\mathrm{d}t} = S|\Psi(t)\rangle, \tag{6.61}$$

where S is the 'evolution operator' (more precisely, the generator of the process). If S is time-independent, equation (6.61) has the formal solution

$$|\Psi(t)\rangle = \mathrm{e}^{St}|\Psi(0)\rangle, \tag{6.62}$$

where $|\Psi(0)\rangle$ is the initial probability distribution.

Although $A_{\mathbf{x}}$ and $A_{\mathbf{x}}^{\dagger}$ are essentially bookkeeping devices, it is instructive to express the basic events in terms of these operators. Thus the component of S describing removal of a particle at a randomly chosen site is

$$W = \sum_{\mathbf{x}} (1 - A_{\mathbf{x}}^{\dagger}) A_{\mathbf{x}}. \tag{6.63}$$

The combination $(1 - A_{\mathbf{x}}^{\dagger})A_{\mathbf{x}}$ serves to transfer probability from a configuration with $\mathbf{x}$ occupied to the corresponding one having $\mathbf{x}$ vacant, ensuring that W conserves probability. Suppose for a moment that $S = W$. Then

$$\frac{d}{dt} |\Psi(t)\rangle = \sum_{\{\sigma_{\mathbf{x}}\}} p(\{\sigma_{\mathbf{x}}\}, t) \sum_{\mathbf{x}} \delta_{\sigma_{\mathbf{x}},1} \left(|\{\sigma_{\mathbf{x}}\}_{\mathbf{x}}\rangle - |\{\sigma_{\mathbf{x}}\}\rangle \right) \tag{6.64}$$

where $|\{\sigma_{\mathbf{x}}\}_{\mathbf{x}}\rangle$ denotes configuration $|\{\sigma_{\mathbf{x}}\}\rangle$ with the particle at $\mathbf{x}$ removed. Thus if $S = W$,

$$\frac{d}{dt}\text{Prob}(\sigma_{\mathbf{x}} = 1) = \frac{d}{dt} \langle \quad |A_{\mathbf{x}}| \Psi(t)\rangle = -\text{Prob}(\sigma_{\mathbf{x}} = 1), \tag{6.65}$$

so that W is, as expected, the evolution operator for removing particles at unit rate, independently at each site.

Similarly, addition of a particle at a randomly chosen site is represented by

$$U = \sum_{\mathbf{x}} (1 - A_{\mathbf{x}}) A_{\mathbf{x}}^{\dagger}, \tag{6.66}$$

and NN hopping corresponds to

$$\mathcal{D} = \sum_{\mathbf{x}} \sum_{\hat{\mathbf{e}}} (1 - A_{\mathbf{x}+\hat{\mathbf{e}}} A_{\mathbf{x}}^{\dagger}) A_{\mathbf{x}+\hat{\mathbf{e}}}^{\dagger} A_{\mathbf{x}}, \tag{6.67}$$

where the second sum is over NNs of $\mathbf{x}$. To represent the CP we require an operator that places a particle at $\mathbf{x}$ (vacant) at a rate proportional to the number of occupied NNs. This takes the form:

$$V = \frac{1}{q} \sum_{\mathbf{x}} \sum_{\hat{\mathbf{e}}} (1 - A_{\mathbf{x}}) A_{\mathbf{x}}^{\dagger} A_{\mathbf{x}+\hat{\mathbf{e}}}^{\dagger} A_{\mathbf{x}+\hat{\mathbf{e}}}. \tag{6.68}$$

The evolution operator for the basic CP is then $S_{CP} = W + \lambda V$, that for the diffusive process, $S_{CP} + D\mathcal{D}$, and so on. For multi-species models we may introduce operators analogous to $A_{\mathbf{x}}^{\dagger}$ and $A_{\mathbf{x}}$ for each type of particle. Thus it is straightforward to construct an S corresponding to a given model. The corresponding (quantum) spin model in some instances admits an exact solution (Grynberg, Newman and Stinchcombe 1994).

Besides providing a compact representation of the master equation, the operator description serves to organize the task of deriving perturbation expansions for various physical properties. The simplest such expansion is in powers of t — the Taylor series for equation (6.62). This is particularly useful for a simple initial state, for example, the CP with only a single particle; through transformation and analysis of the resulting series, useful estimates of the exponents δ and η may be obtained (Jensen & Dickman 1993b). A more powerful approach for the CP is to expand the ultimate survival probability (obtained *via* Laplace transformation of $|\Psi(t)\rangle\rangle$ in powers of λ or λ^{-1} (Dickman & Jensen 1991). In series analysis of the steady state, the stationary solution $|\overline{\Psi}\rangle$ is expanded by decomposing the evolution operator into a *noninteracting* part, S^0, for which we can find an exact solution, and treating the interaction perturbatively (Dickman & Burschka 1988, Dickman 1989a). In each case a series for a specific observable is derived by means of a computer algorithm embodying the effects of operators like W and V on configurations. (The length of the series is generally memory-limited; for example, 24 terms have been obtained for P_∞ in the CP.) The analysis of series to yield critical properties uses methods such as ratio analysis and Padé approximants, familiar from equilibrium critical phenomena (Stanley 1971, Baker 1990, Guttmann 1989). Further applications of series analysis in nonequilibrium problems are to multi-particle processes (Poland 1993), to an interacting monomer–monomer reaction (Zhou, Redner, & Park 1993), and to random sequential adsorption (Baram & Kutasov 1989, Dickman, Wang, & Jensen 1991, de Oliveira, Tomé, & Dickman 1992, Bonnier, Hontebeyrie, & Meyers 1993).

Operator formulations have also been devised for expressing conditional probabilities as path integrals (Doi 1976, Grassberger & Scheunert 1980, Peliti 1985). This approach, which allows multiple occupancy of lattice sites, is a convenient starting point for introducing an effective Hamiltonian density, and for field-theoretic renormalization group analysis. (Relaxing the rule against multiple occupancy and taking the continuum limit means that we are no longer dealing with precisely the same model, but given the appropriate birth and saturation terms, one is assured of studying a field theory in the same universality class as the original lattice model.) An important feature of the path-integral approach is that it provides a route to renormalization group analysis without the need to formulate a Langevin equation, and to stipulate the associated noise autocorrelation (Cardy & Täuber 1996). The same is true of the Poisson transform method (Elderfield & Vvedensky 1985, Droz & McKane 1994). Finally we note that series analysis has also been applied to probabilistic cellular automata (Ruján 1987), in particular, to DP (Adler & Duarte 1987, Adler *et al.* 1988, Essam *et al.* 1986, Jensen 1996).

7
Models of disorder

The ordinary kinetic versions of the Ising model[1] may be modified to exhibit steady nonequilibrium states. This is illustrated in chapter 4 where a conflict between two canonical mechanisms (*diffusion* and *reaction*) drives the configuration away from equilibrium. A more systematic investigation of this possibility, when the conflict is between different reaction processes only, is described here. We focus on spin systems evolving by a superposition of independent local processes of the kind variously known as spin flips, birth/death or creation/annihilation.[2] The restriction to spin flip dynamics does not prevent the systems in this class from exhibiting a variety of nonequilibrium phase transitions and critical phenomena. Their consideration may therefore help in developing nonequilibrium theory. In addition, they have some practical interest, e.g., conflicting dynamics may occur in disordered materials such as dilute magnetic systems, and some of these situations can be implemented in the laboratory.

The present chapter describes some exact, mean-field and MC results that together render an intriguing picture encouraging further study. It is argued in §7.1 that some of the peculiar, emergent, macroscopic behavior of microscopically disordered materials may be related to diffusion of disorder. This provides a physical motivation for the *nonequilibrium random-field Ising model* (NRFM). This system is then used throughout

[1] The kinetic magnetic model (with nonconserved magnetization) was introduced by Glauber (1963) — see also Stanley (1971), for instance —; Kawasaki (1966, 1972) has studied the kinetic lattice gas (in which density is conserved). We refer the reader to §4.2 for a precise statement of these systems.

[2] The superposition of spin flips has been investigated systematically in Garrido & Marro (1989) and subsequent work quoted below; see also chapter 8. The present chapter is mainly based in López-Lacomba, Garrido, & Marro (1990), Alonso & Marro (1992b), López-Lacomba & Marro (1992), and Garrido & Marro (1994).

the chapter to illustrate concepts and techniques. Further nonequilibrium conditions that are induced by (fast) diffusion of disorder are defined in §7.2, and §7.3 introduces a generalized class of systems with conflicting dynamics. These involve diffusion of disorder in such a way that a changeover from the *nonequilibrium impure models* in §7.2 to the ordinary (equilibrium) models of quenched disorder is described as one varies a parameter. §7.4 concerns the elementary transition rates used to construct specific models. In §7.5, the concept of an *effective Hamiltonian* is introduced, and §7.6 is devoted to reviewing some exact results for one-dimensional lattices. §7.7 considers the relation between nonequilibrium, quenched and annealed models of disorder. In §7.8, exact and mean-field results for arbitrary dimension are reviewed. Some interesting examples of the class of systems introduced in §7.2 and §7.3 are studied in chapter 8.

7.1 Diffusion of microscopic disorder

The early study of cooperative phenomena focused on ideal pure systems such as perfect crystals. Impurities were regarded as producing only high-order quantitative effects. In fact, much experimental effort was put into reducing sample irregularities, and a common theoreticians' strategy consisted of mapping an impure system onto some *effective* pure one. There is now widespread recognition, however, that in addition to being the rule in nature, the presence of disorder (however small) can dramatically alter many observable properties. In particular, a new world of fascinating phenomena has emerged concerning macroscopic ordering in microscopically disordered systems.[3]

The present understanding of this topic has benefited significantly from the study of magnetic systems. This is due to the fact that *good*, i.e., close to ideally homogeneous, examples of disordered magnetic systems are experimentally accessible. Furthermore, several impure lattice models have been proposed following the success of the Ising model to capture the essential physics in complex systems. These models involve a periodic lattice whose sites, for example, $\mathbf{r} \in \mathbf{Z}^d$, are occupied by spin(-$\frac{1}{2}$) variables with two states, $s_\mathbf{r} = \pm 1$. Every configuration, $\mathbf{s} = \{s_\mathbf{r}\}$, has a potential energy given by the Ising Hamiltonian,

$$H(\mathbf{s}) = -\sum_{|\mathbf{r}-\mathbf{r}'|=1} J_{\mathbf{r}\mathbf{r}'}\, s_\mathbf{r} s_{\mathbf{r}'} - \sum_\mathbf{r} h_\mathbf{r}\, s_\mathbf{r} , \tag{7.1}$$

i.e., any pair of NN *spins* contributes $\pm J_{\mathbf{r}\mathbf{r}'}$, and each spin interacts also with a magnetic field of local intensity $h_\mathbf{r}$. Disordered magnetic systems

[3] Ziman (1979), Grinstein (1985), and Mézard, Parisi, & Virasoro (1987), for instance.

based on (7.1) are the site and the bond *dilute models*;[4] in these cases, the spin variables at a fraction of randomly distributed sites are set to zero, or $J_{\mathbf{rr}'} = 0$ at a fraction of NN pairs selected at random. Another outstanding example is the *random exchange model*, where all couplings are ferromagnetic (alternatively, antiferromagnetic), i.e., $J_{\mathbf{rr}'} > 0$ (< 0) for any $|\mathbf{r} - \mathbf{r}'| = 1$, while their strength varies randomly from bond to bond. Another important example is the *Edwards–Anderson model* for which $J_{\mathbf{rr}'}$ is a random variable symmetrically distributed around zero.[5] In this model, competition between positive and negative couplings may preclude any configuration minimizing simultaneously all the exchange interactions in (7.1); the system is said to suffer *frustration* — see Parisi (1980), and Liebman (1986), for instance. This may induce uncommon macroscopic behavior which is perhaps shared, at least in part, by a class of materials known as *spin glasses* — Cannella & Mydosh (1972); see also, for instance, Binder & Young (1986), Mézard *et al.* (1987). Frustration has a different origin in the *random field model*,[6] where only the local field $h_{\mathbf{r}}$ is a random variable in (7.1).

Systems based on (7.1) or similar Hamiltonians are called *models of quenched disorder* (MQD) hereafter. They are characterized by a frozen-in spatial distribution of disorder, i.e., $J_{\mathbf{rr}'}$ and/or $h_{\mathbf{r}}$ vary at random with $\mathbf{r}$ but remain fixed with time. The study of this conceptually simple situation is difficult (see §6.8). It has led to the development of new concepts and techniques that apply to a large class of phenomena. However, exact results are scarce, and some open problems remain despite great effort; in particular, the comparison between the behavior of MQD and real materials is not quite satisfactory. Therefore, even though further study of MQD is advisable, one may argue they require modifications to fit actual cases in nature. Concerning spin glasses, for example, the MQD neglect diffusion of magnetically active ions. Diffusion constantly modifies the distance between each specific pair of spin ions in substances and, consequently, one should probably allow for variations both in space and time of $J_{\mathbf{rr}'}$ in a model. The same argument applies to other cases of

4 Site dilution has been studied in §6.8. See also Labarta, Marro, & Tejada (1986), where concentration-dependent *effective* critical exponents for the three-dimensional site-diluted ferromagnetic Ising model induced by crossover phenomena are reported, Chowdhury & Stauffer (1986), Holey & Fähnle (1990), and Heuer (1993), for instance.

5 Edwards & Anderson (1975); see also Sherrington & Kirkpatrick (1975). For simplicity, think of this case as an example of (7.1) with $h_{\mathbf{r}} = 0$ for any $\mathbf{r}$.

6 See Imry & Ma (1975); see also Imry (1984), Imbrie (1986), Fisher (1986), Bricmont & Kupiainen (1987), Newman *et al.* (1993), Dotsenko (1994), and Eichhorn & Binder (1996), for instance.

disordered systems, e.g., one may imagine the local field $h_{\mathbf{r}}$ also varies with time in materials. These effects do not seem to be correctly described by *annealed* systems. For example, the change with time of the spatial distribution of Js in the annealed version by Thorpe & Beeman (1976) of the Edwards–Anderson model is constrained by the need to reach equilibrium with the other degrees of freedom. Therefore, impurities tend to be strongly correlated (for example, they may eventually tend to be located at the interfaces), which is not observed in most substances. Instead, one may conceive a situation in which both $\mathbf{s}$ and the spatial distribution of *impurities*, e.g., $J_{\mathbf{rr}'}$, vary with time (perhaps with some correlations within the two sets). That is, one may assume that spins and impurities behave more independently of each other than in the annealed model so that a conflict occurs, and a steady nonequilibrium condition prevails asymptotically. This is consistent with the reported observation of nonequilibrium effects, for example, the influence of the details of the dynamical process on the steady state in some materials.[7]

A simple realization of this idea is by means of kinetic versions of the MQD, to be denoted MKD hereafter, in which impurities move rapidly and at random. More precisely, the MKD are Ising-like systems in which the probability of any configuration $\mathbf{s}$ at time t satisfies the master equation

$$\frac{\partial P(\mathbf{s};t)}{\partial t} = \sum_{\mathbf{s}^{\mathbf{r}} \in \mathbf{S}} \left[c(\mathbf{s} \mid \mathbf{s}^{\mathbf{r}}) P(\mathbf{s}^{\mathbf{r}};t) - c(\mathbf{s}^{\mathbf{r}} \mid \mathbf{s}) P(\mathbf{s};t) \right]. \tag{7.2}$$

Here, $\mathbf{S} = \{\mathbf{s}\}$ is the set of possible configurations, and the r.h.s. describes changes $s_{\mathbf{r}} \rightarrow -s_{\mathbf{r}}$ that generate a new configuration $\mathbf{s}^{\mathbf{r}}$ from $\mathbf{s}$ with probability $c(\mathbf{s}^{\mathbf{r}} \mid \mathbf{s})$ per unit time. Unlike for the ordinary case, however, $c(\mathbf{s}^{\mathbf{r}} \mid \mathbf{s})$ corresponds to the simultaneous competition of independent spin-flip processes each induced by a *Glauber generator* $\mathscr{L}_G$ as defined in §4.2. One may formulate this by writing

$$c(\mathbf{s}^{\mathbf{r}} \mid \mathbf{s}) = \left\langle\left\langle c\left(\mathbf{s};\vec{\zeta};\mathbf{r}\right)\right\rangle\right\rangle \equiv \int_{-\infty}^{+\infty} d\vec{\zeta}\, p\left(\vec{\zeta}\right) c\left(\mathbf{s};\vec{\zeta};\mathbf{r}\right). \tag{7.3}$$

Here, $\vec{\zeta}$ stands for a random variable with normalized distribution $p(\vec{\zeta})$; $\vec{\zeta}$ represents a set of degrees of freedom not included in $\mathbf{s}$, for example, the coupling constants, or the local magnetic fields, or both, or any other parameter in the configurational *energy function* that enters (7.3), $H_{\vec{\zeta}}(\mathbf{s})$.

[7] For comparisons between theory and experiments, see the reviews mentioned above, and Pynn & Skjeltorp (1987), for example; for some specific nonequilibrium effects in disordered systems, see Vincent, Hamman, & Ocio (1992), and Rieger (1995), for instance.

For simplicity, this is taken to be of the form (7.1) in most of the examples below, and each elementary rate $c(\mathbf{s};\vec{\zeta};\mathbf{r})$ is assumed to satisfy individually a detailed balance condition,

$$c\left(\mathbf{s}^{\mathbf{r}};\vec{\zeta};\mathbf{r}\right)=c\left(\mathbf{s};\vec{\zeta};\mathbf{r}\right)\exp\left(-\beta\delta H_{\vec{\zeta}}\right), \tag{7.4}$$

where $\delta H_{\vec{\zeta}} \equiv H_{\vec{\zeta}}(\mathbf{s}^{\mathbf{r}}) - H_{\vec{\zeta}}(\mathbf{s})$, and β is the inverse temperature of the bath involved by assumption (7.2).

Condition (7.4) states the microscopic reversibility of the Markov process (7.2) for $c\left(\mathbf{s}^{\mathbf{r}}|\,\mathbf{s}\right) = c(\mathbf{s};\vec{\zeta};\mathbf{r})$. That is, the system (7.1)–(7.4) reduces to the Glauber–Ising model for $p(\vec{\zeta}) = \delta(\vec{\zeta}-\text{const.})$, where δ is Dirac's delta function. Otherwise, the (dynamic) competition between different values of $\vec{\zeta}$ generates a steady nonequilibrium state in general. The superposition (7.3) differs from the reaction–diffusion one in chapter 4, and also from the anisotropic dynamics in chapter 2, but the result is similar in a sense. That is, the present system behaves, in general, as if acted on by an external non-Hamiltonian agent, and the steady state depends on $H_{\vec{\zeta}}(\mathbf{s})$, $p(\vec{\zeta})$, T, and $c(\mathbf{s};\vec{\zeta};\mathbf{r})$.

Investigating such a dependence may be relevant to understanding some of the emergent properties of substances with microscopic disorder. In fact, dynamics here can have two different interpretations:

(1) the quantity $\vec{\zeta}$ (e.g., J or h, assuming $J_{\mathbf{r}\mathbf{r}'} = J\ \forall\mathbf{r},\mathbf{r}'$ and $h_{\mathbf{r}} = h\ \forall\mathbf{r}$) takes at each step the same value — a random one chosen from distribution $p(\vec{\zeta})$ — all over the system (independent of $\mathbf{r}$), or

(2) only the value of $\vec{\zeta}$ that affects the few spins that are involved at each transition $\mathbf{s}\to\mathbf{s}^{\mathbf{r}}$ is changed at random successively during the time evolution.

Interpretation (1) readily follows from (7.3); the alternative interpretation (2) is allowed as long as the spin-flip process is local and the interactions within $H_{\vec{\zeta}}(\mathbf{s})$ are of short range, for example, restricted to NNs as in (7.1). Even though most *thermodynamic* behavior is the same for the two interpretations (see below), they correspond to different physical situations. For (2), a distribution of the disorder variable is reached asymptotically, which has two interesting properties. On one hand, it is similar *for any given t* to the frozen-in distribution with random spatial variations that characterizes the MQD; for example, any snapshot of a MKD with $p(\vec{\zeta}) \equiv p(J)$ within the asymptotic regime cannot be distinguished in practice from the Edwards–Anderson random spatial distribution $p(J)$ of Js. On the other hand, any MKD essentially differs from the corresponding quenched case due to variations with time of $p(\vec{\zeta})$. As demonstrated in §7.3, the variations of $p(\vec{\zeta})$ that are implied by (7.2)–(7.3) correspond to a sort of very rapid and completely random diffusion of disorder. This may be assumed to represent the diffusion of disorder in materials due to thermally activated atomic migration. This is perhaps too naive a representation of such effects in natural systems but it may be viewed as a first step towards a more realistic treatment, as described in §7.2. Another interesting point is that the MKD exhibit varied ordering phenomena which admit a systematic, mathematically precise study.

Stirring fields

A simple realization of MKD is the NRFM. In this case, $\vec{\zeta}$ in equations (7.3) and (7.4) represents a random applied magnetic field (equivalently, a chemical potential), h, with distribution $p(h)$, and

$$H_{\vec{\zeta}}(\mathbf{s}) \equiv H_h(\mathbf{s}) \equiv -J \sum_{|\mathbf{r}-\mathbf{r}'|=1} s_{\mathbf{r}} s_{\mathbf{r}'} - h \sum_{\mathbf{r}} s_{\mathbf{r}}, \tag{7.5}$$

for each h, where J is a constant. Therefore, the conflict is between independent canonical mechanisms each acting with probability $p(h)$ as if the magnetic field had a (random) value h.

According to interpretation (1) above, this may represent a magnet in a homogeneous field that varies at random. This is equivalent to assuming that the magnetic field varies monotonically with time describing $p(h)$ with a period shorter than the mean time between successive transitions that modify the spin configuration. (If one estimates this mean time as of the order of magnitude of the Larmor precession of a nucleus in the field of its neighborhood, a typical value is about 10^{-5} seconds.) This version is called the (Ising) *system under random field* (SURF) hereafter.[8] The model admits an alternative interpretation corresponding to (2) above. Assuming that the elementary dynamical process, i.e., $c(\mathbf{s};h;\mathbf{r})$ in (7.3), only involves a small domain of the lattice, and restricting oneself to the energy function (7.5), one is free to regard $h_{\mathbf{r}}$ as the only field that changes at a given step. Therefore, starting from (almost) any arbitrary spatial distribution for the fields, the dynamics soon establishes a random one which is a realization of the given $p(h)$. It follows that the system is described *at each time* by

$$\tilde{H}_{\mathbf{h}}(\mathbf{s}) \equiv -J \sum_{|\mathbf{r}-\mathbf{r}'|=1} s_{\mathbf{r}} s_{\mathbf{r}'} - \sum_{\mathbf{r}} h_{\mathbf{r}} s_{\mathbf{r}}, \tag{7.6}$$

where $\mathbf{h} = \{h_{\mathbf{r}}\}$ with $h_{\mathbf{r}}$ spatially distributed according to $p(h)$. This *instantaneous Hamiltonian* corresponds to the one that characterizes the

[8] The situation here differs essentially from the case studied in Lo & Pelcovits (1990), Acharyya & Chakraborti (1995), and Acharyya (1997) in which a system is periodically driven by the action of a field between two ordered phases.

ordinary, quenched random-field Ising model. However, the local field $h_{\mathbf{r}}$ also changes here at random with time at each site according to $p(h)$. Consequently, a sort of *dynamic* conflict or frustration occurs, which differs fundamentally from the standard quenched and annealed random-field models. It is likely that this situation is related to some of the peculiar behavior exhibited by certain substances.

The similarities and differences between the NRFM and the SURF deserve comment. Consider first the MC algorithm one would employ to implement the kinetic process (7.2)–(7.3). This is an important detail to be determined given that steady states may depend on kinetics. Therefore, choosing at random a rate $c(\mathbf{s};h;\mathbf{r})$ with probability $p(h)$ is not equivalent *a priori* to proceeding directly with the effective rate $c(\mathbf{s}^{\mathbf{r}}|\,\mathbf{s})$ computed according to (7.3). The latter procedure is more appropriate for the SURF version of the model. The NRFM, however, would require the probability of flipping $s_{\mathbf{r}}$ to be computed as a function of T and, consequently,

$$2Js_{\mathbf{r}}\left(\sum_{\{\mathbf{r}',\,|\mathbf{r}-\mathbf{r}'|=1\}} s_{\mathbf{r}'}+\frac{1}{2}h\right),$$

where h is to be sampled from $p(h)$.

Only the energy is biased in practice by the applied field. Therefore, significant differences between the NRFM and the SURF refer to the magnitude of energy fluctuations, which are expected to be anomalously large for the latter given that variations of the field then affect the whole system. In fact, the energy may be defined naturally for the SURF as a double average of (7.5),

$$U \equiv [[H_h(\mathbf{s})]] \equiv \sum_{\mathbf{s}\in\mathbf{S}} P_{st}(\mathbf{s})\,\langle\langle H_h(\mathbf{s})\rangle\rangle = \sum_{\mathbf{s}\in\mathbf{S}} P_{st}(\mathbf{s}) \int_{-\infty}^{+\infty} \mathrm{d}h\, p(h)\, H_h(\mathbf{s}), \quad (7.7)$$

where $P_{st}(\mathbf{s})$ is the stationary solution of (7.2). That is, the energy for each configuration, $H_h(\mathbf{s})$, is random, and one needs to average with respect to *two* random variables. Equation (7.7) states how to compute any macroscopic quantity in this case; in particular, one has $\sigma_U^2 \equiv [[\{H_h(\mathbf{s}) - U\}^2]]$ for the mean square fluctuations of the energy. For the NRFM, the energy and its fluctuations are defined similarly except that $\mathrm{d}h\,p(h)$ and (7.5) are replaced by $\prod_k \mathrm{d}h_k\, p_t^s(h_k)$ and (7.6), respectively. It follows that the energy is the same for both interpretations, while the term $[[h^2(\sum_{\mathbf{r}} s_{\mathbf{r}})^2]]$ which enters σ_U^2 for the SURF is replaced by

$$\left[\left[\left(\sum_{\mathbf{r}} h_{\mathbf{r}} s_{\mathbf{r}}\right)^2\right]\right] = \left[\left[h^2\left(\sum_{\mathbf{r}} s_{\mathbf{r}}\right)^2\right]\right] + 2\left(\mu^2 - \langle\langle h^2\rangle\rangle\right)\left\langle\sum_{\mathbf{r}\neq\mathbf{r}'} s_{\mathbf{r}} s_{\mathbf{r}'}\right\rangle$$

for the NRFM; here, $\langle\cdots\rangle$ represents the stationary average, and $\mu = \langle\langle h \rangle\rangle$. Generalizing this argument indicates that the two interpretations differ with respect to any function that is nonlinear in h. Otherwise, most behavior is expected to be the same.

As indicated above, both NRFM and SURF have simple limits for $p(h) = \delta(h \pm h_0)$, where h_0 is a positive constant. In any of these limits, and for rates satisfying (7.4), the model reduces to the well-known equilibrium case. The situation is less simple, however, for more general distributions $p(h)$. In general, the conflict between several values of the field impedes detailed balance, and leads the system asymptotically towards a steady nonequilibrium state whose dependence on $p(h)$, T, J, and $c(\mathbf{s};h;\mathbf{r})$ is unknown. This occurs even for the simplest field distribution, i.e., $p(h) = q\delta(h - h_0) + (1 - q)\delta(h + h_0)$. Therefore this simple model may allow one to analyze a large variety of situations, as illustrated below.[9]

7.2 Generalized conflicting dynamics

The NRFM may be generalized naturally along two different lines thus producing additional interesting cases. On one hand, one may consider further types of disorder or *impurities*. This leads to other examples of MKD that are defined immediately below in analogy with the NRFM. Alternatively, or in addition to this, one may assume that the spatial distribution of disorder changes with time according to a dynamical process that generalizes the one for the MKD. This possibility is introduced next in the present section. It is convenient, in particular, to clarify the significance of the MKD. That is, the MQD and the MKD follow as two cases of such a generalized system under different limiting conditions; these involve the relation between the time scales for the evolution of the impurity and of the spin subsystems.

We first consider further examples of MKD. They always follow the Markovian evolution (7.2) with effective rates (7.3); for the sake of simplicity, one may assume that (7.4) holds for the elementary process $c(\mathbf{s};\vec{\zeta};\mathbf{r})$. Different systems correspond to different interpretations of the random variable $\vec{\zeta}$. That is, one may think of $\vec{\zeta}$ as representing the temperature and/or any of the parameters that enter the expression for the energy function. For example, assuming (7.1), $\vec{\zeta}$ may represent not only the applied field h as in the NRFM but also the exchange energy J, or both. Some interesting cases are:

[9] We mention some interesting related work, mainly on the zero-temperature random-field Ising model, described in Dahmen & Sethna (1996), and references therein, focusing on the detailed study of nonequilibrium effects such as hysteresis and power-law distribution of avalanches — see also Marro & Vacas (1997).

- For $\vec{\zeta} \equiv J$ and, for example, $p(J) = q\delta(J - J_0) + (1 - q)\delta(J + J_0)$, where J_0 is a constant, one has a sort of nonequilibrium spin glass model.

- $p(J) = q\delta(J - J_0) + (1 - q)\delta(J)$ may be interpreted as a nonequilibrium dilute magnetic model.

After a transient regime, these systems (Garrido & Marro 1989, 1992) will exhibit a spatial distribution of Js similar *for any t* to the one characterizing the corresponding MQD in §7.1. The disorder configuration changes with time in the present nonequilibrium cases, however, to represent ideally (i.e., in a completely random manner) the diffusion of disorder that may be caused by atomic migration.

- López-Lacomba & Marro (1994a) have studied the consequences of assuming that $\vec{\zeta}$ represents both h and J, and $p(\vec{\zeta}) = p'(J)\,p''(h)$ with appropriate functions for p' and p'' in (7.3). According to the above philosophy, this may represent a spin-glass material subject to a random magnetic field, and has some interesting applications (for example, the so-called proton glasses).

- The same case with the appropriate distribution $p(\vec{\zeta})$ may be of interest in the description of neural networks (Sompolinski 1988), for example, systems with neural activity noise (Ma, Zhang, & Gong 1992, for instance).

- Alternatively, one may think of $\vec{\zeta}$ as symbolizing the bath temperature by inserting the inverse temperature β into definition (7.1). The dynamics then involves the action of a heat bath whose temperature is a random variable of distribution $p(\beta)$. This corresponds to a nonequilibrium system with locally competing temperatures, first studied in Garrido, Labarta, & Marro (1987) as a simple representation of the action of an external agent inducing nonequilibrium phenomena.

- One may also generalize the expression for the energy function (7.5). As an illustration, consider

$$H_{J_{NNN}}(\mathbf{s}) = -J_{NN} \sum_{|\mathbf{r}-\mathbf{r}'|=1} s_{\mathbf{r}} s_{\mathbf{r}'} - J_{NNN} \sum_{|\mathbf{r}-\mathbf{r}'_z|=2} s_{\mathbf{r}} s_{\mathbf{r}'_z}, \tag{7.8}$$

  where the last sum is over the *next* NN of $\mathbf{r} = (x, y, z)$ along the $\hat{\mathbf{z}}$-direction, denoted here $\mathbf{r}'_z$. From a formal point of view, this is identical to the Hamiltonian that characterizes the axial next-NN Ising (ANNNI) model.[10] López-Lacomba & Marro (1994b) studied

[10] Elliot (1961); see Selke (1992), for a review.

the case in which J_{NN} =const. and J_{NNN} is a variable changing at random during the evolution. Some of the properties of this *kinetic ANNNI model* and of the other cases just mentioned are described in chapter 8.

Next, as a further generalizing step, consider the case in which the probability at time t of the whole system configuration, $P_t(\mathbf{s}, \vec{\zeta})$, satisfies the master equation

$$\frac{\partial P_t\left(\mathbf{s},\vec{\zeta}\right)}{\partial t} = \left[\mathscr{L}_{\mathbf{s}} + \Gamma \mathscr{L}_{\vec{\zeta}}\right] P_t\left(\mathbf{s},\vec{\zeta}\right), \tag{7.9}$$

where $\mathbf{s}$ is the spin configuration, and $\vec{\zeta} = \{\zeta_i \in \Re\}$ is a set of degrees of freedom not included in $\mathbf{s}$, for example, $\{J_{\mathbf{rr'}}\}$, $\{h_{\mathbf{r}}\}$, etc. Furthermore,

$$\mathscr{L}_{\mathbf{s}} g\left(\mathbf{s};\vec{\zeta}\right) = \sum_{\mathbf{s^r}\in\mathbf{S}} \left[c\left(\mathbf{s^r};\vec{\zeta};\mathbf{r}\right) g\left(\mathbf{s^r};\vec{\zeta}\right) - c\left(\mathbf{s};\vec{\zeta};\mathbf{r}\right) g\left(\mathbf{s};\vec{\zeta}\right)\right], \tag{7.10}$$

where $g(\mathbf{s};\vec{\zeta})$ stands for an arbitrary function of its arguments. That is, $\mathscr{L}_{\mathbf{s}}$ is the Glauber generator in (4.9), and $c(\mathbf{s};\vec{\zeta};\mathbf{r})$ is the (elementary) transition rate in (7.3) for the process $s_{\mathbf{r}} \to -s_{\mathbf{r}}$ when the other variables, including $\vec{\zeta}$, are given. The operator $\mathscr{L}_{\vec{\zeta}}$ in (7.9) is defined as

$$\mathscr{L}_{\vec{\zeta}}\, g\left(\mathbf{s};\vec{\zeta}\right) = \sum_{\vec{\zeta}'} \left[\varpi\left(\vec{\zeta}' \to \vec{\zeta} \mid \mathbf{s}\right) g\left(\mathbf{s};\vec{\zeta}'\right) - \varpi\left(\vec{\zeta} \to \vec{\zeta}' \mid \mathbf{s}\right) g\left(\mathbf{s};\vec{\zeta}\right)\right]. \tag{7.11}$$

This adds to the action of $\mathscr{L}_{\mathbf{s}}$, and induces (simultaneously) changes of $\vec{\zeta}$ with probability per unit time given by $\varpi(\vec{\zeta} \to \vec{\zeta}' \mid \mathbf{s})$ when the spin configuration is $\mathbf{s}$. Equation (7.9) also involves the assumption that the two processes have *a priori* frequencies $\Gamma_{\mathbf{s}}$ and $\Gamma_{\vec{\zeta}}$, and one may speed up one of the processes relative to the other by varying $\Gamma \equiv \Gamma_{\vec{\zeta}}/\Gamma_{\mathbf{s}} \in [0,\infty]$.

It is convenient to simplify (7.9)–(7.11) by assuming that the rate $c(\mathbf{s};\vec{\zeta};\mathbf{r})$ satisfies detailed balance with respect to an energy function of the NN Ising type, as in (7.4). This leads to the standard Glauber model with equilibrium states for $\Gamma \equiv 0$ and $\vec{\zeta} = \{J_{\mathbf{rr'}} = J = \text{const.},\ h_{\mathbf{r}} = h = \text{const.}\}$. By avoiding this choice, rather general and interesting nonequilibrium situations may arise. Two classes of these are:

(i) $\vec{\zeta}$ changes with time independently of $\mathbf{s}$, i.e.,

$$\varpi\left(\vec{\zeta} \to \vec{\zeta}' \mid \mathbf{s}\right) = \varpi\left(\vec{\zeta} \to \vec{\zeta}'\right) \quad \text{for any } \mathbf{s}. \tag{7.12}$$

One may imagine this situation is caused by an external agent.

(ii) The processes (7.10) and (7.11) depend on each other in some way that is specified by the function $\varpi(\vec{\zeta} \to \vec{\zeta}' \mid \mathbf{s})$. This may correspond, for example, to a system in which impurities are canonically driven by a heat bath at temperature T' while spins change due to contact with a heat bath at temperature T — see Garrido & Marro (1997). The case $T' = T$ reproduces the familiar annealed concept.

The next section examines case (i).

7.3 Action of external agents

The consequences of assumption (7.12) are considered now in some detail. It is concluded that, in general, (7.9)–(7.12) lead in the limit $\Gamma \to 0$ to the standard, equilibrium MQD, while $\Gamma \to \infty$ corresponds to the nonequilibrium MKD. It is interesting to describe the crossover from one situation to the other as Γ is varied. Furthermore, it is shown that the time scales characterizing the evolution of $\mathbf{s}$ and of $\vec{\zeta}$ are well differentiated for $\Gamma \to \infty$; this helps to elucidate the physical significance of the nonequilibrium situations defined in the previous section. The formalism is illustrated by considering the explicit relation between a MQD by Mattis (1976) and the corresponding MKD version. An explicit, microscopic realization of the heat bath concept, which is another illustration of (7.12), is described in chapter 8.

The solution of the master equation (7.9) may be written as

$$P_t\left(\mathbf{s},\vec{\zeta}\right) = \Pi_t\left(\mathbf{s} \mid \vec{\zeta}\right) p_t\left(\vec{\zeta}\right), \tag{7.13}$$

with

$$\sum_{\mathbf{s}}\sum_{\vec{\zeta}} P_t\left(\mathbf{s},\vec{\zeta}\right) = 1, \quad \sum_{\mathbf{s}} \Pi_t\left(\mathbf{s} \mid \vec{\zeta}\right) = 1 \quad \forall \vec{\zeta}. \tag{7.14}$$

It follows under assumptions (7.12) and (7.13) that the distribution of disorder at each time, $p_t(\vec{\zeta})$, depends only on the initial distribution and on the intrinsic mechanism $\mathscr{L}_{\vec{\zeta}}$, while $\mathbf{s}$ evolves coupled to $\vec{\zeta}$, in general. The strength of this coupling depends on the value of Γ.

The simplest case occurs for $\Gamma = 0$. Then

$$p_t\left(\vec{\zeta}\right) = p_0\left(\vec{\zeta}\right) \quad \text{for any } t, \tag{7.15}$$

and the conditional probability satisfies

$$\frac{\partial \Pi_t\left(\mathbf{s} \mid \vec{\zeta}\right)}{\partial t} = \mathscr{L}_{\mathbf{s}} \Pi_t\left(\mathbf{s} \mid \vec{\zeta}\right). \tag{7.16}$$

That is, the initial distribution of disorder, $p_0(\vec{\zeta})$, remains quenched or frozen in time for $\Gamma = 0$, and $\mathbf{s}$ is governed by the Glauber equation (7.16). This is the situation that characterizes the MQD. As a matter of fact, the ordinary bond dilute, random field, and spin-glass Ising models are realizations of (7.8)–(7.12) for $\Gamma = 0$ and distributions

$$\left.\begin{aligned} p_t\left(\vec{\zeta}\right) &= \prod_{\mathbf{r},\mathbf{r}'} \left[q\delta\left(J_{\mathbf{rr}'}\right) + (1-q)\,\delta\left(J_{\mathbf{rr}'} - J_0\right)\right], \\ p_t\left(\vec{\zeta}\right) &= \prod_{\mathbf{r}} \left\{q\delta\left(h_{\mathbf{r}}\right) + \tfrac{1}{2}(1-q)\left[\delta\left(h_{\mathbf{r}} - h_0\right) + \delta\left(h_{\mathbf{r}} - h_0\right)\right]\right\}, \\ p_t\left(\vec{\zeta}\right) &= \prod_{\mathbf{r},\mathbf{r}'} \left[\tfrac{1}{2}\delta\left(J_{\mathbf{rr}'} - J_0\right) + \tfrac{1}{2}\delta\left(J_{\mathbf{rr}'} + J_0\right)\right], \end{aligned}\right\} \qquad (7.17)$$

respectively, where $|\mathbf{r} - \mathbf{r}'| = 1$.

Next, consider the limit $\Gamma \to 0$ such that $t \to \infty$ and the time scale $\tau \equiv \Gamma t$ remains finite. On this time scale, the master equation transforms into

$$\frac{\partial \tilde{p}_\tau\left(\vec{\zeta}\right)}{\partial \tau} = \mathscr{L}_{\vec{\zeta}}\, \tilde{p}_\tau\left(\vec{\zeta}\right), \quad \tilde{p}_\tau\left(\vec{\zeta}\right) \equiv p_{\tau/\Gamma}\left(\vec{\zeta}\right). \qquad (7.18)$$

One also obtains on the same time scale that

$$\mathscr{L}_{\mathbf{s}}\tilde{\Pi}_\tau\left(\mathbf{s} \mid \vec{\zeta}\right) \to 0 \text{ as } \Gamma \to 0, \quad \tilde{\Pi}_\tau\left(\mathbf{s} \mid \vec{\zeta}\right) \equiv \Pi_{\tau/\Gamma}\left(\mathbf{s} \mid \vec{\zeta}\right). \qquad (7.19)$$

This implies that

$$\tilde{\Pi}_\tau\left(\mathbf{s} \mid \vec{\zeta}\right) = \Pi_{st}\left(\mathbf{s} \mid \vec{\zeta}\right) \text{ for any } \tau > 0 \qquad (7.20)$$

corresponding to the steady solution of (7.16). Therefore, the general solution of (7.9) as $\Gamma \to 0$ is

$$\tilde{P}_\tau\left(\mathbf{s}, \vec{\zeta}\right) \equiv P_{\tau/\Gamma}\left(\mathbf{s}, \vec{\zeta}\right) = \Pi_{st}\left(\mathbf{s} \mid \vec{\zeta}\right) \tilde{p}_\tau\left(\vec{\zeta}\right). \qquad (7.21)$$

The validity of this simplified description rests upon the existence for $\Gamma \to 0$ of well-defined time scales characterized by time variables t and τ, respectively. More precisely, it is required that $\max_{\vec{\zeta}}\{t_{\mathbf{s}}(\vec{\zeta})\} \ll \Gamma^{-1}$, where $t_{\mathbf{s}}(\vec{\zeta})$ is the relaxation time associated to (7.16).

A similar discussion applies to the limit $\Gamma \to \infty$, $t \to 0$. The solution of the master equation may be written as

$$P_t\left(\mathbf{s}, \vec{\zeta}\right) = \Pi_t^*\left(\vec{\zeta} \mid \mathbf{s}\right) \mu_t(\mathbf{s}), \quad \sum_{\mathbf{s}} \Pi_t^*\left(\vec{\zeta} \mid \mathbf{s}\right) = 1, \quad \forall \vec{\zeta}. \qquad (7.22)$$

Then one substitutes into (7.9), uses condition (7.12), and introduces the time scale $\tau \equiv \Gamma t =$ finite. It follows for $\Gamma \to \infty$ and $t \to 0$ that

$\tilde{\mu}_\tau(\mathbf{s}) \equiv \mu_{\tau/\Gamma}(\mathbf{s}) = \tilde{\mu}_0(\mathbf{s})$ for any τ, and

$$\frac{\partial \Pi_\tau^*\left(\vec{\zeta} \mid \mathbf{s}\right)}{\partial \tau} = \mathscr{L}_{\vec{\zeta}} \Pi_\tau^*\left(\vec{\zeta} \mid \mathbf{s}\right). \tag{7.23}$$

Assuming uniqueness of the solution, this implies that $\Pi_\tau^*(\vec{\zeta} \mid \mathbf{s}) = \tilde{p}_\tau(\vec{\zeta})$ given that $\mathscr{L}_{\vec{\zeta}}$ is independent of $\mathbf{s}$. Therefore,

$$\tilde{P}_\tau\left(\mathbf{s},\vec{\zeta}\right) \equiv P_{\tau/\Gamma}\left(\mathbf{s},\vec{\zeta}\right) = \tilde{p}_\tau\left(\vec{\zeta}\right)\tilde{\mu}_0(\mathbf{s}) \quad \text{as } \Gamma \to \infty \tag{7.24}$$

(for small enough t).

Finally, one may consider the combined limit $\Gamma \to \infty$, $\tau \to \infty$, and the time scale $t = \tau\Gamma^{-1}$ = finite. One finds

$$\Pi_t^*\left(\vec{\zeta} \mid \mathbf{s}\right) = p_{st}\left(\vec{\zeta}\right), \quad \frac{\partial \mu_t(\mathbf{s})}{\partial t} = \sum_{\vec{\zeta}} p_{st}\left(\vec{\zeta}\right) \mathscr{L}_{\mathbf{s}} \mu_t(\mathbf{s}). \tag{7.25}$$

That is, during the time between two consecutive spin flips, the other degrees of freedom generally undergo enough changes when $\Gamma, \tau \to \infty$ to ensure that $p_t(\vec{\zeta})$ has reached the steady distribution $p_{st}(\vec{\zeta})$. One needs to assume that $\mathscr{L}_{\vec{\zeta}}$ is such that this distribution is unique and may be reached from almost any initial condition in a time small enough compared to the characteristic time associated to the evolution of $\mathbf{s}$.

Summing up, there are two interesting cases of (7.9)–(7.12): (a) For $\Gamma \to 0$, spins remain frozen in while impurities evolve as given by (7.18) *within the scale* $\tau = \Gamma t$ for large enough values of t. Within the original time scale t, the limiting case $\Gamma = 0$ corresponds to the ordinary MQD characterized by (7.15) and (7.16). (b) For $\Gamma \to \infty$, one may distinguish a time scale τ in which spins remain essentially frozen in, while impurities evolve according to (7.23). Moreover, a (coarse-grained) time scale t exists well separated from the other one in which impurities have already reached their stationary state, $p_{st}(\vec{\zeta})$. According to (7.25), the spin system evolves within the latter time scale as implied by an *effective Glauber generator*,

$$\mathscr{L}_{\mathbf{s}}^{eff} \equiv \sum_{\vec{\zeta}} p_t\left(\vec{\zeta}\right) \mathscr{L}_{\mathbf{s}}. \tag{7.26}$$

The MKD defined above correspond to this situation. We consider next a model that explicitly describes a changeover between cases (a) and (b).

A generalization of Mattis' model

Consider the one-dimensional Hamiltonian

$$H_{\vec{\zeta}}(\mathbf{s}) = -\sum_x J_x s_x s_{x+1}, \quad J_x = J_0 \zeta_x \zeta_{x+1}, \tag{7.27}$$

where J_0 is a constant, and $s_x = \pm 1$, $\zeta_x = \pm 1$. Here ζ_x is a random variable representing disorder that is distributed spatially between sites according to $p(\zeta_x) = \frac{1}{2}\delta(\zeta_x - 1) + \frac{1}{2}\delta(\zeta_x + 1)$. This system was devised by Mattis (1976); the one-dimensional case of interest here corresponds to the quenched spin-glass model of Edwards & Anderson (1976) (the correspondence breaks down for $d > 1$, however).

One may generalize this as an explicit (one-dimensional) realization of the model (7.9)–(7.12). With this aim, consider

$$c\left(\mathbf{s};\vec{\zeta};x\right) = 1 - \tanh\left[\beta J_0 s_x \zeta_x \left(s_{x+1}\zeta_{x+1} + s_{x-1}\zeta_{x-1}\right)\right] \tag{7.28}$$

and

$$\varpi\left(\vec{\zeta} \to \vec{\zeta}^{\,x} \mid \mathbf{s}\right) = \text{const.} \quad \text{for any } \vec{\zeta},\ \mathbf{s}\,. \tag{7.29}$$

The latter is the simplest realization of (7.12); together with (7.11) it implies that $\vec{\zeta}$ changes by *flipping*, i.e., $\zeta_x \to -\zeta_x$, completely at random — as due to the action of a heat bath at *infinite temperature*. The choice (7.28) reduces the system to a simple case, namely, it is shown below to be Bernoulli in the limit $\Gamma \to \infty$. In this case, the model has no critical point, even at $T = 0$. (Other one-dimensional MKD with more complicated functions $c(\mathbf{s};\vec{\zeta};x)$ do exhibit zero-T critical points.) The only motivation for (7.28) is simplicity (which illustrates nontrivial behavior, however). This choice introduces correlations between $\mathbf{s}$ and $\vec{\zeta}$ (although no correlations exist within either set) for any finite $\Gamma\,(\neq 0)$, and the original Mattis model (where correlations exist in both $\mathbf{s}$ and $\vec{\zeta}$) is recovered for $\Gamma \to 0$.

Let us introduce the variables $\sigma_x = \zeta_x s_x$, $\vec{\sigma} = \{\sigma_x\}$, and let the rate be $\tilde{c}\left(\vec{\sigma} \to \vec{\sigma}^x\right) = 1 - \frac{1}{2}\gamma\sigma_x\left(\sigma_{x+1} + \sigma_{x-1}\right)$ where $\gamma \equiv \tanh(2\beta J_0)$. One may write from (7.9)–(7.11) with (7.28) and (7.29) that

$$\frac{\partial\langle g\rangle_t}{\partial t} = \sum_{\vec{\sigma}^x}\Big\{\left\langle\left[g\left(\vec{\sigma}^x,\vec{\zeta}\right) - g\left(\vec{\sigma},\vec{\zeta}\right)\right]\tilde{c}\left(\vec{\sigma} \to \vec{\sigma}^x\right)\right\rangle_t$$
$$+\Gamma\left\langle g\left(\vec{\sigma}^x,\vec{\zeta}^{\,x}\right) - g\left(\vec{\sigma},\vec{\zeta}\right)\right\rangle_t\Big\}, \tag{7.30}$$

where

$$\left\langle g\left(\vec{\sigma},\vec{\zeta}\right)\right\rangle_t = \sum_{\vec{\sigma},\vec{\zeta}} g\left(\vec{\sigma},\vec{\zeta}\right) P_t\left(\vec{\sigma},\vec{\zeta}\right). \tag{7.31}$$

For example, making $g(\vec{\sigma},\vec{\zeta}) = \zeta_x$ leads to the prediction that

$$\langle\zeta_x\rangle_t = \langle\zeta_x\rangle_0 \exp(-2\Gamma t), \tag{7.32}$$

i.e., exponential relaxation toward a stationary distribution of disorder with $\langle\zeta_x\rangle_{st} = 0$ for $\Gamma > 0$. For $g(\vec{\sigma},\vec{\zeta}) = \zeta_0\sigma_x$, one obtains after Fourier transforming that

$$F_t(0) = \langle s_0 \rangle_t = e^{-2(1+2\Gamma)t} \sum_x F_0(x)\, I_{|x|}(2\gamma t)$$
$$+ 4\Gamma \int_0^t d\tau\, F_{t-\tau}(0)\, e^{-2(1+2\Gamma)\tau}\, I_0(2\gamma\tau), \qquad (7.33)$$

where $F_t(x) \equiv \langle \zeta_0 \sigma_x \rangle_t$ and $I_n(X)$ stands for the modified Bessel function. As expected, this reduces to the solution of the model by Mattis for $\Gamma \to 0$; otherwise, the study of the solutions of (7.33) leads to new kinds of behavior, we refer the reader to Garrido & Marro (1994) for details.

The infinite temperature limit $\gamma \to 0$ leads to $\langle s_0 \rangle_t = \langle s_0 \rangle_0 \exp(-2t)$ which is independent of Γ. Consequently, the spin system relaxes at infinite temperature independent of the other degrees of freedom with a characteristic time $\tau_s = \frac{1}{2}$, while the relaxation time for the disorder is $\tau_b = 1/2\Gamma$, as implied by the result (7.32). For small enough γ, one may expand (7.33); one obtains for $t < \frac{1}{2}\tau_b = 1/4\Gamma$ that the relaxation again becomes independent of Γ, namely,

$$\langle s_0 \rangle_t = e^{-2t} \{F_0(0) + \gamma t\,[F_0(1) + F_0(-1)]$$
$$+ \tfrac{1}{2}\gamma^2 t^2\,[F_0(2) + F_0(-2) + 2F_0(0)] + \mathcal{O}(\gamma^3)\}. \qquad (7.34)$$

This corresponds to the peculiar relaxation that characterizes the quenched spin-glass model. In the present case, (7.34) holds for any t as $\Gamma \to 0$. Only for $t > \frac{1}{2}\tau_b$, where $\tau_b \to 0$ as $\Gamma \to \infty$, is the relaxation influenced by the evolution of the disorder (bond) distribution. In any case, the relaxation depends on F_0, which describes initial correlations between the bond and spin configurations. One may also study the limit $\Gamma \to \infty$ by expanding (7.33) for small Γ^{-1}; one obtains

$$\langle s_0 \rangle_t = e^{-2t} \left\{ \langle s_0 \rangle_0 \left[1 + \frac{2\gamma^2 t}{\Gamma}\right] + \frac{\gamma}{4\Gamma}\,[F_0(1) + F_0(-1)] + \mathcal{O}\left(\frac{1}{\Gamma^2}\right)\right\}, \qquad (7.35)$$

i.e., the bonds evolve as if they were in contact with a bath at infinite temperature, as stated in (7.29). In order to recover the nonequilibrium Ising glass from the present generalization of Mattis' model one needs to consider the limit $\Gamma \to \infty$ as well as condition (7.29). This illustrates a general result cited above: in addition to the lack of correlations between impurities, the MKD involve different time scales for the evolution of **s** and $\vec{\zeta}$.

The function $G(x,y;z,w) = \langle \sigma_x \sigma_y \zeta_z \zeta_w \rangle$ may be obtained from (7.30) as

$$\begin{aligned}4\{\delta_{x,y} - \Gamma\,[2 - \delta_{x,y} - \delta_{x,w} - \delta_{y,w} - \delta_{z,w} - \delta_{x,z}\,(1-2\delta_{z,w}) \\ - \delta_{y,z}\,(1-2\delta_{y,w}) + 2\delta_{x,y}\,(\delta_{y,z} + \delta_{y,w} - 2\delta_{y,z}\delta_{y,w})] - 1\}\; G(x,y;z,w) \\ +\gamma\,(1-2\delta_{x,y})\,[G(x,y+1;z,w) + G(x,y-1;z,w)] \\ +\gamma\;[G(x+1,y;z,w) + G(x-1,y;z,w)] = 0,\end{aligned} \qquad (7.36)$$

where the condition $\langle \sigma_x \sigma_x \zeta_y \zeta_y \rangle = 1$ has been used; $\delta_{x,y}$ is the Kronecker delta. For $\Gamma = 0$, this reduces to

$$G(x,y;z,w) = \zeta_z \zeta_w\,[\tanh(\beta J)]^{|x-y|}\,, \qquad (7.37)$$

which is precisely the behavior in the Mattis case; otherwise, one may study (7.36) numerically. This has been done by performing up to 10^3 iterations on a 20^4 hypercube, with the boundaries fixed to $G = 0$, using Mattis' equilibrium correlations as initial conditions. In this way, one obtains

$$G(x,y;x,y) = \langle s_x s_y \rangle = \delta_{x,y}\,, \quad G(x,x;z,w) = \langle \zeta_z \zeta_w \rangle = \delta_{z,w}\,, \qquad (7.38)$$

which conform to the comment above that (7.28) induces correlations which are exceptionally simple within a physical context.. The correlations between the two sets of degrees of freedom, i.e., **s** and $\vec{\zeta}$, as measured by

$$G(r) \equiv G(x,y;z,z) = \langle s_x s_y \rangle, \quad r \equiv |x-y|\,, \qquad (7.39)$$

are nontrivial, however. In particular, one observes that:

(1) $G(r)$ decreases from the Mattis value with increasing Γ for any given $r > 0$. This is in agreement with the expectation that **s** and $\vec{\zeta}$ should become essentially uncorrelated (except at very short distances) for the nonequilibrium case, i.e., in the limit $\Gamma \to \infty$, for the simple choice (7.28).

(2) $G(r)$ increases with decreasing temperature for any $r > 0$.

(3) $G(r)$ decays exponentially with r for any value of Γ and T. Figure 7.1 depicts the behavior with T and Γ of the correlation length ξ defined by requiring that $G(r) \sim \exp(-r/\xi)$ for large enough r. This turns out to be practically constant if $T < 0.6$ (for $\Gamma = 0.1$) revealing that the case $\Gamma \neq 0$ does not have a zero-T critical point (for the special choice (7.28)).

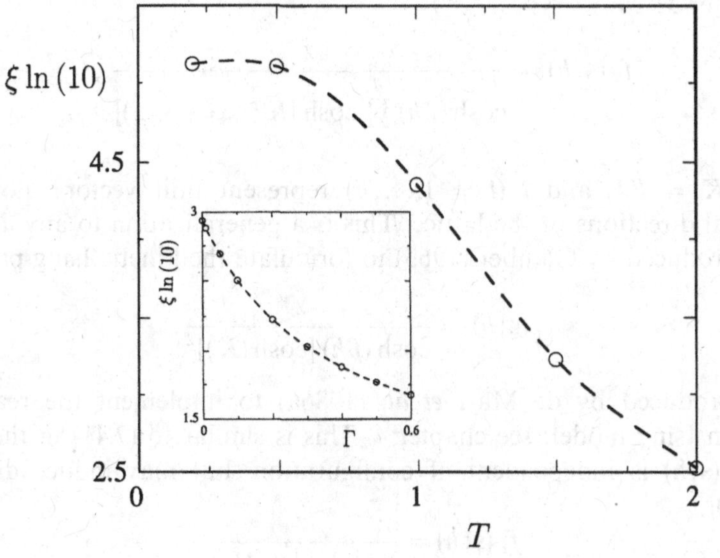

Fig. 7.1. The correlation length ξ, as defined in the text, for the generalized Mattis model for $\Gamma = 0.1$ as a function of T (main graph), and as a function of Γ for $T = 2$ (inset). The vertical axis includes a constant factor ln 10. (From Garrido & Marro (1994).)

7.4 Elementary transition rates

The physical situations of interest do not provide sufficient criteria to determine the elementary spin flip transition rates $c(\mathbf{s};\vec{\zeta};\mathbf{r})$ one should use in (7.3) or (7.10). The only restriction that has been introduced so far is the detailed balance condition (7.4) which is assumed for the sake of simplicity only. This implies that

$$c\left(\mathbf{s};\vec{\zeta};\mathbf{r}\right) = f_{\mathbf{r}}\left(\mathbf{s};\vec{\zeta}\right) \exp\left(-\frac{1}{2}\beta\delta H_{\vec{\zeta}}\right). \tag{7.40}$$

Here, $\delta H_{\vec{\zeta}} = H_{\vec{\zeta}}(\mathbf{s}^{\mathbf{r}}) - H_{\vec{\zeta}}(\mathbf{s})$, and the function $f_{\mathbf{r}}(\mathbf{s};\vec{\zeta})$ is analytic, positive, and independent of $s_{\mathbf{r}}$. For the NRFM, $H_{\vec{\zeta}}$ is given in (7.5); for concreteness, specific model expressions refer to this case henceforth, unless otherwise indicated.

The following functions correspond to familiar examples of rates that satisfy detailed balance.

$$f_{\mathbf{r}}(\mathbf{s};h) = \alpha = \text{const.} \tag{7.41}$$

has been used by van Beijeren & Schulman (1984) and others in different problems, for example, to set up a fast rate limit of the DLG as described in chapter 3:

$$f_{\mathbf{r}}(\mathbf{s};h) = \frac{\alpha}{\cosh(\beta h) \prod_{i=1}^{d} \cosh\left[K\left(s_{\mathbf{r}+\hat{\imath}} + s_{\mathbf{r}-\hat{\imath}}\right)\right]}, \tag{7.42}$$

where $K = \beta J$, and $\hat{\imath}$ $(i = 1,\ldots,d)$ represent unit vectors along the principal directions of the lattice. This is a generalization to any d of the rate introduced by Glauber (1963) to formulate the kinetic Ising problem.

$$f_{\mathbf{r}}(\mathbf{s};h) = \frac{\alpha}{\cosh(\beta h)\left[\cosh(K)\right]^{2d}} \tag{7.43}$$

was introduced by de Masi *et al.* (1986a) to implement the reaction–diffusion Ising model; see chapter 4. This is similar to (7.41) in the sense that $f_{\mathbf{r}}(\mathbf{s};h)$ is independent of configuration, but may induce different behavior.

$$f_{\mathbf{r}}(\mathbf{s};h) = \frac{\alpha}{2\cosh\left(\frac{1}{2}\beta\delta H_h\right)} \tag{7.44}$$

is the original rate used by Kawasaki (1966, 1972) for the kinetic lattice gas.

$$f_{\mathbf{r}}(\mathbf{s};h) = \exp\left(-\frac{1}{2}\left|\beta\delta H_h\right|\right) \tag{7.45}$$

corresponds to the Metropolis algorithm. For illustrative purposes, it turns out convenient to consider the case

$$f_{\mathbf{r}}(\mathbf{s};h) = \exp\left(-\left|\beta\delta H_h\right|\right), \tag{7.46}$$

even though it involves only an apparently trivial modification of the Metropolis case. We note that, in principle, no motivation exists for these choices other than that they correspond to rates used before in different problems.

It seems natural also to consider rates of the form

$$c(\mathbf{s};h;\mathbf{r}) = \varphi\left(\beta\delta H_h\right). \tag{7.47}$$

Here, $\varphi(X)$ represents a function of the configuration which is arbitrary except that $\varphi(X) = e^{-X}\varphi(-X)$; also, $\varphi(0) = 1$, and $\varphi(X) \to 0$ as $X \to \infty$, to ensure proper normalization and boundary conditions. From this point of view, (7.41), (7.43), (7.44), and (7.45) correspond, respectively, to

$$\varphi(X) = e^{-\frac{1}{2}X}, \tag{7.48}$$

$$\varphi(X) = e^{-\frac{1}{2}X}\left[\cosh(K)\right]^{-2d}, \tag{7.49}$$

$$\varphi(X) = \left(1 + e^{X}\right)^{-1}, \tag{7.50}$$

and

$$\varphi(X) = \min\{1, e^{-X}\}. \tag{7.51}$$

The case (7.42) in which directions are involved cannot be written as in (7.47); (7.49), where $\cosh K$ is not, in general, a constant, cannot be said to be of the form (7.47).

The peculiarity of the one-dimensional case makes it convenient to use a third representation of spin flip rates. This stresses their local nature, i.e., the fact that the rates of interest involve only the (two) NNs of the spin whose flip is being considered, s_r. The most general rate that satisfies detailed balance for $d = 1$ is

$$c\left(\mathbf{s};\vec{\zeta};\mathbf{r}\right) = f_r\left(\mathbf{s};\vec{\zeta}\right)\left[1 - \alpha_h s_r + \alpha\left(\alpha_h - s_r\right)\frac{1}{2}\left(s_{r+1} + s_{r-1}\right)\right], \tag{7.52}$$

where $\alpha \equiv \tanh(2K)$, $\alpha_h \equiv \tanh(\beta h)$, and $f_r\left(\mathbf{s};\vec{\zeta}\right) = f_r\left(\mathbf{s}^r;\vec{\zeta}\right)$, i.e.,

$$f_r\left(\mathbf{s};\vec{\zeta}\right) = a + b\frac{1}{2}\left(s_{r+1} + s_{r-1}\right) + c\, s_{r+1}s_{r-1}, \tag{7.53}$$

where $a > 0$. It follows by comparison that some of the rates of interest can be represented by (7.53) with the parameters given in table 7.1. The effective rate (7.3) may therefore be written rather generally as

$$c\left(\mathbf{s}^{\mathbf{r}} \mid \mathbf{s}\right) = A_r + \frac{1}{2}B_r\left(s_{r+1} + s_{r-1}\right) + C_r\, s_{r+1}s_{r-1}, \tag{7.54}$$

where

$$\left.\begin{aligned} A_r &= \langle\langle a\rangle\rangle + \tfrac{1}{2}\langle\langle \alpha\alpha_h b\rangle\rangle - s_r\left[\langle\langle \alpha_h a\rangle\rangle + \tfrac{1}{2}\langle\langle \alpha b\rangle\rangle\right], \\ B_r &= \langle\langle \alpha\alpha_h a\rangle\rangle + \langle\langle b\rangle\rangle + \langle\langle \alpha\alpha_h c\rangle\rangle - s_r\left[\langle\langle \alpha a\rangle\rangle + \langle\langle \alpha_h b\rangle\rangle + \langle\langle \alpha c\rangle\rangle\right], \\ C_r &= \langle\langle c\rangle\rangle + \tfrac{1}{2}\langle\langle \alpha\alpha_h b\rangle\rangle - s_r\left[\langle\langle \alpha_h c\rangle\rangle + \tfrac{1}{2}\langle\langle \alpha b\rangle\rangle\right]; \end{aligned}\right\} \tag{7.55}$$

$\langle\langle\cdots\rangle\rangle$ is defined in (7.3), and h may be interpreted as any of the parameters in the expression for the energy function. One has $\langle\langle \alpha_h \cdots\rangle\rangle = \alpha_h\langle\langle\cdots\rangle\rangle$ in (7.55) if h is not random but has a constant value, as in some cases below.

It may be remarked that using a specific rate introduces some arbitrariness. That is, the rate function often influences the steady state,[11] which makes it convenient to maintain the general forms (7.40), (7.47), or (7.52) for as long as possible; specific choices are introduced below for concreteness in order to illustrate the variety of emergent behavior the models may exhibit.

---

[11] A classification of rates according to their implications for macroscopic behavior is attempted in §7.6.

Table 7.1. The parameter values in equation (7.53) corresponding to various one-dimensional rates, as indicated. Here, $\pi \equiv \exp\left(-4\left|K\right|\right)$, $\mu \equiv \exp(-2\beta h)$, $y \equiv \cosh(\beta h)$, $z \equiv 1 + \mu$, and $\varepsilon \equiv 1 - \tanh^2(2K)\tanh^2(\beta h)$.

Rate	a	b	c
(7.41)	$y\cosh^2 K$	0	$y\sinh^2 K$
(7.42)	1	0	0
(7.44)	$(1+\varepsilon)/2\varepsilon$	$-\varepsilon^{-1}\sqrt{1-\varepsilon}$	$(1-\varepsilon)/2\varepsilon$
(7.45) for $h > 2J$	$z\dfrac{1+6\pi+\pi^2}{16\pi}$	$-z\dfrac{1-\pi^2}{8\pi}$	$z\dfrac{1-\pi^2}{16\pi}$
(7.45) for $h < 2J$	$z^2\dfrac{1+\pi}{16\mu}+\dfrac{z}{4}$	$(1+\pi)\dfrac{\mu^2-1}{8\mu}$	$z^2\dfrac{1+\pi}{16\mu}-\dfrac{z}{4}$

7.5 Effective Hamiltonian

For some of the models described in the preceding section, there is a systematic method for constructing the steady state. The essence of the method (Garrido & Marro 1989, López-Lacomba *et al.* 1990) is as follows. Let us assume that the master equation (7.2) has a strictly positive stationary solution for any configuration, $P_{st}(\mathbf{s})$. One may define an analytic object $E(\mathbf{s})$,

$$P_{st}(\mathbf{s}) \equiv \frac{\exp[-E(\mathbf{s})]}{\sum_{\mathbf{s}\in\mathbf{S}} \exp[-E(\mathbf{s})]} \tag{7.56}$$

which is expected to have the general structure

$$E(\mathbf{s}) = \sum_{k=1}^{N} \sum_{(\alpha,\ldots,\omega)} J_{k;\alpha,\ldots,\omega}\, s_\alpha \cdots s_\omega \, . \tag{7.57}$$

Here, the second sum is over every set of k spins on sites denoted $\alpha, \epsilon, \ldots, \rho, \sigma, \ldots, \omega$, respectively, and the coefficients $J_{k;\alpha,\ldots,\omega}$ are invariant under permutation of these indexes. In order to be useful, $E(\mathbf{s})$ needs to be short-range. For example, it is convenient to require that $E(\mathbf{s})$ involves only a finite number of coefficients, even if it refers to a macroscopic

$(N \to \infty)$ system. That is,

$$J_{k;\alpha,\ldots,\omega} = 0 \quad \text{for any} \quad k \geqslant k_0, \tag{7.58}$$

where k_0 is independent of N, at least for $N > N_0$. Clearly, a one-dimensional system with a short-range $E(\mathbf{s})$ may be solved by using the relatively simple tools of equilibrium theory; this is illustrated in §7.6. Consequently, if a unique stationary solution $P_{st}(\mathbf{s})$ exists that satisfies (7.56)–(7.58), then the object $E(\mathbf{s})$ that follows from (7.56) may be termed the *effective Hamiltonian* of the system characterized by the functions $c(\mathbf{s};\vec{\zeta};\mathbf{r})$ and $p(\vec{\zeta})$ in (7.3). Of course, the effective Hamiltonian does not represent, in general, the real energy of state $\mathbf{s}$, for example, the number of *up–down* bonds does not need to be related to $E(\mathbf{s})$. As is made explicit below, the effective Hamiltonian, if it exists, should be regarded primarily as a useful tool.

A series of results imply that $E(\mathbf{s})$ may indeed be computed explicitly in some (relatively simple) cases. The following holds: for a finite system driven by any dynamics $c(\mathbf{s}^{\mathbf{r}} \mid \mathbf{s})$, where $\mathbf{s}, \mathbf{s}^{\mathbf{r}} \in \mathbf{S}$, a unique, strictly positive, steady state $P_{st}(\mathbf{s})$ exists as an asymptotic $(t \to \infty)$ solution of (7.2) for (almost) any initial condition $P(\mathbf{s};0)$. This is simply a consequence of the fact that (7.2) describes in the present case a Markov chain which is homogeneous and irreducible (López-Lacomba *et al.* 1990). As a corollary, a system with N sites is characterized by a unique $E(\mathbf{s})$ given by (7.57) with

$$J_{k;\alpha,\ldots,\omega} = \frac{1}{2^N} \sum_{\mathbf{s} \in \mathbf{S}} s_\alpha \cdots s_\omega \ln P_{st}(\mathbf{s}) . \tag{7.59}$$

Two main problems arise. On one hand, the chances are that the $E(\mathbf{s})$ that follows in this way contains an indefinite number of terms as $N \to \infty$. Fortunately, property (7.58) may be seen to hold for many cases of interest, and the restriction of the theorem to finite N does not then imply a real limitation for our purposes, as proved below. On the other hand, even if (7.58) holds, one still needs to compute $P_{st}(\mathbf{s})$ from the master equation and to deal with (7.59) and (7.57), which involves a system of 2^N homogeneous linear equations, a lengthy and, sometimes, futile procedure.

Therefore, a simpler systematic method, even if it involves some restriction, is required to study the general existence of the (short-range) effective Hamiltonian $E(\mathbf{s})$ and to compute it explicitly. It turns out convenient to consider first systems that satisfy *global detailed balance* (GDB). That is, as a condition that simplifies the search for an effective Hamiltonian, it is required in some instances that the effective transition rate has the symmetry

$$c(\mathbf{s}^{\mathbf{r}} \mid \mathbf{s}) \exp[-E(\mathbf{s})] = c(\mathbf{s} \mid \mathbf{s}^{\mathbf{r}}) \exp\left[-E(\mathbf{s}^{\mathbf{r}})\right] \tag{7.60}$$

for any $\mathbf{s} \in \mathbf{S}$ and $\mathbf{r} \in \mathbf{Z}^d$, where $E(\mathbf{s})$ is defined in (7.56). This implies a sort of reversibility in the problem (see below, however). It then follows after combining (7.60) for a given $\mathbf{r}$ with (7.57) that

$$\ln \frac{c(\mathbf{s}^{\mathbf{r}} \mid \mathbf{s})}{c(\mathbf{s} \mid \mathbf{s}^{\mathbf{r}})} = -\delta E = 2s_{\mathbf{r}} \sum_{k=1}^{N} \sum_{(\mathbf{r},\epsilon,\ldots,\omega)} J_{k;\mathbf{r},\epsilon,\ldots,\omega}\, s_{\epsilon} \cdots s_{\omega}\,, \tag{7.61}$$

$\delta E \equiv E(\mathbf{s}^{\mathbf{r}}) - E(\mathbf{s})$. Next, one obtains

$$J_{k;\alpha,\ldots,\mathbf{r},\ldots,\omega} = \frac{1}{2^{N+1}} \sum_{\mathbf{s}\in\mathbf{S}} s_{\alpha} \cdots s_{\mathbf{r}} \cdots s_{\omega} \ln \frac{c(\mathbf{s}^{\mathbf{r}} \mid \mathbf{s})}{c(\mathbf{s} \mid \mathbf{s}^{\mathbf{r}})} \tag{7.62}$$

by identifying coefficients in (7.61). These quantities differ from the unknowns $J_{k;\alpha,\ldots,\omega}$ in which all subscripts play the same role. That is, one needs to repeat the procedure for every $\mathbf{r}$, and this implies that any set of k spins at (say) $\alpha, \epsilon, \ldots, \rho, \sigma, \ldots, \omega$ must satisfy

$$\sum_{\mathbf{s}\in\mathbf{S}} s_{\alpha} \cdots s_{\rho} s_{\sigma} \cdots s_{\omega} \ln \frac{c(\mathbf{s}^{\rho} \mid \mathbf{s})\, c(\mathbf{s} \mid \mathbf{s}^{\sigma})}{c(\mathbf{s} \mid \mathbf{s}^{\rho})\, c(\mathbf{s}^{\sigma} \mid \mathbf{s})} = 0 \tag{7.63}$$

for all sites ρ and σ. It follows that (7.63) is the necessary and sufficient condition for any dynamics $c(\mathbf{s}^{\mathbf{r}} \mid \mathbf{s})$ to satisfy GDB. One also has that $c(\mathbf{s}^{\mathbf{r}} \mid \mathbf{s}) = f_{\mathbf{r}}(\mathbf{s}) \exp(-\frac{1}{2}\delta E)$, where $f_{\mathbf{r}}(\mathbf{s}) = f_{\mathbf{r}}(\mathbf{s}^{\mathbf{r}}) > 0$, i.e., $f_{\mathbf{r}}(\mathbf{s})$ has no dependence on the value of $s_{\mathbf{r}}$. This characterizes a (relatively simple) subclass of the systems of interest, associated with a family of functions $c(\mathbf{s}; \vec{\zeta}; \mathbf{r})$ and $p(\vec{\zeta})$ that conform to the GDB condition (7.60), whose interesting behavior is illustrated below.

Another general result concerns the number of coefficients $J_{k;\alpha\ldots\omega}$ that are involved in (7.57) as N is varied and, in particular, the possibility of having condition (7.58) fulfilled. The following holds: $E(\mathbf{s})$ is the effective Hamiltonian if the effective rate $c(\mathbf{s}^{\mathbf{r}} \mid \mathbf{s})$ satisfies GDB and depends only on a constant number of lattice sites, independent of N. In fact, if the transition rates $c(\mathbf{s}^{\mathbf{r}} \mid \mathbf{s})$ and $c(\mathbf{s} \mid \mathbf{s}^{\mathbf{r}})$ involve (only) n spin variables, it follows that only the coefficients $J_{k;\alpha\ldots\omega}$ that involve these n variables (namely, 2^n coefficients, at most) may differ from zero, given that they follow from (7.62) when GDB holds. This result simplifies the search for an effective Hamiltonian for familiar Ising-like systems and elementary rates, for which n equals a few units, if GDB holds. The problem reduces in practice to the analysis of the latter condition, equivalently of condition (7.63). One conclusion from this analysis is that usually no effective Hamiltonian exists, which suggests more complex behavior. Explicit expressions for the effective Hamiltonian are given in §7.6 for $d = 1$, and in §8.4 for $d > 1$.

Summing up, if the effective rate (7.3) is short-range, as it is for a superposition of the elementary rates in §7.4, and GDB holds, the system has an effective Hamiltonian. This turns out to be so in certain one-dimensional cases (as well as for certain rates when $d > 1$), and the effective Hamiltonian is then of the NN Ising type (perhaps with a field term included) if this is the structure of the original series of *Hamiltonians* involved in the elementary rates by means, for example, of equation (7.47). In spite of canonical symmetry, which seems to be implied by (7.60), this situation is far from trivial. That is, an effective Hamiltonian obtained from (7.60) does not imply that a single parameter such as an *effective temperature* exists, and the corresponding emergent behavior is shown below to be complex in general. On the other hand, the GDB condition does not hold in many cases, as illustrated below and in chapter 8. The theorems above do not, however, exclude the existence of an effective Hamiltonian defined via equations (7.56)–(7.58). Several paths may be followed to investigate such a possibility. One method consists of replacing (7.60) by a weaker condition. López-Lacomba *et al.* (1990) have described a perturbative theory for small departures from equilibrium. It is remarkable that a perturbative approach for systems in which GDB does not hold may take condition (7.60) as the *unperturbed state*. A method of successive approximations to determine the effective Hamiltonian in the absence of GDB has been considered by de Oliveira (1994) — see also Künsch (1984); this neglects coupling parameters corresponding to clusters of spins larger than a certain size. The concept of effective Hamiltonian is also invoked in the numerical renormalization group method of Swendsen (1984). In this, as implemented by Browne & Kleban (1989), one computes correlation functions in actual configurations obtained by the MC method using the given kinetic rules. The coupling parameters are then obtained from the correlation functions by using correlation identities. These methods are not described here (they would, however, be interesting to apply to some of the models in this book). Instead, we illustrate below the use of approximate methods such as mean-field theory and standard MC simulations to deal with some complex cases.

7.6 Thermodynamics in one dimension

The general results in the preceding section imply the existence of an effective Hamiltonian for several one-dimensional versions of the systems of interest. This leads in turn to the explicit *exact* computation of steady-state properties by the methods of equilibrium theory. This program is illustrated in the present section for the one-dimensional NRFM.

The notation becomes simpler if one introduces the definitions

$$A \equiv \ln \frac{\langle\langle \phi_0 \exp(\beta h) \rangle\rangle}{\langle\langle \phi_0 \exp(-\beta h) \rangle\rangle}, \tag{7.64}$$

$$B \equiv \ln \frac{\langle\langle \phi_+ \exp(\beta h + 2K) \rangle\rangle}{\langle\langle \phi_+ \exp(-\beta h - 2K) \rangle\rangle}, \tag{7.65}$$

$$C \equiv \ln \frac{\langle\langle \phi_- \exp(\beta h - 2K) \rangle\rangle}{\langle\langle \phi_- \exp(-\beta h + 2K) \rangle\rangle}. \tag{7.66}$$

Here, ϕ_0 and $\phi_\pm$ denote the function $f_{\mathbf{r}}(\mathbf{s};h)$ defined in (7.40) when $s_{\mathbf{r}+\hat{\imath}} + s_{\mathbf{r}-\hat{\imath}}$ equals zero and ± 2, respectively. The necessary and sufficient condition for GDB reduces to

$$2A = B + C. \tag{7.67}$$

Consider a one-dimensional system for which this holds. It follows by algebra from the formalism in the previous section that an effective Hamiltonian exists, given by

$$E(\mathbf{s}) = -K_e \sum_{|\mathbf{r}-\mathbf{r}'|=1} s_{\mathbf{r}} s_{\mathbf{r}'} - \beta h_e \sum_{\mathbf{r}} s_{\mathbf{r}} \tag{7.68}$$

with

$$\beta h_e = \tfrac{1}{2} A, \quad K_e = \tfrac{1}{8}(C - B). \tag{7.69}$$

That is, the effective Hamiltonian has the same structure as the NN Ising Hamiltonian appearing in the definition of the system.

This result, which is a consequence of GDB, implies a canonical structure in some cases. Nevertheless, the *effective parameters* in $E(\mathbf{s})$ involve a rather complex dependence on T, $p(h)$, J, and $c(\mathbf{s};h;\mathbf{r})$. That is, thermodynamics depends on the specific realizations of the rates and of the distribution characterizing dynamical disorder, and this is reflected in macroscopic properties, as made explicit below. In other words, noncanonical features are involved. This *quasi canonical* situation corresponds to having the spin system subject to an external agent that represents the effect of competing kinetics by modifying the parameters of the original Hamiltonian. That is, even though the model is relatively simple when GDB holds, nontrivial constraints exist on the spin system. As indicated in §7.5, the GDB condition (for the effective rate) holds only exceptionally, so that most of the models of interest have an even *more complex*, fully nonequilibrium behavior rather than the quasi canonical one described in this section.

A macroscopic classification of rates

The theory in the preceding section suggests a meaningful classification of elementary rates based on their influence on emergent behavior. We first define two classes which are convenient to describe the NRFM.[12]

Soft kinetics. This denotes hereafter rates such as (7.41), (7.42), and (7.43) studied by van Beijeren & Schulman, Glauber, and de Masi *et al.*, respectively. In this case, GDB is satisfied for any distribution $p(h)$, and it always follows that $K_e = K = \beta J$. The fact that kinetics has a competing nature, and the only practical differences between the rates in this class are reflected in the existence of a nonzero effective field given by

$$\tanh(\beta h_e) = \frac{\langle\langle \sinh(\beta h)\rangle\rangle}{\langle\langle \cosh(\beta h)\rangle\rangle} \tag{7.70}$$

for (7.41), and by

$$\tanh(\beta h_e) = \langle\langle \tanh(\beta h)\rangle\rangle \tag{7.71}$$

for both (7.42) and (7.43).

Hard kinetics. This identifies hereafter rates such as (7.44), (7.45), and (7.46) corresponding, respectively, to Kawasaki and Metropolis rates, and a modification of the latter. They may induce a more intricate situation than soft rates. In fact, GDB is satisfied for any distribution $p(h)$ only if the original Hamiltonian has $J = 0$ (and one obtains $K_e = 0$, and h_e given by (7.71) for (7.44), for instance). For the more interesting case with $J \neq 0$, GDB is satisfied only for field distributions such that $p(h) = p(-h)$, and one gets $h_e = 0$, and either

$$\tanh(2K_e) = \langle\langle \tanh(2K + \beta h)\rangle\rangle \tag{7.72}$$

for (7.44), or else

$$\tanh(2K_e) = \frac{\langle\langle \exp[-|n(2K + \beta h)|] \sinh(2K + \beta h)\rangle\rangle}{\langle\langle \exp[-|n(2K + \beta h)|] \cosh(2K + \beta h)\rangle\rangle} \tag{7.73}$$

for (7.45) and (7.46) with, respectively, $n = 1$ and 2.

The differences between *soft* and *hard* rates are due to the fact that the function that characterizes the former factorizes, $f_{\mathrm{r}}(\mathrm{s}; h) = f_{\mathrm{r}}^{(1)}(s_{\mathrm{r}+1} + s_{\mathrm{r}-1}) f^{(2)}(h)$, where $f^{(2)}(h)$ has no dependence on the variables $s_{\mathrm{r}\pm 1}$, while this is not an attribute of *hard kinetics.* The latter consequently requires a condition on $p(h)$ in order to satisfy the (strong) property of GDB.

---

[12] See a related discussion in §7.8 from a more general perspective, and further interesting consequences of this in chapter 8.

Effective interactions and fields

Next, we work out the effective parameters K_e and h_e for some specific field distributions of interest. The simplest case is *soft kinetics* with $p(h) = p(-h)$. In a sense, this reduces to the Ising model under zero field, which justifies the name given to these rates. More precisely, one has $K_e = K$, and both (7.70) and (7.71) reduce to $h_e = 0$. However, energy fluctuations differ from the standard equilibrium ones; see below. A more complex behavior when the competition is *soft* occurs, for example, for any symmetric distribution $p(\mu + h) = p(\mu - h)$ with $\mu \neq 0$. Interestingly enough, this produces the 'canonical' behavior $h_e = \mu$ and $K_e = K$ for (7.41), while one has $K_e = K$, and certain rather complicated expressions for h_e, when the rate is either (7.42) or (7.43). This fact may be made explicit by considering the distribution $p(h) = \frac{1}{2}q\delta\,(h - (\mu + \kappa)) + \frac{1}{2}q\delta\,(h - (\mu - \kappa)) + (1 - q)\,\delta\,(h - \mu)$, for instance. This transforms (7.71) into

$$\beta h_e = \tfrac{1}{2}\ln\frac{\cosh^2(\beta\mu)\,[1 + \tanh(\beta\mu)] + \sinh^2(\beta\kappa)\,[1 + (1 - q)\tanh(\beta\mu)]}{\cosh^2(\beta\mu)\,[1 - \tanh(\beta\mu)] + \sinh^2(\beta\kappa)\,[1 - (1 - q)\tanh(\beta\mu)]}. \tag{7.74}$$

For the same (soft) rate (7.41), it seems interesting *a priori* to consider $p(h) = q\delta\,(h - \mu_1) + (1 - q)\,\delta\,(h - \mu_2)$ with $\mu_2 > \mu_1$. Then, in addition to $K_e = K$, one gets from (7.70) that

$$\tanh(\beta h_e) = \frac{\tanh\dfrac{\mu_1 + \mu_2}{2} + (1 - 2q)\tanh\dfrac{\mu_2 - \mu_1}{2}}{1 - (1 - 2q)\tanh\dfrac{\mu_1 + \mu_2}{2}\tanh\dfrac{\mu_2 - \mu_1}{2}}. \tag{7.75}$$

The other cases, i.e., (7.42) and (7.43), also imply $K_e = K$ but (7.70) gives

$$\tanh(\beta h_e) = q\tanh(\beta\mu_1) + (1 - q)\tanh(\beta\mu_2) \tag{7.76}$$

instead. These changes in the effective field as one varies the system parameters, even for the relatively simple one-dimensional case with soft kinetics, illustrate the richness exhibited by this family of models.

The emergent properties are even more varied for *hard kinetics.* Here only an even field distribution $p(h) = p(-h)$ satisfies GDB in nontrivial cases, as noted above. In order to make this more explicit, while still focusing on general facts, one may study the case $p(h) = \frac{1}{2}q\delta\,(h - \kappa) + \frac{1}{2}q\delta\,(h + \kappa) + (1 - q)\,\delta\,(h)$ which is characterized by $h_e = 0$. The situation may be summarized as follows. For rate (7.44), one gets from (7.72) that

$$K_e = K + \tfrac{1}{4}\ln\frac{1 - (1 - q)\tanh^2(2K)\tanh^2(\beta\kappa) - q\tanh(2K)\tanh^2(\beta\kappa)}{1 - (1 - q)\tanh^2(2K)\tanh^2(\beta\kappa) + q\tanh(2K)\tanh^2(\beta\kappa)}. \tag{7.77}$$

For rate (7.46), (7.73) with $n = 2$ leads to

$$\left.\begin{aligned} K_e &= K + \tfrac{1}{4}\ln\frac{q\cosh(\beta\kappa) + (1-q)}{q\cosh(3\beta\kappa) + (1-q)} && \text{for} \quad 2J > \kappa > 0 \\ K_e &= K + \tfrac{1}{4}\ln\frac{q\cosh(2K) + (1-q)}{q\cosh(6K) + (1-q)} && \text{for} \quad 2J = \kappa \\ K_e &= K + \tfrac{1}{4}\ln\frac{q\,\mathrm{e}^{-\beta\kappa} + q\mathrm{e}^{-3\beta\kappa+8K} + 2(1-q)}{q\,\mathrm{e}^{-3\beta\kappa} + q\mathrm{e}^{-\beta\kappa+8K} + 2(1-q)} && \text{for} \quad 2J < \kappa\,. \end{aligned}\right\} \tag{7.78}$$

Finally, (7.73) with $n = 1$ gives a distinct behavior, namely,

$$K_e = K - \tfrac{1}{4}\ln\left[q\cosh(2\beta\kappa) + (1-q)\right] \quad \text{for} \quad 2J \geqslant \kappa > 0\,, \tag{7.79}$$

and an involved expression for $2J < \kappa$.

Emergent behavior

The energy U and its fluctuations are defined in §7.1 in terms of $p(h)$ and $P_{st}(\mathbf{s})$, and the latter distribution may now be written as a function of K_e and h_e for some of the model versions. The *specific heat* may then be obtained as $C = \partial U/\partial T$. On the other hand, the magnetization M and its mean square fluctuations may also be defined in the usual way, i.e., $\sigma_M^2 = \langle[\,(\sum_{\mathbf{r}} s_{\mathbf{r}}) - M\,]^2\rangle$, where $\langle\cdots\rangle \equiv \sum_{\mathbf{s}\in\mathbf{S}} P_{st}(\mathbf{s})\cdots$, and the magnetic susceptibility is $\chi_T = \partial M/\partial\mu$, where μ stands for the mean of distribution $p(h)$. One also has, for instance, that $M = \partial\ln Z/\partial(\beta h_e)$, where $Z = \sum_{\mathbf{s}\in\mathbf{S}}\exp[-E(\mathbf{s})]$, and $U = -J\partial\ln Z/\partial K_e - \mu M$. The transfer matrix method — see Stanley (1971), Thompson (1972), for instance — may then be used to obtain the quantities of interest in terms of K_e, h_e, K, and μ. Some of the relevant expressions are:

$$M = N\frac{\sinh(\beta h_e)}{\sqrt{\sinh^2(\beta h_e) + \mathrm{e}^{-4K_e}}}\,, \tag{7.80}$$

$$U = -JN\left[1 + \frac{2}{\mathrm{e}^{4K_e} - 1}\left(1 - f_{\frac{1}{2}}(K_e, h_e)\right)\right] - \mu M\,, \tag{7.81}$$

$$\sigma_M^2 = Nf_{\frac{1}{2}}(K_e, h_e)\left[1 - \left(\frac{M}{N}\right)^2\right]\,, \tag{7.82}$$

$$C = \frac{JN\,\mathrm{e}^{4K_e}}{\mathrm{e}^{4K_e}-1}\left\{\frac{8}{\mathrm{e}^{4K_e}-1}\left(1 - f_{\frac{1}{2}}(K_e,h_e)\right) + 4\,\mathrm{e}^{-4K_e} f_{\frac{3}{2}}(K_e,h_e)\right\}\frac{\partial K_e}{\partial T}$$
$$-\mu N f_{\frac{1}{2}}(K_e,h_e)\left[1-\left(\frac{M}{N}\right)^2\right]\frac{\partial(\beta h_e)}{\partial T}$$
$$-2N\tanh(\beta h_e) f_{\frac{3}{2}}(K_e,h_e)\left(J\frac{\partial(\beta h_e)}{\partial T} + \mu\frac{\partial K_e}{\partial T}\right), \tag{7.83}$$

and

$$\chi_T = \sigma_M^2\frac{\partial(\beta h_e)}{\partial\mu} + 2N\tanh(\beta h_e) f_{\frac{3}{2}}(K_e,h_e)\frac{\partial K_e}{\partial\mu}\,. \tag{7.84}$$

Here,

$$f_{\frac{1}{2}}(K_e,h_e) \equiv \frac{\cosh(\beta h_e)}{\sqrt{\sinh^2(\beta h_e) + \mathrm{e}^{-4K_e}}} \tag{7.85}$$

and

$$f_{\frac{3}{2}}(K_e,h_e) \equiv \frac{\mathrm{e}^{-4K_e}\cosh(\beta h_e)}{\left[\sinh^2(\beta h_e) + \mathrm{e}^{-4K_e}\right]^{\frac{3}{2}}}. \tag{7.86}$$

Given that one has $K_e = K$ and $\partial K_e/\partial\mu = 0$ for soft kinetics, and that hard kinetics requires even distributions $p(h)$ implying $\mu = 0$ and $h_e = 0$, the above formulas simplify significantly for specific cases; we refer the reader to López-Lacomba & Marro (1992) for details.

The existence and nature of a critical point is an interesting question here. One may investigate this by considering the correlation length, $\xi = \xi(h,T)$, assuming that the spin–spin correlation function behaves as $g(r) \sim \exp(r/\xi)$ at large distances r.

Using standard notation, $\xi(0,\epsilon) \sim \epsilon^{-\nu}$ as $\epsilon \to 0^+$, and $\xi(0,\epsilon) \sim (-\epsilon)^{-\nu'}$ as $\epsilon \to 0^-$ for the Ising model. More precisely, the NN one-dimensional Ising model corresponding to $H(\mathbf{s}) = -\sum_r s_r(Js_{r+1} + h)$ with h =const. is characterized by

$$g(r) = \sin^2\left[2\tau\left(\frac{\psi_-}{\psi_+}\right)^r\right], \quad \xi = \left(\ln\frac{\psi_-}{\psi_+}\right)^{-1}, \tag{7.87}$$

where

$$\psi_\pm = \mathrm{e}^K\cosh(\beta h) \pm \sqrt{\mathrm{e}^{2K}\sinh^2(\beta h) + \mathrm{e}^{-2K}} \tag{7.88}$$

and $\cot\tau = \mathrm{e}^{2K}\sinh(\beta h)$. This suggests that one use $\epsilon \equiv \mathrm{e}^{-2K}$ as the relevant temperature parameter, and the existence of a critical point at

$T = 0$ and $h = 0$ with $\nu = 1$ is implied. It is convenient to introduce the scaling assumption

$$m = \theta\,|\theta|^{1/\delta-1}\,\Phi\left(\epsilon\,|\theta|^{-1/\beta\delta}\right); \tag{7.89}$$

here m is the magnetization for small values of ϵ, $\theta \equiv \beta h$, and Φ is an undetermined *scaling function;* see Baxter (1982). For $|\theta| \ll 1$, (7.89) holds for the (pure) one-dimensional Ising model with $\Phi(X) = \left(1 + X^2\right)^{-\frac{1}{2}}$, and $\beta\delta = 1$ where δ is infinite and $\beta = 0$. Moreover, $\alpha = 1$ follows from the scaling relation $2 - \alpha = \gamma = \nu = 1$.

These equations may be applied to quasi canonical systems with competing kinetics. That is, one may expect they have critical behavior for $h_e = 0$ and $T \to 0^+$ when GDB is satisfied. This limit is not as simple as it may appear at first glance, however. In fact, the models involve a distribution $p(h)$ whose specific form is known to affect both GDB and the condition $h_e = 0$ strongly, and the fact that h_e and K_e depend on temperature compels one to perform a careful study of the limit $T \to 0^+$. It follows from (7.87) that ξ diverges as $\psi_-\psi_+^{-1} \to 1$, which occurs for $\exp(-4K_e)\left[1 - \tanh^2(\beta h_e)\right] + \tanh^2(\beta h_e) \to 0$. Given that $\tanh^2(\beta h_e) \geqslant 0$, and $\exp(-4K_e)\left[1 - \tanh^2(\beta h_e)\right] \geqslant 0$, two conditions are required: $\tanh^2(\beta h_e) \to 0$, which implies $\beta h_e \to 0$, and $\exp(-4K_e) \to 0$, which implies $K_e \to \infty$. The simplest case in which this occurs is when $h_e \to 0$ as the mean of $p(h)$ goes to zero (assuming this limit causes no extra problems in K_e). This corresponds to the existence of a proper critical point at $T = 0$. One may expect more complex situations in general, however, as is made clear below.

Impure critical behavior

The thermodynamic formulas above are now considered for specific rates and disorder distributions. We focus here on the NRFM near the zero-temperature limit.

The soft case with any generic even distribution, $p(h) = p(-h)$ of mean $\mu = 0$ and variance σ_h^2, has been shown to correspond to $K_e = K$ and $h_e = 0$. Furthermore, one gets $M = 0$, $U = -2JN\sinh(2K)\sinh^{-2}K$, and $C = \left(4J^2N/k_BT^2\right)\sinh^2 K\sinh^{-2}(2K)$. The noncanonical nature of the system is reflected in fluctuations. One obtains $\sigma_M^2 = N\exp(2K)$, and

$$\sigma_U^2 = k_BT^2C + \sigma_e^2\,. \tag{7.90}$$

Here, $\sigma_e^2 = \sigma_h^2\sigma_M^2$ is an *excess* fluctuation associated with the nonzero variance of $p(h)$. Consequently, a fluctuation-dissipation relation does not hold in this simple case in the sense that σ_U^2 differs from k_BT^2C.

Next, consider $p(\mu + h) = p(\mu - h)$ with $\mu \neq 0$. The hard case satisfies GDB only for $J = 0$ which corresponds to a rather trivial system. The soft case is more interesting; one may distinguish two subcases:

(i) For rate (7.41), $K_e = K$ and $h_e = \mu$. This corresponds to the pure Ising model under field μ except for the presence of excess fluctuations given by $\sigma_e^2 = \sigma_h^2 (\sigma_M^2 + M^2)$.

(ii) For rates (7.42) and (7.43), there is no general result valid for any symmetric distribution with $\mu \neq 0$. One may consider, however, the example $p(h) = \frac{1}{2}q\delta(h - (\mu + \kappa)) + \frac{1}{2}q\delta(h - (\mu - \kappa)) + (1 - q)\delta(h - \mu)$. This leads to $K_e = K$, and h_e given by (7.71). As long as $q \neq 1$, one obtains as $T \to 0$:

$$\left.\begin{array}{lll} U \approx -N(J + \mu), & \sigma_U^2 \approx N\sigma_h^2, & C \to 0, \\ M \approx N, & \sigma_M^2 \to 0, \text{ and} & \chi_T \to 0. \end{array}\right\} \tag{7.91}$$

However, some other quantities may exhibit important differences near zero T depending on the relation between the parameters μ and κ:

$$\tanh(\beta h_e) \to \begin{cases} 1 & \text{for } \mu > \kappa \\ 1 - \frac{1}{2}q & \text{for } \mu = \kappa \\ 1 - q & \text{for } \mu < \kappa, \end{cases} \tag{7.92}$$

or

$$h_e \approx \begin{cases} \mu - \kappa + \frac{1}{2\beta}\ln\left(\dfrac{2}{q}\right) & \text{for } \mu > \kappa \\ \frac{1}{2\beta}\ln\left(\dfrac{4-q}{q}\right) & \text{for } \mu = \kappa \\ \frac{1}{2\beta}\ln\left(\dfrac{2-q}{q}\right) & \text{for } \mu < \kappa, \end{cases} \tag{7.93}$$

and

$$\frac{\partial(\beta h_e)}{\partial \mu} \approx \begin{cases} \beta & \text{for } \mu > \kappa \\ 0 & \text{for } \mu \leqslant \kappa, \end{cases} \tag{7.94}$$

$$\frac{\partial(\beta h_e)}{\partial T} \approx \begin{cases} \beta^2(\kappa - \mu) & \text{for } \mu > \kappa \\ 0 & \text{for } \mu \leqslant \kappa. \end{cases} \tag{7.95}$$

The most interesting situation for soft kinetics occurs for $\mu < \kappa$ and $q = 1$, i.e., $p(h) = \frac{1}{2}\delta(h - (\mu + \kappa)) + \frac{1}{2}\delta(h - (\mu - \kappa))$. A dependence on the sign of $(\kappa - \mu) - J$ is found in this case:

$$\chi_T \approx \begin{cases} 2N\beta\, e^{4\beta(\kappa - \mu - J)} \to 0 & \text{for } \kappa - \mu < J \\ N\beta \to \infty & \text{for } \kappa - \mu = J \\ 2N\beta\, e^{-2\beta(\kappa - \mu - J)} \to 0 & \text{for } \kappa - \mu > J \end{cases} \tag{7.96}$$

and

$$\sigma_M^2 \approx \begin{cases} N\,\mathrm{e}^{\beta[6(\kappa-\mu)-4J]} & \text{for} \quad \kappa-\mu<J \\ N\,\mathrm{e}^{2\beta J} \to \infty & \text{for} \quad \kappa-\mu=J \\ N\,\mathrm{e}^{2\beta J} \to \infty & \text{for} \quad \kappa-\mu>J\,. \end{cases} \tag{7.97}$$

That is, the situation depends on whether the exchange interaction J is strong enough to compensate the action of the fields, namely $\mu+\kappa$ and $\mu-\kappa<0$ (there is even a dependence on the relation between $3(\kappa-\mu)$ and $2J$ reflected in σ_M^2), and the critical behavior varies accordingly. One finds two classes of critical points for the Glauber rate (7.42):

(1) One occurs for $\mu\to 0$ and $T\to 0^+$. The situation for $q\neq 1$ resembles the one for the equilibrium case, except that the magnetization m scales now according to (7.89) with $\theta=\beta\mu(1-q)$. More unexpected is the case $q=1$ for which $\beta\delta=1-\kappa/J$, and spontaneous magnetization is exhibited for $\kappa<J$ which scales with $\theta=4\beta\mu$, while $m=0$ for $\kappa>J$.

(2) The situation differs for $q=1$ with $\mu\neq 0$ as $T\to 0^+$. A line of critical points exists for $\mu<\kappa$, in which ξ diverges with $\nu=\min\left[1,(\kappa-\mu)J^{-1}\right]$. This is to be compared with $\nu=1$ that follows for $q=1$ by taking $T\to 0^+$ after $\mu\to 0$, i.e., the two limits do not commute in general. ξ does not diverge for $\mu>\kappa$.

The hard case with even field distribution, for example, $p(h)=\frac{1}{2}q\delta(h-\kappa)+\frac{1}{2}q\delta(h+\kappa)+(1-q)\,\delta(h)$ with $q\neq 0$, presents interesting properties as well, as anticipated above. One obtains

$$\left.\begin{aligned} &U=-2NJ\frac{\sinh^2 K_e}{\sinh(2K_e)}, \quad M=0, \quad \sigma_M^2=N\,\mathrm{e}^{2K_e}, \\ &C=-4NJ\frac{\sinh^2 K_e}{\sinh^2(2K_e)}\frac{\partial K_e}{\partial T}, \quad \sigma_h^2=q\kappa^2, \\ &\sigma_U^2=4NJ^2\frac{\sinh^2 K_e}{\sinh^2(2K_e)}+\sigma_h^2\sigma_M^2 \end{aligned}\right\} \tag{7.98}$$

(χ_T cannot be defined for $\mu=0$). On the other hand, it follows from (7.77) that as $T\to 0^+$ for the Kawasaki rate (7.44),

$$K_e \approx \begin{cases} K\left(1-\dfrac{\kappa}{2J}\right)+\frac{1}{4}\ln\left(\dfrac{2}{q}\right) & \text{for} \quad 2J>\kappa \\ \frac{1}{4}\ln\left(\dfrac{4-q}{q}\right) & \text{for} \quad 2J=\kappa \\ \frac{1}{4}\ln\left(\dfrac{2-q}{q}\right) & \text{for} \quad 2J<\kappa. \end{cases} \tag{7.99}$$

This exhibits two distinct classes of behavior at low temperature. One gets

$$\xi \approx \sqrt{\frac{q}{8}}\epsilon^{-\nu}, \quad \nu = 1 - \frac{\kappa}{2J}, \tag{7.100}$$

for $2J > \kappa$, while $2J = \kappa$ leads to

$$\xi \approx \frac{1}{2}\sqrt{\frac{4-q}{q}}. \tag{7.101}$$

That is, the (dynamical) disorder makes the system so *hot* at $T = 0$ in the latter case that the usual (thermal) critical point is washed out, but there is a sort of (*percolative,* i.e., having rather a configurational or geometrical origin) critical behavior as $q \to 0$ which may be characterized by an exponent $\nu_q = \frac{1}{2}$ for the correlation length. This occurs also for $2J < \kappa$ and, given that interactions cannot balance out the peaks of the field distribution in this case, it follows that $K_e \to 0$, $U \to 0$, $C \to 0$, $\sigma_U^2 \to N\sigma_h^2$, and $\sigma_M^2 \to N$ for $q = 1$. Figure 7.2 illustrates these situations. The implications of (7.73) for the Metropolis rate (7.45) are also rather complex, as one may work out by oneself; see equation (7.79). Moreover, as illustrated in figure 7.3, one finds from (7.73) for $n = 2$ a situation corresponding to (7.46) which differs essentially from that for (7.45), in spite of the fact that the two rates are quite similar.

Some of the foregoing analysis of the NRFM may be extended to a closely related *impure system,* namely, the case in which $p(h)$ involves such strong fields that the spin direction is *frozen* with some probability (during the evolution). (Alternatively, one may imagine that the chemical potential has some probability of fixing the occupation variable at one of the two possible states.) This occurs for $p(h) = q\delta(h - \mu_1) + (1 - q)\delta(h - \mu_2)$ if $\mu_2 \to \infty$, for instance. GDB then holds for soft kinetics only, as shown above, and one may distinguish two cases. The first one occurs for rate (7.41) for which $\psi_-\psi_+^{-1} \to 0$ as $\mu_2 \to \infty$. The second occurs for rates (7.42) and (7.43); then, $h_e = \mu_1$ for $\mu_2 \to \infty$ and $q \to 1$ in such a way that $\mu_2(1 - q) = \kappa$ remains finite. Further details may be found in López-Lacomba & Marro (1992).

Nonuniversal behavior

Summing up, the great diversity of behavior a simple one-dimensional system may exhibit is notable — as also occurs in nature; see Dagotto (1996), for instance. Figure 7.4, with different graphs for the magnetization of the NRFM is a direct evidence of this fact, as is figure 7.5, which illustrates the *specific heat* for the same system. The differences from the pure case are important. In particular, it is noteworthy that the

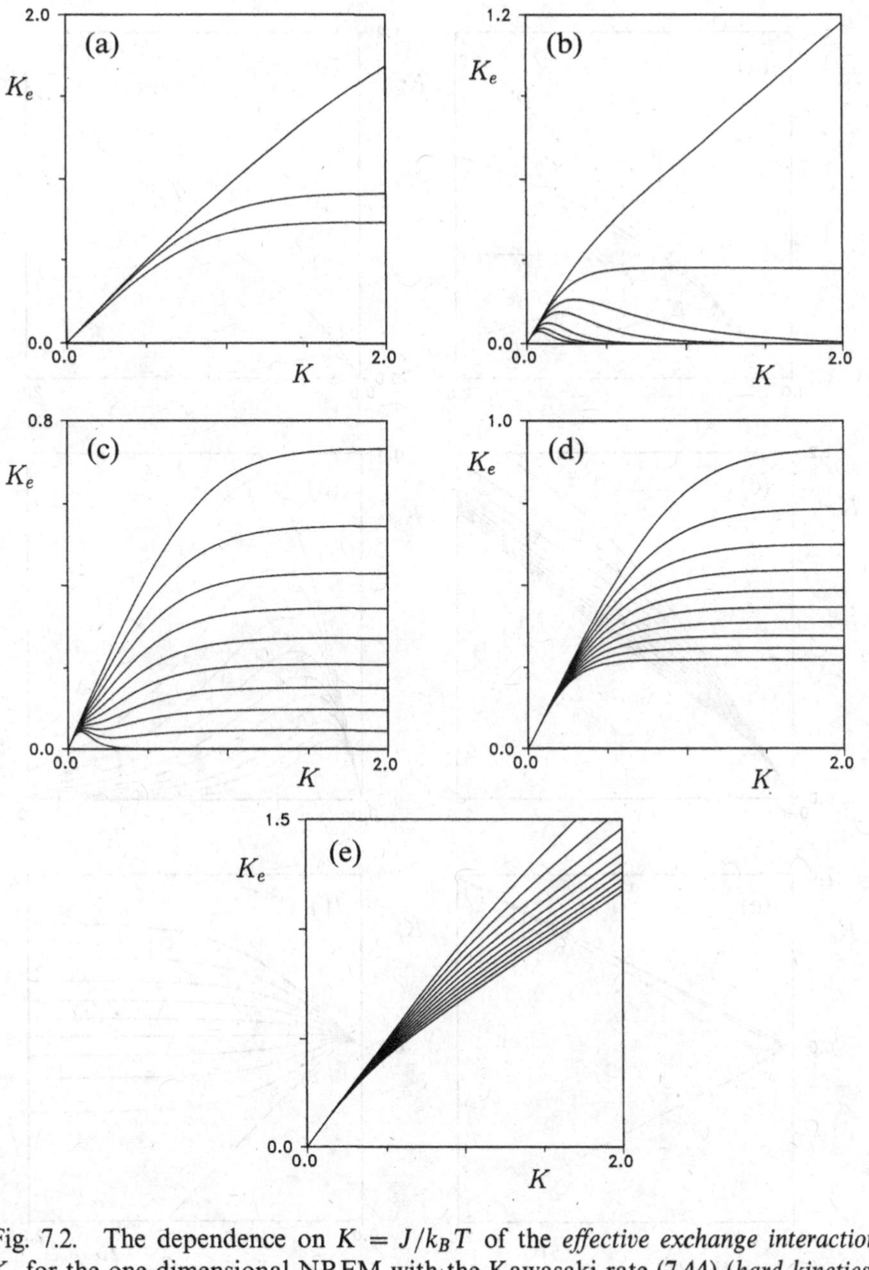

Fig. 7.2. The dependence on $K = J/k_B T$ of the *effective exchange interaction* K_e for the one-dimensional NRFM with the Kawasaki rate (7.44) (*hard kinetics*) with $p(h) = \frac{1}{2}\delta(h-\kappa) + \frac{1}{2}\delta(h+\kappa) + (1-q)\,\delta(h)$, for different values of q and $\lambda \equiv \kappa/2J$: (a) $q = 0.1$ and $\lambda = 0.5$, 1, and 5 from top to bottom. (b) The same for $q = 1$ and $0.5 \leqslant \lambda \leqslant 5$ increasing from top to bottom. A continuous changeover is observed between the cases illustrated in (a) and (b). (c) The case $\lambda = 5$ for $q = 0.1, 0.2, \ldots, 1$ from top to bottom. (d) The same for the *critical value* $\lambda = 1$. (e) The same for $\lambda = 0.5$. (From López-Lacomba & Marro (1992).)

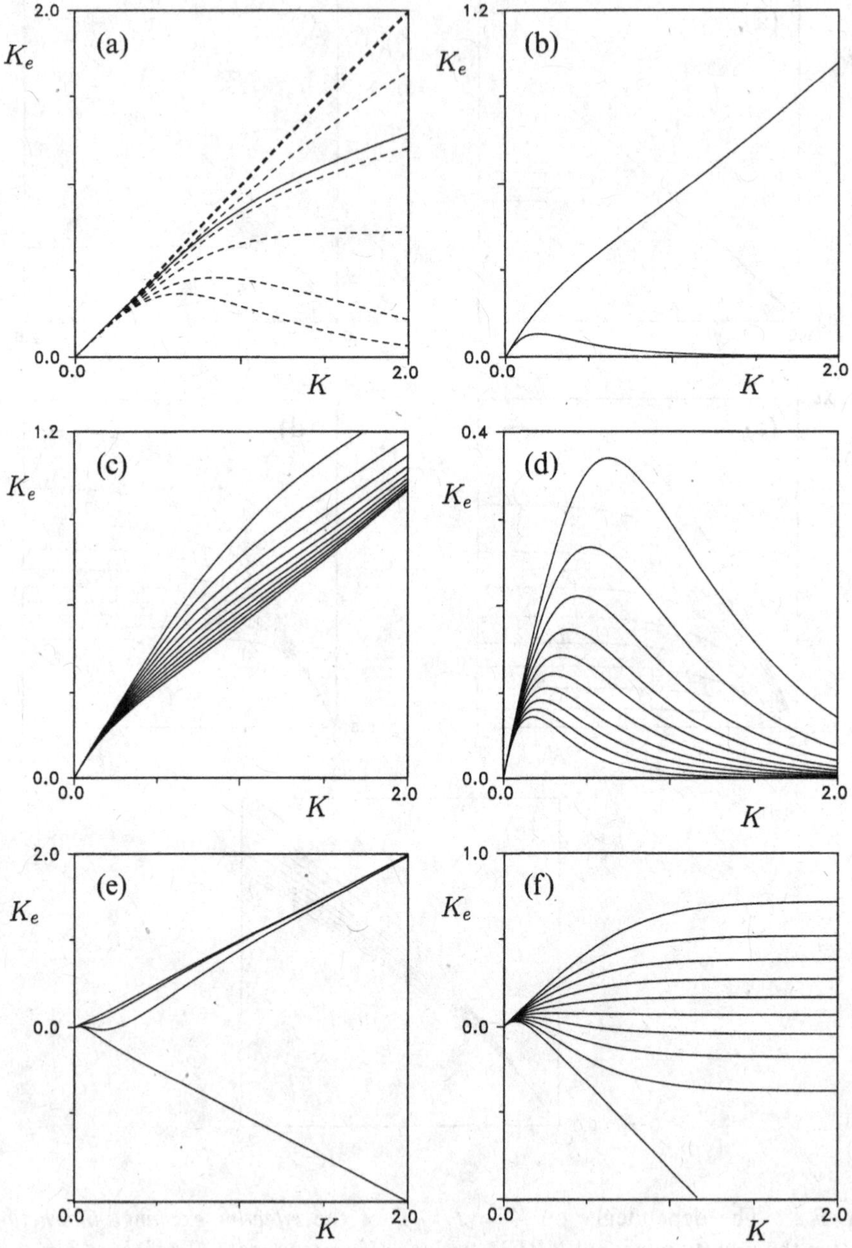

Fig. 7.3. The same as figure 7.2 but for the Metropolis rate (7.46) (*hard kinetics*): (a) $q = 0.1$ and $\lambda = 0.5$ (solid line) and, from bottom to top, increasing λ, $1 \leqslant \lambda \leqslant 5$. (b) $q = 0.9$ and $\lambda = 0.5$ (upper curve) and 1. (c) $\lambda = 0.5$ and $q = 0.1$, $0.2, \ldots, 1$, from top to bottom. (d) The same as (c) but for $\lambda = 1$. (e) $\lambda = 5$ and, from top to bottom, $q = 0.1$, 0.5, 0.9, and 1. (f) The same as (c) but for $\lambda = 2$.

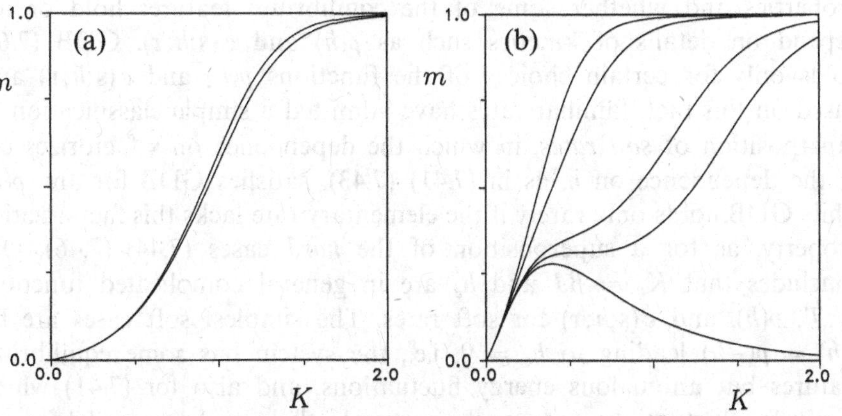

Fig. 7.4. The (inverse) temperature dependence of the magnetization m for the NRFM with the (*soft*) Glauber rate (7.42): (a) For $p(h) = \frac{1}{2}\delta(h-4\mu) + \frac{1}{2}\delta(h+2\mu) + (1-q)\delta(h-\mu)$ with $\mu = 0.1J$ and $q = 0.1$ (upper curve) and $q = 1$ (lower curve) (b) The same for $\mu = J$ and, from top to bottom, $q = 0.1, 0.5; 0.9, 0.95$, and 1. (From López-Lacomba & Marro (1992).)

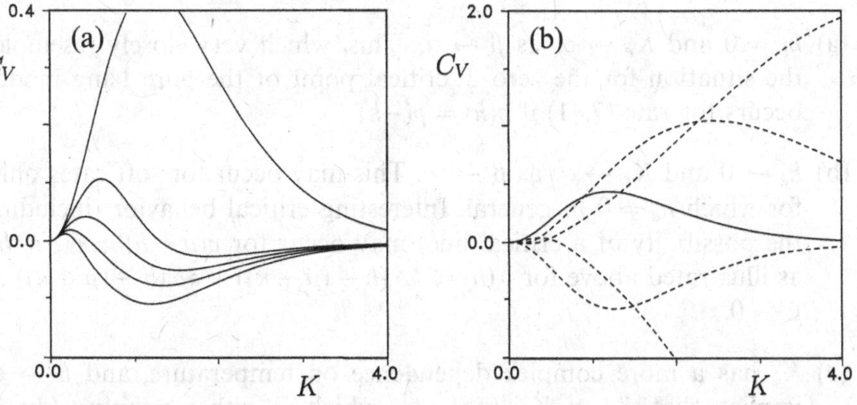

Fig. 7.5. The (inverse) temperature dependence of the *specific heat* C (normalized to Nk_B) for the NRFM with *hard kinetics*. (a) For the Kawasaki rate (7.44), $\lambda = 1.5$ and, from top to bottom, $q = 0, 0.25, 0.5, 0.75$, and 1. (b) For the modification (7.46) of the Metropolis rate; the solid line is for $q = 0$, while the dashed lines are for ($q = 0.75$ and $\lambda = 3$), ($q = 0.75$ and $\lambda = 2.5$), ($q = 0.25$ and $\lambda = 1.5$), and ($q = 1$ and $\lambda = 2$), respectively, from top to bottom along the vertical line $K = 1$.

magnetization in some cases (see figure 7.4(b)) reaches a maximum, and then decreases towards zero as the temperature is decreased; we refer the reader to §7.8 for further discussion of this phenomenon.

A general conclusion from the above is that emergent macroscopic properties and whether some of the equilibrium features hold or not depend on details of kinetics such as $p(h)$ and $c(\mathbf{s};h;\mathbf{r})$. GDB (7.60) holds only for certain choices of the functions $p(h)$ and $c(\mathbf{s};h;\mathbf{r})$ and, based on this fact, familiar rates have admitted a simple classification. A superposition of *soft rates,* in which the dependence on $\mathbf{s}$ factorizes out of the dependence on h, as in (7.41)–(7.43), satisfies GDB for any $p(h)$, while GDB holds only rarely if the elementary rate lacks this factorization property, as for a superposition of the *hard* cases (7.44)–(7.46). One concludes that $K_e = \beta J$ and h_e are in general complicated functions of T, $p(h)$, and $c(\mathbf{s};h;\mathbf{r})$ for soft rates. The simplest soft cases are for $p(h) = p(-h)$ leading to $h_e = 0$, i.e., the system has some equilibrium features but anomalous energy fluctuations, and also for (7.41) which, except for fluctuations, reduces the system to the pure Ising model for any $p(\mu + h) = p(\mu - h)$ with $\mu \neq 0$.

Such diversity is naturally reflected in the critical properties. If GDB holds, the effective Hamiltonian has the structure of the NN Ising energy function that enters the model definition, and a critical point (at which the correlation length diverges) is exhibited under one of the following conditions:

(a) $h_e = 0$ and $K_e \to \infty$ as $\beta \to \infty$. This, which very closely resembles the situation for the zero-T critical point of the pure Ising model, occurs for rate (7.41) if $p(h) = p(-h)$.

(b) $h_e \to 0$ and $K_e \to \infty$ as $\beta \to \infty$. This may occur for soft rates only, for which $h_e \neq 0$ in general. Interesting critical behavior (including the possibility of a critical line) may occur for $p(\mu + h) = p(\mu - h)$, as illustrated above for $p(h) = \frac{1}{2}\delta(h - (\mu + \kappa)) + \frac{1}{2}\delta(h - (\mu - \kappa))$ as $\mu \to 0$.

(c) K_e has a more complex dependence on temperature, and $\beta \to \infty$ implies that $\beta_e \equiv K_e/J \to \beta^0$, which is either positive (disorder is strong enough at $T = 0$ to impede any critical behavior), negative (the dynamic competition of fields induces an effective antiferromagnetic situation), or infinite (corresponding to a zero-T critical point), depending on the model. This occurs typically for hard rates and even distributions, for example, $p(h) = \frac{1}{2}\delta(h - \kappa) + \frac{1}{2}\delta(h + \kappa) + (1 - q)\,\delta(h)$. It is remarkable that the typical deviation κ, or the relation between κ, the mean μ, and the exchange interaction J, can produce dramatic changes in macroscopic properties.

Several points are worth mentioning. One is that a zero-T critical point may occur in the NRFM for $d = 1$ under some circumstances, unlike for the quenched counterpart. Another is the conclusion that *relevant* and *marginal* parameters (in the renormalization group sense) are the rule here, implying nonuniversal behavior. Finally, we note again the absence of a fluctuation–dissipation relation.

7.7 Comparison with related systems

An interesting question raised above is the relation between the NRFM and the quenched and annealed versions of the random field Ising system (Stinchcombe 1983). An explicit comparison for $d = 1$ is described next. The few similarities one may draw in this way are isolated but interesting. For example, the comparison stresses the influence of dynamics on the steady state, and the possible relevance of the NRFM to understanding disordered systems. We emphasize that the basic microscopic difference between the three systems concerns the nature of local fields. The distribution of fields is fixed for the quenched system, changes with time as driven by a very fast and completely random (i.e., infinite-T) process for the NRFM, and maintains itself in equilibrium at temperature T with the other degrees of freedom for the annealed case. This case, in which disorder may, for example, move to the interface (thus being highly correlated) trying to lower the free energy, seems to be rare in nature. A nonequilibrium version of a dilute antiferromagnet under a uniform field which, in the light of an equilibrium result (Fishman & Aharoni 1979), one might have expected to be closely related to the NRFM is also considered in this section.

Quenched and annealed random fields

The quenched random-field model (§7.1) is an equilibrium ferromagnetic Ising system in which spins at different sites experience random local magnetic fields h that are independent and have a distribution $p(h)$. This situation has been actively investigated both theoretically and experimentally (Derrida, Vannimenus, & Pomeau 1978, Bruinsma & Aepli 1983, Weher 1989). An early, exact solution of the one-dimensional spin-glass model under a uniform field (Fan & McCoy 1969) may be adapted to write an implicit solution for the one-dimensional quenched random-field Ising model. This has been analyzed by Grinstein & Mukamel (1983) among others. These authors studied the case (denoted GMM hereafter) in which the field is (spatially) distributed according to a distribution of mean μ and variance $\sigma_h^2 = q\kappa$, namely,

$$p(h) = \tfrac{1}{2}q\delta(h-\mu_+)+\tfrac{1}{2}q\delta(h-\mu_-)+(1-q)\,\delta(h-\mu),\, \mu_\pm = \mu\pm\kappa, \quad (7.102)$$

in the limit $\kappa \to \infty$.

This is similar to one of the cases in §7.6 and, in fact, Grinstein & Mukamel reported that the noted relevance of the parameter κ/J was a main motivation for their study. Consider the spin–spin correlation function $g(r) \equiv [[s_R s_{R+r}]]$ where $[[\cdots]]$ is the double average in (7.7), i.e., the ensemble average plus a disorder average with respect to $p(h)$. Assuming that $z \equiv \tanh(\beta J) \neq 1$ (so that $T \neq 0$) and $rz^r \ll 1$, one has

$$g(r) \simeq (qr+1)(1-q)^r z^r \tag{7.103}$$

for the GMM, neglecting terms of order $r(1-q)^r z^{3r}$ or smaller. In order to reduce the NRFM result, $g(r) = [\tanh K_e]^r$, to a comparable one, the hard rate (7.44) and distribution (7.102) with $\mu = 0$ may be used; it follows in the same limit for any $\kappa > 2J$ that:

$$g(r) \simeq (1-q)^r z^r, \tag{7.104}$$

neglecting terms of order $r(1-q)^r z^{r+2}$ or smaller. That is, high-T nonequilibrium correlations have pure exponential behavior for $d = 1$, unlike for the GMM. The correlation length, however, behaves similarly in both cases, i.e., $\xi^{-1} \approx -\ln z - \ln(1-q)$. For $T = 0$, on the other hand, one obtains $g(r) \sim (1-q)^r$ for the GMM and $g(r) \sim [r(1-q)]^r$ for the NRFM. In any case, it is to be remarked that the similarities here cannot be extrapolated beyond the conditions stated, for example, the quenched and nonequilibrium models differ fundamentally when the latter is implemented with a rate other than (7.44). In fact, the same $p(h)$ leads for rate (7.46) to $K_e = K$ for $\kappa > 8J$ which does not have a quenched counterpart.

The situation is similar if the NRFM is compared to a random-field Ising model whose impurities are *annealed*, i.e., have reached equilibrium with the other degrees of freedom instead of remaining frozen in (Thorpe & Beeman 1976). One may characterize this equilibrium situation (for $d = 1$) by the partition function

$$\tilde{Z}_N = [[Z_N(T,h)]] \equiv \int \prod_{j=1}^{N} \mathrm{d}h_j\, p(h_j) \sum_{\mathbf{s}\in\mathbf{S}} \exp\left[\beta \sum_{i=1}^{N} s_i (J s_{i+1} + h_i)\right]. \tag{7.105}$$

Except for the additional factor

$$\left\{[[\cosh(\beta h)]]^2 - [[\sinh(\beta h)]]^2\right\}^{N/2},$$

which is independent of $\mathbf{s}$, this is the partition function for the ordinary case under field $\eta = h_e$ given in (7.70). Consequently, the two systems have the same stationary solution $P_{st}(\mathbf{s})$ which is also the one for the NRFM with the soft rate (7.41). Therefore, annealed *configurational quantities*, such as the magnetization and its fluctuations and correlation functions, are identical to those for the nonequilibrium system, while thermal quantities, such as U and C, for the latter may reflect additional dependence on $p(h)$ besides the one involved by η. For instance, if δU denotes the difference in energy between the two systems, it follows for (7.102) that $\delta U = \kappa \tanh(\beta\kappa)$ for $q = 1$, $\delta U \to 0$ as $T \to \infty$, and $\delta U \to \kappa$ as $T \to \infty$. The comment at the end of the previous paragraph also applies here.

Nonequilibrium dilute antiferromagnets

The random field model is believed to describe several interesting practical cases. In particular, specific realizations of it seem to represent phase transitions in porous media[13] and certain dilute antiferromagnets.[14] Therefore, one may ask whether such a relation also holds for nonequilibrium systems; this has been investigated for $d = 1$ as follows.

A relation was established in Fishman & Aharony (1979) between two quenched cases, the random-field model and the dilute antiferromagnet, whose spins are present at each lattice site with a probability which is independent of other spins, in a uniform field. A nonequilibrium dilute antiferromagnet under a constant field may be modeled as a case of MKD. Consider a superposition (7.3) of elementary rates (7.4) defined according to the energy function (7.5), where h is a constant and J is a random variable with distribution $g(J) = q\delta(J) + (1-q)\delta(J - J_0)$, $J_0 > 0$. That is, the rates differ in the exchange interaction, and there is an external uniform magnetic field. For $d = 1$ and $h \neq 0$, an effective Hamiltonian exists under GDB that has the Ising structure (7.52) only if the rate involved is the soft one (7.41). The corresponding effective parameters are

$$K_e = \tfrac{1}{4} \ln \frac{q + (1-q)\,e^{-2\beta J_0}}{q + (1-q)\,e^{2\beta J_0}}, \quad h_e = h. \tag{7.106}$$

This is precisely the case of the NRFM above for $J = -\frac{1}{2}J_0$, rate (7.41), and $p(\mu + h) = p(\mu - h)$ with $\mu \neq 0$. It may be checked that no other simple correspondence exists between these two nonequilibrium models. That is, the only relation occurs for rates (7.41) which, as discussed above, are a very simple case for which the kinetics is almost identical to the (equilibrium) one considered by Fishman & Aharony. Therefore, the conclusion is that the relation which is familiar for the equilibrium MQD does not, in general, hold for nonequilibrium MKD.

[13] de Gennes (1984), Goh, Goldburg, & Knobler (1987), Wiltzius *et al.* (1989), Wong & Chan (1990), Lee (1993).

[14] Birgeneau *et al.* (1982), Cardy (1984), Hill *et al.* (1991), Pontin, Segundo, & Pérez (1994).

7.8 Lattices of arbitrary dimension

This section is devoted to reviewing briefly the results that may be obtained for MKD of arbitrary dimension, i.e., for the lattice $\mathbf{Z}^d$, $d \geqslant 1$. Two kinds of approaches are described: the derivation of exact bounds that locate the region of the phase diagram in which the system is ergodic, and kinetic cluster mean-field theory. As in the rest of the chapter, we concentrate on the NRFM as an example. We discuss again the properties of the two classes of dynamics (*hard* and *soft*) defined above, and describe the existence of effective Hamiltonian for $d > 1$ for a certain class of rates.

Some exact results

The method of §7.5 gives the following information for $d > 1$ (Garrido & Marro 1989): if an effective Hamiltonian exists, and the rate involved is local and has certain symmetry properties, which is the case for the rates enumerated in §7.4, then the effective Hamiltonian necessarily has the NN Ising structure (7.52). The problem is that evaluation of the effective Hamiltonian by this method involves the GDB condition, and this is very rarely satisfied by the MKD for $d > 1$. This has two main consequences. On one hand, two- and three-dimensional systems cannot, in general, have the quasi canonical property discussed in §7.6; in fact, the GDB condition has been shown to fail in some cases even for $d = 1$. This seems to guarantee that the systems with $d > 1$ have, in general, more intriguing behavior and should be investigated by other methods.

Before attempting a more general method, we mention that a class of rates exists for which the MKD have a simple effective Hamiltonian, and that this holds for rather general cases of MKD, for example, one may generalize the energy function (7.1) in a nontrivial way. In order to incorporate this important result into the formalism of the present chapter, it is convenient to specify the classification of dynamics in §7.6 as follows:

Hard dynamics. The elementary rate generally has the property that $\varphi(-X)$ (equation (7.47)) goes to a positive constant as $X \to \infty$, as for the cases $\varphi(X) = 1 - \tanh\left(\frac{1}{2}X\right)$, (7.44), and $\varphi(X) = \min\left\{1, \mathrm{e}^{-X}\right\}$, (7.45). This produces a more complex macroscopic behavior in many of the cases investigated (but not always; see end of §8.1).

Soft dynamics. The elementary rate is such that

$$\varphi(X) = \varphi\left(\{\lambda_D\}\right) = \prod_D \varphi_D(\lambda_D) \tag{7.107}$$

where each factor satisfies detailed balance, i.e.,

$$\varphi_D(\lambda_D) = \exp(-\lambda_D)\,\varphi_D(-\lambda_D). \tag{7.108}$$

Furthermore, it seems that $\varphi(-X) \to e^{\Delta X}$, where $\Delta < 1$, for $X \to \infty$, in general. For the NN Ising energy function (7.1), $X = \sum_D \lambda_D s_D$, and

$$\varphi\left(\sum_D \lambda_D s_D\right) = \left[\prod_D \varphi_D(\lambda_D)\exp\left(\tfrac{1}{2}\lambda_D\right)\right] \exp\left(-\tfrac{1}{2}\sum_D \lambda_D s_D\right). \tag{7.109}$$

The cases $\varphi(X) = e^{-X/2}$, (7.41), and $\varphi(X) = e^{-X/2}[\cosh(\beta J)]^{-2d}$, (7.43), belong to this class.

An important fact is that, as a consequence of this factorization property of soft rates, the probability of flipping any spin is the superposition of independent processes, one for each bond ending at the given spin. That is, some correlations are *suppressed* by the dynamics. Therefore, simple behavior ensues in some cases (but not always) for soft rates. In particular, a simple effective Hamiltonian may exist for $d > 1$. This result is illustrated in §8.4 for specific systems.

Further information for $d \geqslant 1$ may be derived from the following result (Liggett 1985, Garrido & Marro 1991). The system is said to be *exponentially ergodic* when

$$\varpi \equiv \inf_{\mathbf{r}} \left\{ Q(\mathbf{r}) - \sum_D |P_D(\mathbf{s})| \right\} > 0. \tag{7.110}$$

Here the sum is over all possible sets D of spins, and

$$Q(\mathbf{r}) \equiv \sum_{\mathbf{s}\in\mathbf{S}} c(\mathbf{s}^{\mathbf{r}}|\,\mathbf{s}) > 0 \tag{7.111}$$

and

$$P_D(\mathbf{r}) \equiv \sum_{\mathbf{s}\in\mathbf{S}} \left(\prod_{\mathbf{r}'\in D} s_{\mathbf{r}'}\right) s_{\mathbf{r}}\, c(\mathbf{s}^{\mathbf{r}}|\,\mathbf{s}) \tag{7.112}$$

characterize the (effective) rate of interest, (7.3). In practice, (7.110) is readily checked, given that only contributions from sets D consisting of the spin at $\mathbf{r}$ and/or the spin(s) that interact with $s_{\mathbf{r}}$ are involved. This is a sort of positivity property of the rate (for all $\mathbf{r}$) which implies that, for almost any initial condition, the system relaxes exponentially quickly in time towards a single phase. In other words, a phase transition may occur solely for those parameter values that violate (7.110).

Consider first the NRFM for any $p(h)$ with soft rate (7.41). It ensues that no phase transition may occur for $\beta < \beta_0$ where β_0 satisfies

$$[1+\tanh(\beta_0 J)]^{2d} = \frac{2}{1+[[\sinh(\beta_0 h)]]\,[[\cosh(\beta_0 h)]]}. \tag{7.113}$$

For distributions such that $p(h) = p(-h)$, one obtains $\beta_0 J = 0.44$, 0.19, and 0.123 for $d = 1$, 2, and 3, respectively, as for the ordinary Ising model with zero field. For $p(\mu+h) = p(\mu-h)$ with $\mu > 0$, one obtains $\beta_0 J = 0.266$, 0.149, and 0.104 for $d = 1$, 2, and 3, respectively, as for the ordinary Ising model under field μ. This confirms that (7.41) generally corresponds to a rather trivial case of the NRFM problem. For example, (7.102) with $\mu = 0$ and the hard rate (7.44) imply instead that

$$\left|q\left(T_2^+ - T_2^-\right) + 2(1-q)\,t_2\right| = 2 \tag{7.114}$$

for $d = 1$, and

$$\begin{aligned}&\left|q\left(T_4^+ + 2T_2^+ - T_4^- - 2T_2^-\right) + 2(1-q)(t_4+2t_2)\right|\\ &\quad+\left|q\left(T_4^+ - 2T_2^+ - T_4^- + 2T_2^-\right) + 2(1-q)(t_4-2t_2)\right| = 4\end{aligned} \tag{7.115}$$

for $d = 2$, etc., where $T_n^\pm \equiv \tanh[\beta_0(\kappa \pm nJ)]$ and $t_n \equiv \tanh[n\beta_0 J]$.

It may also be remarked that these systems have a simple realization at $T = 0$. For example, the last system mentioned then corresponds to

$$c\left(\mathbf{s}^{\mathbf{r}}\middle|\,\mathbf{s}\right) = 1 - s_r\sigma_1\left[1+\tfrac{1}{2}q\phi(\lambda_1)\right] \tag{7.116}$$

if $d = 1$,

$$c\left(\mathbf{s}^{\mathbf{r}}\middle|\,\mathbf{s}\right) = 1 - s_{\mathbf{r}}\left[\alpha_1(\sigma_1+\sigma_2) + \alpha_2(\sigma_1\pi_2+\sigma_2\pi_1)\right] \tag{7.117}$$

if $d = 2$, and

$$c\left(\mathbf{s}^{\mathbf{r}}\middle|\,\mathbf{s}\right) = 1 - s_{\mathbf{r}}\left[\alpha_3\sum_{\ell}\sigma_\ell + \alpha_4\left(\sum_{\ell,m}\sigma_\ell\pi_m + \sigma_1\sigma_2\sigma_3\right) + \alpha_5\sum_{\ell,m,n}\sigma_\ell\pi_m\pi_n\right] \tag{7.118}$$

with $\ell, m, n = 1, 2, 3$, $\ell \neq m \neq n$, if $d = 3$. Here,

$$\begin{aligned}&\sigma_1 \equiv \tfrac{1}{2}\left(s_{\mathbf{r}+\hat{\imath}} + s_{\mathbf{r}-\hat{\imath}}\right), \quad \sigma_2 \equiv \tfrac{1}{2}\left(s_{\mathbf{r}+\hat{\jmath}} + s_{\mathbf{r}-\hat{\jmath}}\right), \quad \sigma_3 \equiv \tfrac{1}{2}\left(s_{\mathbf{r}+\hat{\mathbf{k}}} + s_{\mathbf{r}-\hat{\mathbf{k}}}\right),\\ &\pi_1 \equiv s_{\mathbf{r}+\hat{\imath}}\,s_{\mathbf{r}-\hat{\imath}}, \quad \pi_2 \equiv s_{\mathbf{r}+\hat{\jmath}}\,s_{\mathbf{r}-\hat{\jmath}}, \quad \pi_3 \equiv s_{\mathbf{r}+\hat{\mathbf{k}}}\,s_{\mathbf{r}-\hat{\mathbf{k}}},\end{aligned}$$

where $\mathbf{r} \pm \mathbf{r}'$ $\left(\mathbf{r}' = \hat{\imath}, \hat{\jmath}, \text{ or } \hat{\mathbf{k}}\right)$ represent the NNs of $\mathbf{r}$, and

$$\left.\begin{aligned}\alpha_1 &\equiv \tfrac{1}{8}\left\{4 + q\left[2+2\phi(\lambda_1)+\phi(\lambda_2)\right]\right\},\\ \alpha_2 &\equiv \tfrac{1}{8}\left\{-2 - 2q\phi(\lambda_1) + q\phi(\lambda_2)\right\},\\ \alpha_3 &\equiv \tfrac{1}{8}\left\{4 + \tfrac{1}{4}q\left[3\phi(\lambda_1)+4\phi(\lambda_2)+\phi(\lambda_3)\right]\right\},\\ \alpha_4 &\equiv \tfrac{1}{8}\left\{-4 - 3q\phi(\lambda_1) + q\phi(\lambda_3)\right\},\\ \alpha_5 &\equiv \tfrac{1}{8}\left\{1 + \tfrac{1}{4}q\left[5\phi(\lambda_1)-4\phi(\lambda_2)+\phi(\lambda_3)\right]\right\},\end{aligned}\right\} \tag{7.119}$$

where

$$\phi(X) = \theta(X)\left[\theta(-X) - 2\right], \qquad \theta(X) = \begin{cases} 0 & \text{for } X < 0 \\ 1 & \text{for } X \geqslant 0, \end{cases} \tag{7.120}$$

and $\lambda_n \equiv \kappa/2J - n$. It follows from (7.110) that this zero-T system with $\kappa > 6J$ is ergodic for any $q > 0$, $1/3$, and $23/31$ for $d = 1$, 2, and 3, respectively.

Kinetic mean-field theory

Kinetic cluster theory in the zeroth- and first-order approximations has been developed for the d-dimensional NRFM by Alonso & Marro (1992b) who consider

$$p(h) = \frac{1}{2}\delta(h - h_0) + \frac{1}{2}\delta(h + h_0) \tag{7.121}$$

and various rates. A key result is the existence of a nonequilibrium tricritical point for $d > 2$, and *reentrance phenomena* (which has already been anticipated in §7.6 for $d = 1$).

The zeroth-order approximation turns out to hide some of the nonequilibrium features of the model. Nevertheless, some interest in it follows from the fact that Aharony (1978) has evaluated the partition function for the quenched random-field Ising system within the same order of approximation, and comparison reveals some significant differences between the two cases. One may consider the equation for the magnetization, $m \equiv \langle s_{\mathbf{r}} \rangle$ that follows from (7.2):

$$\frac{\partial m}{\partial t} = F(m) \equiv -2\left\langle s_{\mathbf{r}}\, c\left(\mathbf{s}^{\mathbf{r}} \mid \mathbf{s}\right)\right\rangle, \tag{7.122}$$

where $\langle \cdots \rangle$ is the thermal average. No global (thermodynamic) stability criterion is available but steady solutions, $F(m) = 0$, need to fulfill $\left(\partial F/\partial m\right)_{st} < 0$, and one may investigate the trajectories $m(t)$ for different initial conditions.

The zeroth-order description corresponds to the Bragg–Williams approximation in equilibrium. That is, one deals with a cluster consisting of $s_{\mathbf{r}}$ only, and assumes that the influence of the rest of the system on $s_{\mathbf{r}}$ occurs through a self-consistent mean field. The argument that enters the rate in (7.40) or (7.47) is $\beta\delta H_h = 2\beta\,(qJm + h)\,s_{\mathbf{r}}$, where q is the lattice coordination number, and consistency simply requires that the mean magnetization around site $\mathbf{r}$ equals $\langle s_{\mathbf{r}} \rangle$. Therefore, one has $2c\left(\mathbf{s}^{\mathbf{r}} \mid \mathbf{s}\right) = \varphi\left(2\beta\left[qJm + h_0\right]\right) + \varphi\left(2\beta\left[qJm - h_0\right]\right)$ to be used in (7.122). If the function φ here is implemented as for (7.48), one obtains the same behavior as for $h = 0$. The same holds within the first-order approximation

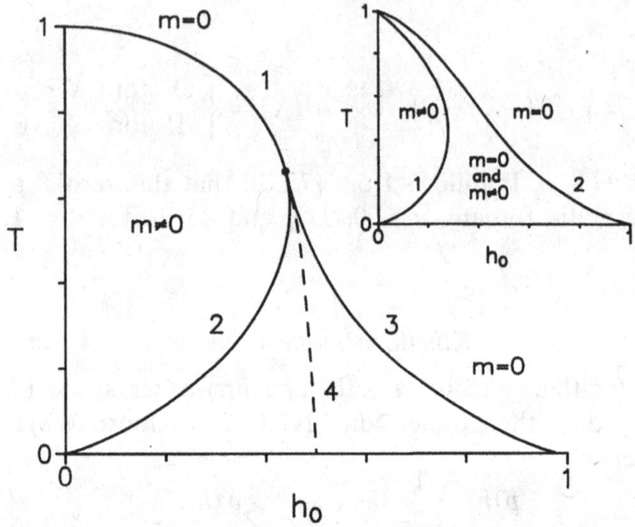

Fig. 7.6. The lower (lines 1 and 2) and upper (lines 1 and 3) limits of local stability of $m = 0$ solutions (both $m = 0$ and $m \neq 0$ solutions may occur in between) as obtained from (7.123) with (7.121) for the NRFM. Lines 2 and 3 join at a tricritical point, (T^*, h_0^*). Lines 1 and 4 are the solution for the corresponding quenched system. The inset depicts the lower and upper limits of local stability (lines 1 and 2, respectively) for the NRFM with rate (7.51); the phase transitions in this case are of first order for any T and h_0. (From Alonso & Marro (1992b).)

below. This is related to the peculiar nature of the rate which, as noted above, admits a factorization that cancels out the contributions from the fields for even distributions. This rate is also atypical in that it lacks a proper normalization, and strongly favors the states of low energy.

Equations (7.50) and (7.51), for example, are more realistic in a sense. The former leads to

$$m = [[\tanh(2\beta\,[qJm + h])]] \tag{7.123}$$

which is the solution derived by Aharony (1978). This solution, however, corresponds to equilibrium and, consequently, is rate-independent, while modifying the rate in the present case produces a different behavior, in general. Figure 7.6 illustrates the consequences of (7.123) when the average involved is evaluated using (7.121). The most notable fact is the existence of a tricritical point, while the rate (7.51) induces phase transitions that are of first order for any T and h_0.

The NRFM for arbitrary dimension has *within this zeroth-order approximation* some of the quasi canonical features described above for $d = 1$.

That is, assuming that GDB holds, one obtains

$$\ln P_{st}(\mathbf{s}) = \tfrac{1}{2} \sum_{r=1}^{N} s_r \ln \frac{\psi_+(m)}{\psi_-(m)}, \tag{7.124}$$

where

$$\psi_\pm(qm) \equiv [[\varphi(\pm 2\beta [qJm + h])]] .$$

Consequently, an *effective free energy* may be defined as

$$f_e = -\beta^{-1} \ln \sum_{\mathbf{s}} P_{st}(\mathbf{s}). \tag{7.125}$$

For large N, one may approximate the sum by the largest term; one is led to the self-consistency condition

$$\frac{\psi_-(qm) - \psi_+(qm)}{\psi_-(qm) + \psi_+(qm)} = m \tag{7.126}$$

which gives

$$\beta f_e = \tfrac{1}{2} \ln \frac{1-m^2}{4} - \tfrac{1}{4} m \ln \frac{\psi_+(qm)}{\psi_-(qm)}. \tag{7.127}$$

This allows for a global stability criterion (which depends on the rate, however); see figure 7.6, and Alonso & Marro (1992b) for further details.

Next, one may consider a larger cluster consisting of spin $s_{\mathbf{r}}$ and its q NNs. The condition of GDB is no longer required. The cluster interacts with the rest of the system via a self-consistent mean field only. Consequently, correlations other than NN ones are not allowed, and $\delta H_h = 2(nJ + h)s_r$, where $n = -q, -q+2, \ldots, q$ stands for the number of spins at NN sites of $\mathbf{r}$ which have state *up* minus those having state *down*. Configurations are then characterized by two independent variables only, for example, $m = \langle s_r \rangle$ and $e = \langle s_r s_{r+1} \rangle$. Equivalently, one may use the density of *up* spins, denoted here $x = \frac{1}{2}(1+m)$, and the density of *up–up* pairs of NN spins, $z = \frac{1}{4}(1 + 2m + e)$. The standard technique (in this first-order approximation) leads to

$$\frac{\mathrm{d}x}{\mathrm{d}t} = F(x,z;T) \equiv \sum_{n=0}^{q} (x-z)^n \binom{q}{n} (\alpha_1 - \alpha_2)\ \psi_+(q-2n) \tag{7.128}$$

and

$$\frac{1}{q}\frac{\mathrm{d}z}{\mathrm{d}t} = G(x,z;T) \equiv \sum_{n=0}^{q} (x-z)^n \binom{q}{n} [n\alpha_1 - (q-n)\alpha_2]\ \psi_+(q-2n), \tag{7.129}$$

where $\alpha_1 \equiv (1+z-2x)^{q-n}(1-x)^{1-q}$, and $\alpha_2 \equiv z^{q-n}x^{1-q}$. The condition for the disordered states that are expected at high enough temperature is $G\left(x=\frac{1}{2}, z; T\right)=0$, and stability breaks down for $(\partial F/\partial x)_{x=\frac{1}{2}, z}=0$. This leads to

$$\sum_{n=0}^{q}\binom{q}{n}(q-2n)\left[\frac{q-2}{q}\right]^{n}\psi_{+}(q-2n)=0, \quad z=\frac{q}{4(q-1)}, \qquad (7.130)$$

which gives a transition temperature, $T_1(h_0)$. One has that $T_C \equiv T_1(h_0) \to 2Jk_B^{-1}\ln[(q-2)/q]$ as $h_0 \to 0$ which corresponds to the Bethe–Peierls equilibrium critical temperature, as expected. In order to distinguish between transitions of first and second order, one defines the upper limit of local stability for $m \neq 0$ solutions, $T_2(h_0)$, as the solution of $(\partial F/\partial x)_{x\neq\frac{1}{2}, z}=0$, where x and z are the solution of $F(x, z; T)=0$ and $G(x, z; T)=0$.

The main results may be summarized as follows. For $q=2$ (corresponding to a line), (7.130) reduces to $2+\eta^{-2}(\zeta^2+\zeta^{-2})=0$, where $\zeta \equiv \exp(-\beta h_0)$ and $\eta \equiv \exp(-2\beta J)$, for rate (7.50). A solution $T_1(h_0)=0$ exists for any $h_0 < 2J$. This means that the system exhibits the familiar critical point at zero temperature if the field strength is small enough, while it disappears for $h_0 > 2J$ in (7.121). For rate (7.51), $\eta^2(\zeta^2+\zeta^{-2})=0$ and $1+\eta^2\zeta^2=0$, respectively, for $h_0 < 2J$ and $h_0 > 2J$. The solution of the former is $T_1=0$ for any $h_0 < 2J$ while no real solution exists for the latter equation. That is, the present approximation essentially reproduces the main *exact* results for $d=1$.

The case $q>2$ is illustrated in figures 7.7–7.9. In general, the system tends to become macroscopically disordered as h_0 is increased, the phase transition is always of second order for $q=3$ and 4 (for example, the square lattice), while (nonequilibrium) tricritical points occur for $q>4$ (for example, the sc lattice) so that the phase transition is of first order for large enough values of h_0. The metastable region extends over relatively small values of h_0 for $q=6$, but not for $q=5$ where T_1 and T_2 go to zero at the same value of h_0. Figure 7.8 reveals the existence of metastable states at low temperature for a given value of the field if $q=6$. On the other hand, figure 7.7 indicates that the slope of the $T_1(h_0)$ curve also changes sign for $q=3$ (but not for $q=4$). This is illustrated in greater detail in figure 7.9; the only qualitative difference between rates (7.50) and (7.51) in the present approximation seems to be this change of sign, which does not occur for the latter. It is interesting to interpret the bottom curve for $d=1$ in figure 7.4(b) in the light of the reentrance phenomena exhibited in the present case in figure 7.9 for $h_0 > 1$.

Summing up, qualitative features of the NRFM within the first-order approximation are the dependence of the phase diagram on q, the tendency

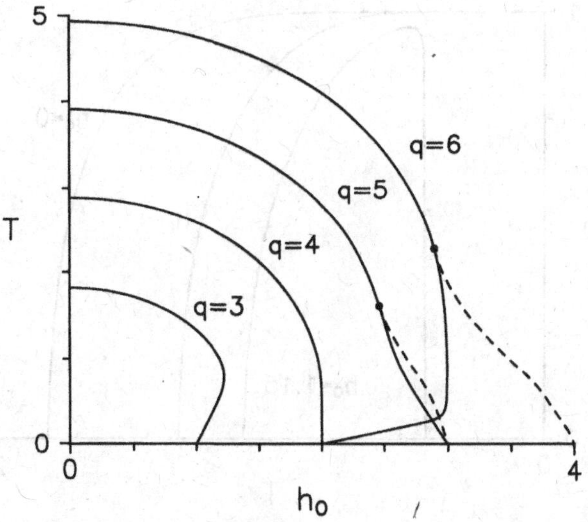

Fig. 7.7. The same as figure 7.6 but in the first-order approximation for rate (7.50) and different values of the lattice coordination number q, as indicated. T is in units of J/k_B; h_0 is in units of J. The solid lines are for T_1; the dashed lines are for T_2. Tricritical points are indicated at $(T, h_0) = (1.62, 2.46)$ and $(2.28, 2.90)$ for $q = 5$ and 6, respectively; rate (7.51) induces qualitatively similar behavior with the tricritical points at (1.85, 2.12) and (3.15, 1.95), respectively.

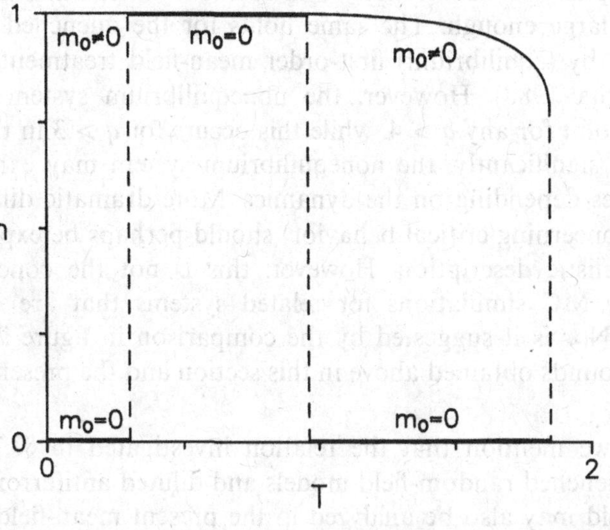

Fig. 7.8. The stationary solutions of equations (7.128) and (7.129) for lattice coordination number $q = 6$ (for example, the sc lattice), rate (7.51), and $h_0 = 2.75J$ for initial conditions $m_0 \equiv m(t = 0) = 0$ and 1, as indicated. This reveals the existence of metastable states for both low and high (but not intermediate) temperatures. (From Alonso & Marro (1992b).)

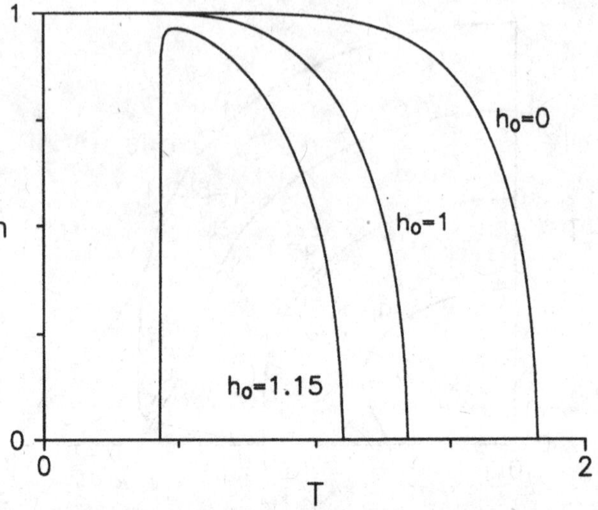

Fig. 7.9. The same as figure 7.8 but for coordination number $q = 3$, rate (7.50), and different values of h_0, as indicated. The existence of disordered states at low temperatures for some values of h_0 is indicated.

of the system to become disordered at low values of T and h_0 as q is decreased, and the occurrence of metastable states near $T = 0$ if h_0 and q are large enough. The same holds for the quenched counterpart as revealed by (equilibrium) first-order mean-field treatments (Bruinsma 1984, Yokota 1988). However, the nonequilibrium system only has a tricritical point for any $q > 4$, while this occurs for $q > 3$ in the quenched case. More significantly, the nonequilibrium system may exhibit various steady states depending on the dynamics. More dramatic differences (for example, concerning critical behavior) should perhaps be expected within a more realistic description. However, this is not the conclusion from preliminary MC simulations for related systems that are described in chapter 8. Nor is it suggested by the comparison in figure 7.10 between the exact bounds obtained above in this section and the present mean-field results.

Finally, we mention that the relation investigated in §7.7 for $d = 1$ between quenched random-field models and diluted antiferromagnets in a uniform field may also be analyzed in the present mean-field description for $d \geq 1$. It follows that the two nonequilibrium systems behave quite differently, in general, independent of the order of the approximation investigated. In particular, the nonequilibrium dilute antiferromagnet has $m \neq 0$ solutions at any temperature in the presence of any uniform external field, and no broken symmetry from $m = 0$ to $m \neq 0$ occurs, in

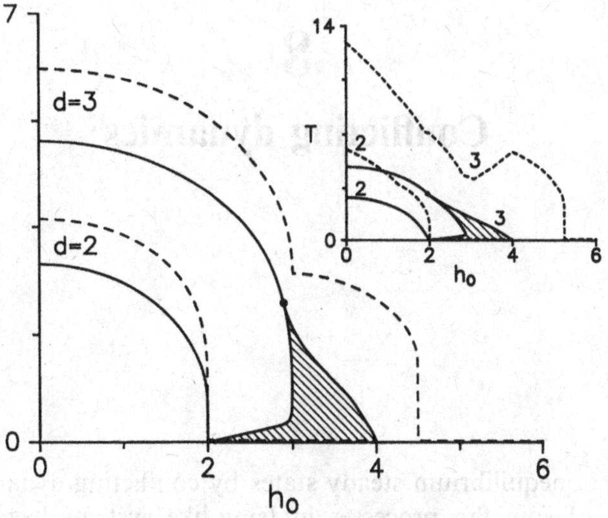

Fig. 7.10. Exact bounds are compared with mean-field results for the transition lines for rates (7.50) (main graph) and (7.51) (inset) and $d = 2$ and 3, as indicated by the numbers near the curves. This reveals that exact bounds are relatively accurate and, consequently, may be useful in practice, especially for rate (7.50). The shaded regions represent metastable behavior. (From Alonso & Marro (1992b).)

contrast to the NRFM. Together with the exact result above for $d = 1$, this suggests that such a relation is an equilibrium feature which holds for quenched disorder but breaks down if the systems are away from equilibrium.

8
Conflicting dynamics

Generating nonequilibrium steady states by conflicting dynamics, i.e., superposition of spin flip processes in Ising-like systems has a precedent in the consideration of spatially nonuniform distributions of temperature (Garrido & Marro 1987). It was introduced as a drastic simplification of more complex practical situations such as systems with different temperatures at opposing boundaries (Creutz 1986, Harris & Grant 1988). This has in turn suggested studying spins that suffer competing action of several baths, e.g., two independent thermal baths with respective probabilities p and $1-p$,[1] and the case in which different sites of the lattice, either a sublattice or else a set of sites chosen at random, are at different temperatures.[2] After some effort, it appears that these systems involving several temperatures probably have the critical behavior that characterizes the original system from which they derive, at least when all the involved temperatures are finite, and no essential symmetry is broken. In spite of such apparent simplicity, phase diagrams and other thermodynamic properties are varied and interesting, and some questions remain unresolved.

Our concern in this chapter is a class of closely related systems that are less well understood. We discuss a (rather arbitrary) selection of stochastic, interacting particle systems of the kind introduced in chapter 7, namely, a nonequilibrium Ising glass (§8.1 and §8.2), and the cases of bond dilution and very strong bonds (§8.1), invasion and voting processes (§8.3), and kinetic versions of systems in which interactions extend beyond nearest

[1] Garrido *et al.* (1987), Marqués (1989, 1993), Tomé, de Oliveira, & Santos (1991), Garrido & Marro (1992), Muñoz (1994), Bassler & Zia (1994, 1995), Garrido & Muñoz (1995), Achahbar, Alonso & Muñoz (1996).

[2] Garrido & Marro (1987), Labarta *et al.* (1988), Heringa, Blöte, & Hoogland (1994), Rácz & Zia (1994), Bakaev & Kabanovich (1995).

neighbors (§8.4). We also discuss a class of dynamic competition that bears some relevance to understanding proton glasses and neural networks (§8.5), and the microscopic interpretation of thermal baths (§8.6). The aim is to guide the reader through intriguing nonequilibrium phenomena, and to illustrate the applicability of several methods in systems that exhibit a varied physical motivation.

8.1 Impure systems with diffusion

One realization of MKD in §7.1 corresponds to the master equation (7.2) with competing rate (7.3), $\vec{\zeta} \equiv J$, and

$$p(J) = q\delta(J) + (1-q)\,\delta(J - J_0). \tag{8.1}$$

This is a *kinetic* bond-dilute system, i.e., bonds have strength $J = J_0$ in general but are *broken* with probability q. For $d = 1$, the method in §7.5 implies the effective Hamiltonian (7.68) with $h_e = 0$ and

$$K_e = -\tfrac{1}{4}\ln\frac{q + (1-q)\,\varphi(4\beta J_0)}{q + (1-q)\,\varphi(-4\beta J_0)}, \tag{8.2}$$

where $\varphi(X)$ is the elementary rate; see equations (7.3) and (7.47). For any of the rates that satisfies detailed balance, $\varphi(X) = \varphi(-X)\,\mathrm{e}^{-X}$, and is such that $\varphi(-X) \to \mu > 0$ as $X \to \infty$ (e.g., the Metropolis case (7.51) is of this class with $\mu = 2$),

$$K_e \approx -\frac{1}{4}\ln\frac{q}{q + (1-q)\,\mu}, \quad \beta \to \infty. \tag{8.3}$$

Therefore, K_e diverges as $q \to 0$ only, i.e., the zero-T critical point which is familiar in other one-dimensional problems does not occur for a competition of this kind. This is reminiscent of the equilibrium case in which the critical point of the *pure* one-dimensional system is washed away by any nonzero concentration of nonmagnetic impurities. In contrast, a critical point occurs for (7.48) or for any other rate such that $\varphi(-X) \approx \mathrm{e}^{\Delta X}$ with $\Delta < 1$; in this case, critical exponent values differ by a factor of Δ from ordinary ones.

These two classes of behavior correspond to *hard* and *soft* dynamics, respectively, as defined in §7.8. It is remarkable that the behavior in the neighborhood of the critical point in parameter space depends so strongly on kinetics. For the present system, this may be understood in part as a consequence of the fact that *disorder*, i.e., broken bonds, is introduced through (8.1) on a time scale proportional to q^{-1}, as compared to the scale proportional to $[(1-q)\,\varphi(-4\beta J_0)]^{-1}$ which is to be associated with the pure background in which bonds have strength J_0. Therefore,

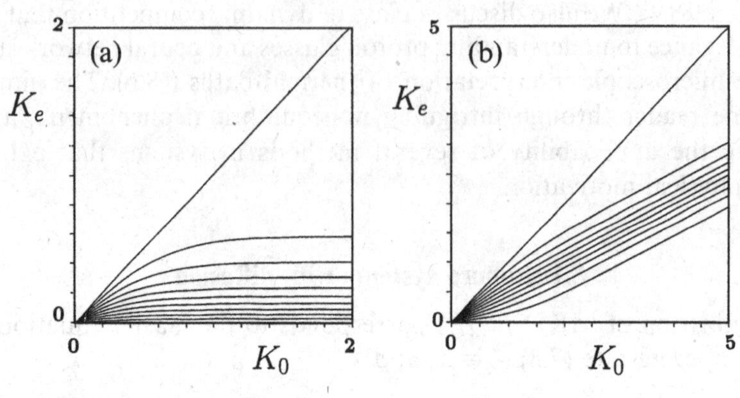

Fig. 8.1. The effective bond strength (8.2) as a function of $K_0 \equiv \beta J_0$, i.e., inverse temperature, for q increasing from $q = 0$ to $q = 0.9$ in steps of 0.1from top to bottom, and for (a) the Metropolis rate (7.51), and (b) the rate of van Beijeren & Schulman (7.48). The curvature of the K_e-versus-K_0 function changes in (b) at $q = \frac{1}{2}$ which marks the onset of different qualitative behavior. (From Garrido & Marro (1992).)

$(1-q)\,\varphi(-4\beta J_0) \gg q$ guarantees stability of the ordinary zero-T critical point. This condition holds for the soft case while both terms are of comparable magnitude for hard rates. The two cases are illustrated in figure 8.1.

The correlation length ξ may be obtained from

$$\xi \equiv -\frac{r}{\ln \langle s_0 s_r \rangle}, \qquad \langle s_0 s_r \rangle = (\tanh K_e)^r . \tag{8.4}$$

One obtains $\xi \approx -1/\ln(\beta J_0)$ as $\beta \to 0$ for any rate and for any $q \neq 1$ (as long as T is high enough). As $\beta \to \infty$, hard rates lead to $\xi \approx \frac{1}{2}\sqrt{\mu/q}$ as $q \to 0$, i.e., a divergence occurs (the nature of which is not thermal but geometrical) which is characterized by the critical exponent $\nu_q = \frac{1}{2}$. In contrast, soft rates imply that $\xi \approx \frac{1}{2}\exp(2\Delta\beta J_0)$ as $\beta \to \infty$, so that the divergence is thermal with $\nu = 2\Delta$, for example, $\nu = 1$ for (7.48); figure 8.2 illustrates these differences. For the pure system in equilibrium one has instead $\xi_0 \approx \frac{1}{2}\exp(2\beta J_0)$, i.e., $\nu = 2$, and $\xi_q^{-1} = -\ln(1-q) + \xi_0^{-1}$ for (quenched) bond dilution. The susceptibility $\chi = \sum_r \langle s_0 s_r \rangle$ has similar properties. That is, $\chi \approx q^{-1}$ as $q \to 0$ in equilibrium, while one obtains $\chi \approx \sqrt{\mu/q}$ for hard rates, and $\chi \approx \exp(2\Delta\beta J_0)$, i.e., $\gamma = \Delta\gamma_{pure} = 2\Delta$ for soft rates. As in §7.6, critical behavior is not universal but depends on asymptotic properties of the rate.

Our discussion in §7.1 about different realizations of the NRFM also applies to the present case. That is, one may imagine that J is changed all over the system at each kinetic step, so that each specific configuration has

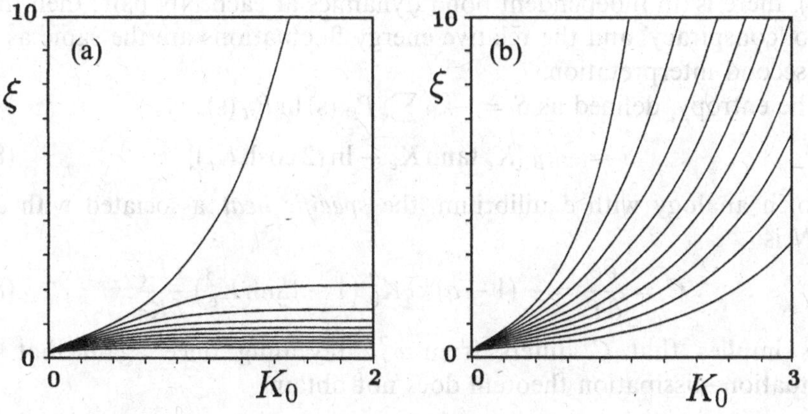

Fig. 8.2. The (inverse) temperature dependence of the correlation length, (8.4), for (a) the Glauber rate (7.49), and (b) the rate (7.48).

energy $E_1(\mathbf{s}) = -J\sum_{NN} s_{\mathbf{r}}s_{\mathbf{r}'}$; this is to be interpreted as the instantaneous energy, with J representing the value of a random variable sampled independently (every time step) from $p(J)$. Alternatively, given that the basic dynamical process consists of attempting to flip a given spin, one may imagine that only the Js connecting this spin to its NNs change at each step; consequently, the energy tends asymptotically to $E_2(\mathbf{s}) = -\sum_{NN} J_{\mathbf{rr}'}s_{\mathbf{r}}s_{\mathbf{r}'}$ with $J_{\mathbf{rr}'}$ spatially distributed at random according to $p(J)$. The relevant energy needs, therefore, to involve averages with respect to both $p(J)$ and $P_{st}(\mathbf{s})$, because J is also random with time for each NN pair $(\mathbf{r}, \mathbf{r}')$. The two interpretations are thus characterized by the same energy, namely,

$$U = [[E_1]] = [[E_2]] = -\langle\langle J\rangle\rangle \tanh K_e = -(1-q)\, J_0 \tanh K_e. \tag{8.5}$$

However, energy fluctuations differ fundamentally in the two interpretations. The mean square fluctuations, σ_U^2, are

$$\langle\langle J^2\rangle\rangle \left(\langle\tanh^2 K_e\rangle - \langle\tanh K_e\rangle^2\right) + \left(\langle\langle J^2\rangle\rangle - \langle\langle J\rangle\rangle^2\right)\langle\tanh K_e\rangle^2 \tag{8.6}$$

and

$$\langle\langle J\rangle\rangle^2 \left(\langle\tanh^2 K_e\rangle - \langle\tanh K_e\rangle^2\right) + 2N\left(\langle\langle J^2\rangle\rangle - \langle\langle J\rangle\rangle^2\right), \tag{8.7}$$

respectively, for the two versions. The relative magnitude $U^{-1}\sqrt{\sigma_U^2}$ vanishes for (8.7) as $N \to \infty$, as in equilibrium, while the last term in (8.6) is of order N^2. This indicates that dynamics may induce (anomalous) changes of order JN within the first model mentioned. For some (soft) rates (see

§8.4), there is an independent bond dynamics at each NN pair; then there is no 'conspiracy' and the relative energy fluctuations are the same as for the second interpretation.

The entropy, defined as $S = -k_B \sum_{\mathbf{s}} P_{st}(\mathbf{s}) \ln P_{st}(\mathbf{s})$, is

$$S = -k_B \left[K_e \tanh K_e - \ln(2\cosh K_e) \right]. \tag{8.8}$$

Also in analogy with equilibrium, the *specific heat* associated with $e = U/N$ is

$$C = \frac{\partial e}{\partial T} = -(1-q)\, k_B K_0^2 \left(1 - \tanh K_e^2\right) \frac{\partial K_e}{\partial K_0}. \tag{8.9}$$

This implies that C differs from σ_U^2, revealing once again that the fluctuation–dissipation theorem does not obtain.

Nonequilibrium Ising glass

Spin-glass materials, such as the metallic alloy CuMn with only a few percent of magnetic Mn ions, provide a further motivation to study MKD. The unusual properties exhibited by these substances have been attributed to *frustration*, i.e., the structural impossibility of satisfying all interactions (§7.1). Despite much effort and quite interesting results, the present understanding of the *spin-glass problem* is not totally satisfactory, however. In particular, no solvable microscopic model is regarded as capturing all the essential features of spin glasses, and there is not even full agreement on the meaning of macroscopic *spin-glass behavior*. It is quite likely that this is not an equilibrium problem, due to the occurrence of frustration and competition. In fact, reaching a unique state asymptotically is often difficult or impossible, and the nature of final states is observed to be influenced by the dynamics (history), which is a feature of nonequilibrium phenomena.[3]

It was argued above that the simplest MKD that contains some of the physics of spin glasses corresponds to (7.2) and (7.3) with $\vec{\zeta} \equiv J$, $h = 0$, and some distribution $p(J)$ for the bonds. The bond distribution may be chosen to obtain a kinetic version of the Edwards–Anderson model. Interesting properties ensue in $d = 1$ for the rather general distribution

$$p(J) = \frac{J_2}{J_1 + J_2}\delta(J - J_1) + \frac{J_1}{J_1 + J_2}\delta(J + J_2), \tag{8.10}$$

$J_1, J_2 \geq 0$, which is nontrivial in spite of giving $\langle\langle J \rangle\rangle = 0$. The cases $\lambda \equiv J_2/J_1 = 0, 1$ correspond to $p(J) = \delta(J)$ and $p(J) = \frac{1}{2}\delta(J - J_1) +$

[3] See Canella & Mydosh (1972) for an early report of unusual macroscopic properties, and Edwards & Anderson (1975, 1976), Toulouse (1977), and Parisi (1980) for early interpretations; some reviews are mentioned in §7.1.

$\frac{1}{2}\delta(J+J_1)$, respectively. It follows for the latter with $J_1 > 0$ that

$$K_e = -\tfrac{1}{4}\ln\frac{1-q+q\,e^{4\beta J_1}}{q+(1-q)\,e^{4\beta J_1}}, \tag{8.11}$$

independent of the dynamics. Taking the limit $T \to 0$ here reveals that the pure zero-T critical point is washed away by this symmetric competition of $\pm J_1$; however, one has $\xi \propto \sqrt{1/q}$ (and a similar behavior for the susceptibility) as $T, q \to 0$.

More interesting is the situation for $\lambda \neq 0, 1$. At high enough T one obtains that

$$K_e \approx -\tfrac{8}{3}\left[\tfrac{1}{2} - 3\ddot{\varphi}(0)\right]\lambda(1-\lambda)(\beta J_1)^3 + \mathcal{O}(\beta J_1)^4. \tag{8.12}$$

That is, rather unexpectedly, the high-temperature properties are strongly influenced by the curvature, $\ddot{\varphi}$, of the rate function near the origin, and the system possesses an effective temperature which is proportional to T^3. This is to be compared with the equilibrium and $\lambda = 1$ cases characterized by universality (with respect to dynamics) and linear behavior. Assuming that φ is differentiable at the origin, one obtains as $T \to 0$ that $K_e \approx \Delta(1-\lambda)\beta J_1 + \frac{1}{4}\ln\lambda$ for soft rates, which are characterized by constant $\Delta < 1$, while hard rates preclude any critical point. Hard rates induce an *effective* ferromagnetic (antiferromagnetic) ground state for $\lambda > 1$ ($\lambda < 1$). The situation is similar for $J_2 \to \infty$: the smallest coefficient in (8.10), which represents the probability of J_2, dominates in this case. For soft rates, however, the zero-T critical point occurs, and the ground state is antiferromagnetic or ferromagnetic depending on the sign of $J_1 - J_2$. That is, the impurity strength dominates over the probability coefficient. In any case, (8.12) reveals that K_e changes sign at a finite T which marks a *continuous* changeover from ferromagnetic to antiferromagnetic emergent properties. The critical point for soft rates is characterized by exponents differing from those of the pure model by a factor of $(1-\lambda)\Delta$, for example, $\nu = 2(1-\lambda)\Delta$ and $\alpha = 2(1-\lambda)\Delta$. That is, these critical exponents not only depend on asymptotic features of the dynamics, such as Δ, but also on the parameter λ that one might have expected to be irrelevant; in fact, the symmetry measured by λ is so important that a continuum of impure values emerges that changes sign at $\lambda = 1$. It would be interesting to investigate whether any of these properties occur in natural one-dimensional substances.[4]

[4] One-dimensional chains in which neighboring (1/2-)spins seem to be coupled by Heisenberg interactions that are randomly ferromagnetic or antiferromagnetic may be realized in the laboratory (Nguyen, Lee, & Loye 1996).

Very strong bonds

Consider next

$$p(J) = q\delta(J - J_0) + (1 - q)\,\delta(J + J_1) \tag{8.13}$$

for q close to 1 and $J_1 \gg J_0 > 0$, say $J_1 \to \infty$. This is the MKD version of an impure system whose equilibrium states have been studied by Labarta *et al.* (1988). Soft rates imply that the effective temperature goes to zero as $J_1 \to \infty$ for any T and q. More intriguing is the case of hard rates for which

$$K_e \approx \begin{cases} -\frac{1}{4}\ln\left(1 + \mu\dfrac{1-q}{q}\right) & \text{as} \quad T \to \infty \\ -\frac{1}{4}\ln\dfrac{1-q}{q} & \text{as} \quad T \to 0 \end{cases} \tag{8.14}$$

as $J_1 \to \infty$. Therefore, the rate influences any state, including the one for $T \to \infty$, except the ground state, which is independent of the kinetics, because spins always need to point in one of the two possible directions. On the other hand, the effective interaction K_e is negative at infinite temperature for any q, and it remains so when $T \to 0$ insofar as $q < \frac{1}{2}$. One may define K_0^C, by means of

$$q\left[\varphi\left(-4K_0^C\right) - \varphi\left(4K_0^C\right)\right] = (1 - q)\,\mu\,, \tag{8.15}$$

which locates the transition between two regions of the phase diagram whose behavior is predominantly ferromagnetic and antiferromagnetic, respectively. One obtains $K_0^C = -\frac{1}{4}\ln\left[(2q-1)/q\right]$ and $K_0^C = -\frac{1}{4}\ln(2q-1)$ for rates (7.51) and (7.49), respectively; in both cases,

$$K_0^C \to \begin{cases} 0 & \text{as} \quad q \to 1 \\ \infty & \text{as} \quad q \to \frac{1}{2}. \end{cases} \tag{8.16}$$

This transition is smooth for $d = 1$, i.e., the correlation length and other quantities remain analytic. It would be interesting to study this situation in $d > 1$.

Some related systems

As for the NRFM, one may compare the above with equilibrium disordered systems, i.e., Ising models with either quenched or annealed impurities.

A quenched impure Ising model may be defined for $d = 1$ and $H_N(\mathbf{s}) = -\sum_r J_r s_r s_{r+1}$, with J_r spatially distributed according to a given (normalized) distribution $p(J_r)$; see Fan & McCoy (1969). The transfer matrix method gives the energy and free energy per site, respectively, as

$$\left.\begin{aligned} e &= N^{-1}\sum_{r=1}^{N}\langle s_r s_{r+1}\rangle = \langle\langle \tanh(\beta J)\rangle\rangle\ , \\ f &= N^{-1}\sum_{r=1}^{N}\ln\left[2\cosh(\beta J_r)\right] = \langle\langle \ln\left[2\cosh(\beta J)\right]\rangle\rangle\ , \end{aligned}\right\} \tag{8.17}$$

where the last terms involve the limit $N \to \infty$, and $\langle\langle\cdots\rangle\rangle$ represents an average with $p(J_r)$. Consider (8.13) with $J_1 = 0$, i.e., broken bonds; one obtains $\xi^{-1} \approx \ln q$ as $K_0 \to \infty$, i.e., any impurity destroys the pure zero-T critical point.

The quenched impure Ising model is thus rather different from the corresponding MKD. The energy e is the same in both cases for $T \to \infty$ if the nonequilibrium system is implemented with hard rates. However, one has $e = q$ for the quenched system as $T \to 0$, while there is a complex expression for the nonequilibrium case which reduces to $e \approx 1 - 2\sqrt{(1-q)/\mu}$ for $q \to 1$. A closer similarity occurs assuming (8.13) with $J_1 \to \infty$, which leads to $e \approx q\tanh K_0 - (1-q)$. Therefore, the limit of e as $T \to 0$ changes sign at $q = \frac{1}{2}$, and one may define a temperature locating this change, namely, $K_0^C = -\frac{1}{2}\ln(2q-1)$, $q > \frac{1}{2}$.

The MKD is also quite distinct from the corresponding annealed case defined (Thorpe & Beeman 1976) via the partition function

$$Z_N = \sum_{\mathbf{s}}\int\prod_{r=1}^{N} \mathrm{d}J_r\, p(J_r)\, \mathrm{e}^{\,\beta J_r s_r s_{r+1}} = \sum_{\mathbf{s}}\prod_{r}\left\langle\left\langle \mathrm{e}^{\,\beta J_r s_r s_{r+1}}\right\rangle\right\rangle . \tag{8.18}$$

Further details may be found in Garrido & Marro (1992); see also §7.7.

8.2 Impure ordering for $d \geq 1$

We first illustrate how the method of §7.8 bounds the ergodic regions of the phase diagram. Consider first the glass case, (8.13) with $J_0 = J_1$. Condition (7.110) indicates that ergodicity may break down in $d = 1$ only for $T < T_0$ such that

$$T_0^{-1} = \tfrac{1}{2}\ln\frac{1 + 2\sqrt{\frac{1}{2} - q + q^2}}{|2q-1|} \tag{8.19}$$

for rate (7.48), and $T_0 = 0$ for (7.50) and (7.51). For the latter rate and $d = 2$, one obtains

$$T_0^{-1} = -\tfrac{1}{4}\ln\left\{\tfrac{1}{3}\sqrt{4(4q-1)^2 - 3(9-16q)} - \tfrac{2}{3}(4q-1)\right\} \tag{8.20}$$

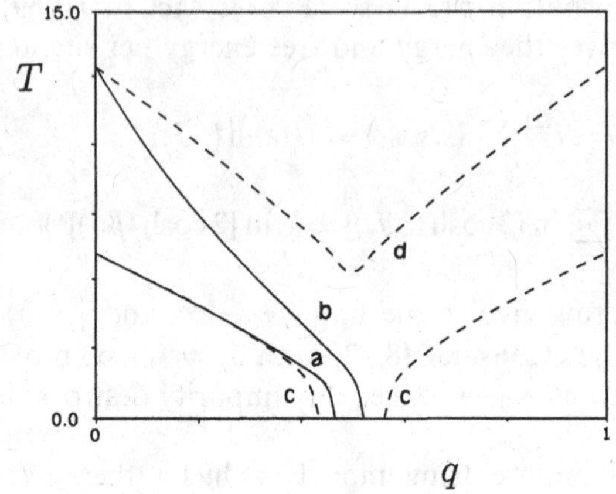

Fig. 8.3. Bounds (7.110) for the transition temperature as a function of the parameter q for a d-dimensional system which evolves by Metropolis rates: (a) $d = 2$ for bond dilution, $p(J) = q\delta(J) + (1-q)\delta(J - J_0)$, with $J_0 > 0$; (b) same $p(J)$ but $d = 3$; (c) $d = 2$ for glass competition, $p(J) = q\delta(J + J_0) + (1-q)\delta(J - J_0)$; (d) same $p(J)$ but $d = 3$. The system is necessarily ergodic above the curve for each case. (From Garrido & Marro (1992).)

for $q \notin \left[\frac{7}{16}, \frac{9}{16}\right]$; $T_0 = 0$ otherwise. For $d = 3$ and the same rate, T_0 is the solution of

$$(5 - 8q)\kappa^3 + 30\kappa^2 + (8q - 3)(15\kappa - 14) = 0, \quad \kappa \equiv \exp\left(\frac{-2}{T_0}\right) \tag{8.21}$$

for any q. For rate (7.48) and $d = 2$, T_0 is the solution of

$$(8q - 7)\kappa^4 + 4\kappa^3 + 6\kappa^2 + 4\kappa + 1 = 8q \tag{8.22}$$

for any q. For the bond-dilute case, (8.13) with $J_0 = 0$, a phase transition may occur for $\frac{7}{15} < q < 0$ below T_0 such that

$$T_0^{-1} = -\tfrac{1}{4} \ln \left[\sqrt{\frac{19 - 27q}{3(1-q)}} - 2 \right] \tag{8.23}$$

for the Metropolis rate (7.51) and $d = 2$. These bounds are illustrated in figures 8.3 and 8.4.

Results from simulations

The above suggests that one should investigate MKD by the MC method. So far, only the nonequilibrium Ising glass has been studied numerically

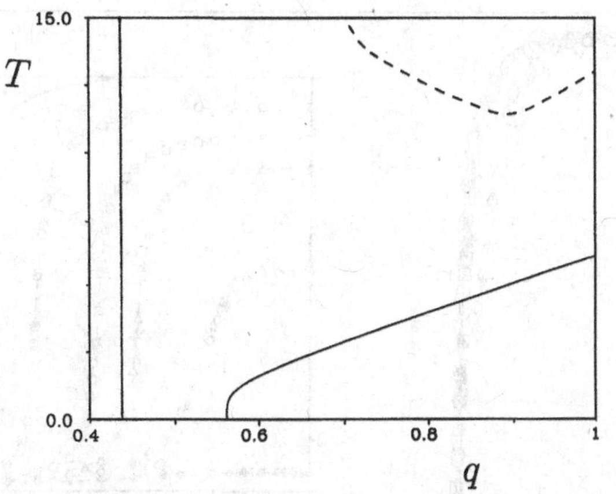

Fig. 8.4. The same as figure 8.3 when the competition includes very strong bonds, namely, $p(J) = q\delta(J - J_0) + (1-q)\delta(J + J_1)$ with $J_1 \to \infty$, for $d = 2$ (solid lines) and $d = 3$ (broken line). The latter goes asymptotically to $q = 0.701$ as $T \to \infty$ (not shown).

(González-Miranda *et al.* 1994). The available data are for the dynamical rule

$$c(\mathbf{s^r} \mid \mathbf{s}) = p \min\left\{1, \exp\left(-\beta\delta H_{J_0}\right)\right\} + (1-p)\min\left\{1, \exp\left(-\beta\delta H_{-J_0}\right)\right\}, \tag{8.24}$$

where $\delta H_J = 2J\theta(\mathbf{s})$ with

$$\theta(\mathbf{s}) \equiv s_{\mathbf{r}} \sum_{\{\mathbf{r}',\, |\mathbf{r}-\mathbf{r}'|=1\}} s_{\mathbf{r}'} . \tag{8.25}$$

This corresponds to a superposition of two Metropolis rates with probabilities p and $1-p$ assuming a NN Ising Hamiltonian with interaction strength $-J_0$ and $+J_0$, respectively. The interest is in the phase diagram and critical behavior for $d > 1$ (detailed balance holds only for $d = 1$).

The magnetization m for the square lattice is illustrated in figure 8.5. This reveals that a phase transition to a ferromagnetic state occurs at $T_C(p)$ for p large enough. (Similarly, antiferromagnetic states occur at low T for small p, but we only consider $p \geq \frac{1}{2}$ below because the phase diagram is symmetric about $p = \frac{1}{2}$.) The degree of saturation and the transition temperature both decrease with p, and there is no low-T ferromagnetic state for small enough p, e.g., for $p \leqslant 0.91$ if $d = 2$. The phase transition is qualitatively similar to the one in the pure system for $p = 1$. In particular, the transition is continuous, and analysis of data gives consistency with equilibrium critical exponents. This is demonstrated, for instance, by the fact that all data for $d = 2$ and $0.94 \lesssim p < 1$ collapse

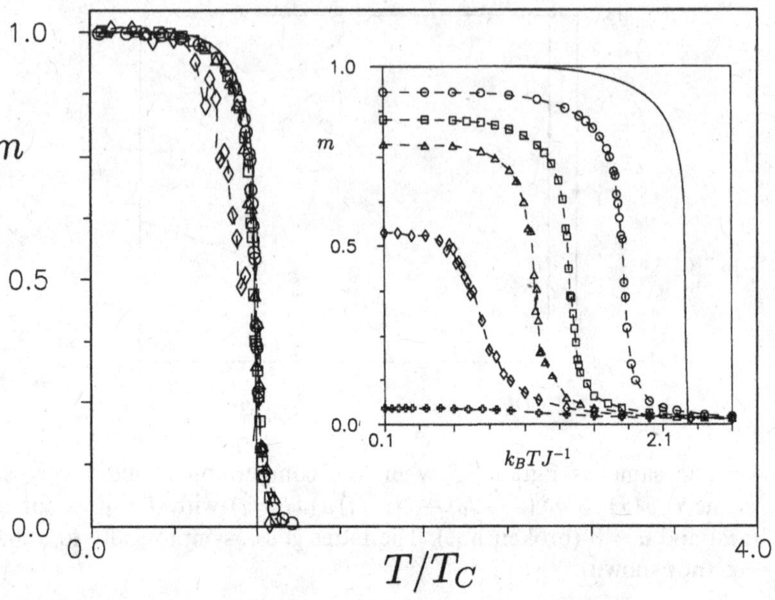

Fig. 8.5. The magnetization versus temperature for the 128×128 lattice for different values of p: $p = 0.97$ (o), 0.95 (□), 0.94 (△), 0.93 (◇), and 0.91 (×). The full line corresponds to the equilibrium result for $p = 1$. The main graph shows the data scaled vertically with $\alpha_1(p)$ given in table 8.1, and temperature in units of the corresponding transition temperature, $T_C(p)$. The inset shows the same data with the temperature in units of J_0/k_B. (From González-Miranda *et al.* (1992).)

within experimental error on the equilibrium curve if one multiplies the corresponding quantity by a parameter that depends on p (see table 8.1), and scales temperatures by $T_C(p)$. The data for $p = 0.93$ $(d = 2)$ do not scale in the same manner, however; in fact, figure 8.5 suggests a departure of this case from the equilibrium solution. This might indicate either that the phase transition becomes of first order as p is decreased, so that a (nonequilibrium) tricritical point exists nearby, or else that critical exponents differ from the Onsager ones for $p \approx 0.93$. This is supported by data for the *energy* (stationary mean number of *up-down* NN pairs) and other measures of short-range order, and its fluctuations. Nevertheless, neither discontinuities nor metastable states have been observed. That is, the case for a continuous transition, perhaps exhibiting crossover, is favored by the data. The deviations shown for $p = 0.93$ in figure 8.5 might reflect the influence of a different (for example, percolative) mechanism competing with thermal fluctuations (§8.3).

A comparison with analytical results helps to clarify this question. Kinetic mean-field theory in the pair approximation (Alonso & Marro

Table 8.1. Variation with p of the parameters that scale all the data for $d = 2$ and $0.94 \leqslant p \leqslant 1$ (but not for $p = 0.93$; see figure 8.5). The magnetization needs to be multiplied by $\alpha_1(p)$, the susceptibility by $1/\alpha_1(p)$, the energy by $\alpha_2(p)$, and the specific heat by $1/\alpha_2(p)$.

	p			
	0.97	0.95	0.94	0.93
α_1	1.111	1.274	1.451	3.074
α_2	1.356	2.014	3.065	11.75

1992a) predicts for $d = 2$ a second-order phase transition for any $p > p_0 = \frac{32}{37}$ (as compared to the MC value $p_0 \simeq 0.928$).[5] However, a tricritical point and phase transitions of first order are predicted if the rate is

$$c(\mathbf{s}^{\mathbf{r}} \mid \mathbf{s}) = p \exp\left(-\tfrac{1}{2}\beta\delta H_{J_0}\right) + (1-p)\exp\left(-\tfrac{1}{2}\beta\delta H_{-J_0}\right) \tag{8.26}$$

instead of (8.24). On the other hand, the bounds for the phase region in which the two-dimensional system is necessarily ergodic indicate that a sharp phase transition may only occur for $T < T_0$, where

$$T_0^{-1} = -\tfrac{1}{4}\ln\left\{\tfrac{1}{3}\left[2 - 8p + \sqrt{4(4p-1)^2 + 3(16p-9)}\right]\right\}, \tag{8.27}$$

if $p > \frac{9}{16}$. Therefore, order might occur for $0.562 < p < p_0$. Inspection of typical spin configurations within this region reveals clustering which suggests that some sort of order (which has not been further qualified) might occur; if real, this would essentially differ from the ferromagnetic order, i.e., the one which is suggested for $p < p_0$ appears to be independent of T (for example, it is rather stable against thermal perturbations). It is difficult to draw more definite conclusions from the available MC data, however.

The behavior of other quantities is also intriguing at low temperature for $p \leqslant 0.93$. This is the case, for example, for the cumulants in figure 8.6, defined by:

$$g_\kappa = 1 - \frac{1}{3}\frac{\langle \kappa^4 \rangle}{\langle \kappa^2 \rangle^2}, \tag{8.28}$$

where κ refers, for example, to the magnetization or to the energy (before any averaging). For an infinite system, $g_\kappa \to \frac{2}{3}$ if $\langle \kappa^n \rangle = \langle \kappa \rangle^n$, i.e., for

[5] This MC value for p_0 is in accordance with the critical value obtained by de Oliveira (1992) for a two-dimensional majority vote process (§8.3).

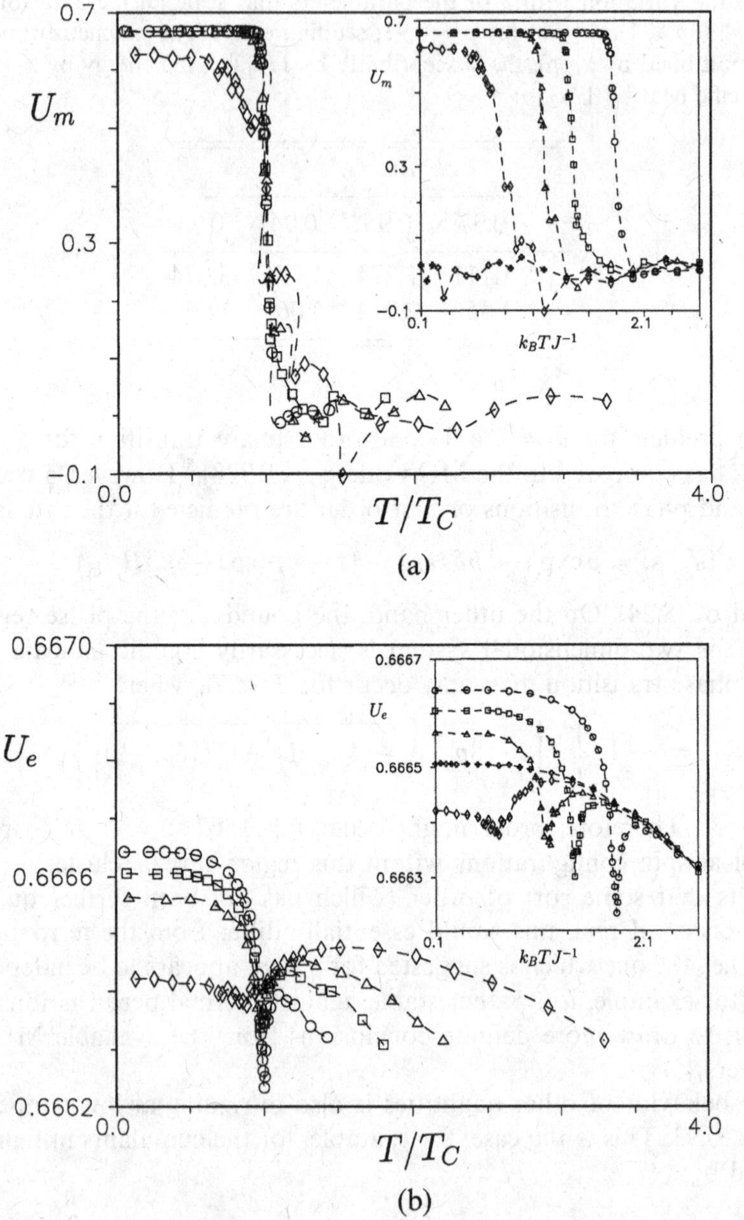

Fig. 8.6. Temperature variation of the normalized cumulants (see Binder 1979) (8.28) for: (a) the magnetization, and (b) the energy. The same system and symbols are used as in figure 8.5. The insets show the raw data while the main graphs involve scaling the temperature by $T_C(p)$.

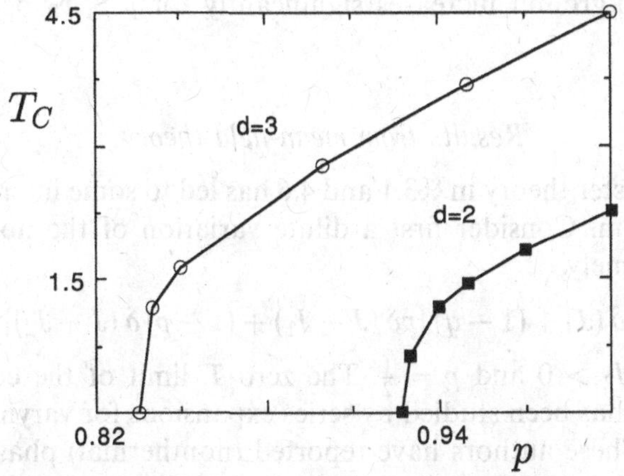

Fig. 8.7. MC estimate of the transition temperature $T_C(p)$ for the nonequilibrium Ising glass for $d = 2, 3$, as indicated. A similar situation occurs for $p < \frac{1}{2}$. (From González-Miranda *et al.* (1992).)

nonpathological distributions, as long as $\langle \kappa \rangle \neq 0$, while $g_\kappa \to 0$ when $\langle \kappa^4 \rangle \to 3 \langle \kappa^2 \rangle^2$, which occurs, for example, for a Gaussian distribution symmetrically distributed around zero. The data for g_m in figure 8.6 are as expected for both $p = 0.91$ (in which case $m = 0$) and $p > 0.93$ for $T < T_C(p)$, while some deviation is observed for $p = 0.93$. The behavior exhibited by the cumulant of the energy is also interesting; there is some indication that the phase transition is not of second order (apparently against all other evidence). Summing up, the whole picture for $d = 2$ and $p \lesssim 0.94$ is not clear cut at the time of writing.

The situation is qualitatively similar for the $L \times L \times L$ lattice with $L \leqslant 64$. The phase diagram for $d = 3$ is indicated in figure 8.7 (which also shows the case $d = 2$). One obtains $p_0 = 0.835$, as compared to the mean-field result $p_0 = 0.786$. The ordinary finite-size scaling analysis leads to exponents that are consistent with the critical indexes of the pure case, but one cannot rule out the possibility of crossover when p decreases as described above for $d = 2$. On the other hand, the exact bounds above do not bar the possibility of an ordered phase for $T < T_0$ with $[1-4(2p-1)]\,e^{-12T_0} + 30\,e^{-8T_0} + [15+60(2p-1)]\,e^{-4T_0} - 56(2p-1) = 14$, $p \geqslant \frac{1}{2}$. As for $d = 2$, but less pronounced here, a sort of order that tends to be washed away as one approaches $p = \frac{1}{2}$ is perhaps suggested by direct inspection of typical configurations outside the ferromagnetic-like region, i.e., for $p < p_0$ at low enough temperature. A probably related *anomaly* is the observation that the peak at $T_C(p)$ of the magnetic susceptibility tends to disappear as p is decreased and one approaches p_0

(while the background increases significantly for $p \lesssim p_0$ between $T_C(p)$ and $T = 0$).

Results from mean-field theory

The kinetic cluster theory in §§3.1 and 4.6 has led to some interesting results for this problem. Consider first a dilute variation of the nonequilibrium Ising glass, namely,

$$p(J) = q\delta(J) + (1-q)\left[p\delta(J-J_1) + (1-p)\delta(J+J_2)\right] , \qquad (8.29)$$

e.g., for $J_1 = J_2 > 0$ and $p = \frac{1}{2}$. The zero-T limit of the corresponding quenched case has been studied by series expansions for varying d by Klein *et al.* (1991). These authors have reported (nonthermal) phase transitions for $T = 0$ at $q = q_{SG} < q_C$, where q_C is the ordinary percolative threshold. The interest in studying this system for any T and d has been discussed from a general perspective in Shapira *et al.* (1994). We present below a mean-field description (Alonso & Marro 1992a) of the steady states that are implied by some cases of (8.29).

The nonequilibrium Ising glass corresponds to $q = 0$ and $J_1 = J_2 > 0$. Hard rates produce a situation consistent with the MC observations above. However, a mean-field study confirms the influence the rate function has on emergent properties, for example, soft rates induce qualitatively different behavior. Some interesting details are as follows. Within the pair approximation, the two-dimensional nonequilibrium Ising glass with any rate function φ exhibits a line, separating paramagnetic from ferromagnetic states, given by

$$\begin{aligned}(17p-1)\,\varphi(8\tau_C) + (20p-4)\,\varphi(4\tau_C) - (20p-16)\,\varphi(-4\tau_C)\\ -(17p-16)\,\varphi(-8\tau_C) = 0, \qquad z_C = \tfrac{1}{3}, \qquad (8.30)\end{aligned}$$

where $\tau_C \equiv J_1\left[k_B T_C(p)\right]^{-1}$ and z represents the density of *up-up* pairs. There is also a line, separating paramagnetic from antiferromagnetic phases, characterized by $z_C = \frac{1}{6}$ and (8.30) (where p is to be replaced by $1-p$). The resulting behavior when φ corresponds to a hard rate is illustrated in figure 8.8 for two cases. The situation for soft rates is illustrated in figure 8.9. This is a very interesting case as it reveals the possible existence of tricritical points and reentrance phenomena. Phase diagrams are presented in figure 8.10. For $d = 3$, the transition lines follow from

$$\begin{aligned}729\phi(3) + 1944\phi(2) + 1620\phi(1) - 720\phi(-1)\\ -384\phi(-2) - 64\phi(-3) = 0, \qquad z_C = \tfrac{3}{10}, \qquad (8.31)\end{aligned}$$

where $\phi(n) = \int dJ\, p(J)\, \varphi\left(4nJ/k_B T_C(p)\right)$. The ensuing picture is qualitatively similar; in fact, one may expect essentially the same properties for any $d > 1$ within the present approximation.

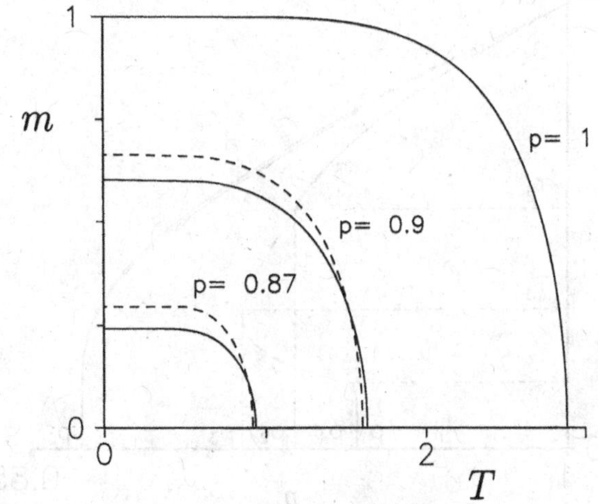

Fig. 8.8. The magnetization for the two-dimensional mean-field nonequilibrium Ising glass, i.e., (8.29) with $J_1 = J_2 > 0$ and $q = 0$, as a function of temperature (in units of J_1/k_B) as one varies p as indicated. The solid lines are for the Metropolis rate (7.51), and the dashed lines are for the Kawasaki rates (7.50). One has zero magnetization for all T for $p < p_0 = 0.865$ in both cases; see figure 8.10. (From Alonso & Marro (1992a).)

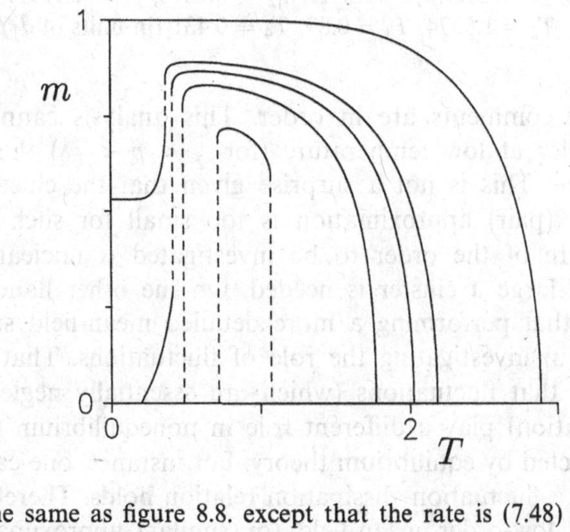

Fig. 8.9. The same as figure 8.8. except that the rate is (7.48) or, equivalently, (7.49). The curves correspond to $p = 1$, 0.95, 0.94, 0.93, and 0.912 from top to bottom. The solid lines represent stable solutions, and the dashed lines correspond to unstable solutions; see figure 8.10 for a different representation of these results.

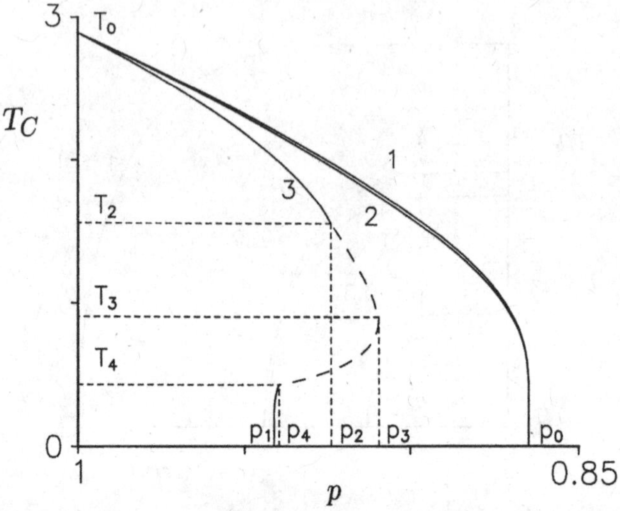

Fig. 8.10. Mean-field phase diagram for a d-dimensional sc lattice which evolves as implied by distribution (8.29) with $J_1 = J_2 > 0$ and $q = 0$ and several rates: curves 1 and 2 are for the *hard* rates by Metropolis (7.51) and Kawasaki (7.50), respectively, and curve 3 is for the *soft* rate (7.48) or (7.49). The solid curves correspond to lines of critical points, and the dashed curves correspond to first-order phase transitions occurring for *soft* rates below a tricritical point at (T_2, p_2). One obtains qualitatively similar curves for both $d = 2$ and 3. It follows for $d = 2$ that $p_0 = 0.8650$, $p_1 = 0.9412$, $p_2 = 0.9240$, $p_3 = 0.9095$, $p_4 = 0.9396$, and $T_0 = 2.8854$, $T_2 = 1.5574$, $T_3 = 0.87$, $T_4 = 0.431$ (in units of J_1/k_B).

Two general comments are in order. This analysis cannot describe the sort of order at low temperature (for $\frac{1}{2} < p < p_0$) that has been suggested above. This is not a surprise given that the cluster involved by the present (pair) approximation is too small for such a purpose. Since the nature of the order to be investigated is unclear, one does not know how large a cluster is needed. On the other hand, it should be mentioned that performing a more detailed mean-field study would also be useful in investigating the role of fluctuations. That is, definite evidence exists that fluctuations (which are essentially neglected in the pair approximation) play a different role in nonequilibrium phenomena than that predicted by equilibrium theory. For instance, one cannot prove in general that a fluctuation–dissipation relation holds. Therefore, it may happen that a low-order mean-field (or similar) approximation yields spurious behavior, as has been shown explicitly in one particular case (Garrido *et al.* 1996).

The dilute nonequilibrium Ising glass, i.e., (8.29) with $q \neq 0$, has been studied for several values of J_1 and J_2. The boundary for $d = 2$ between

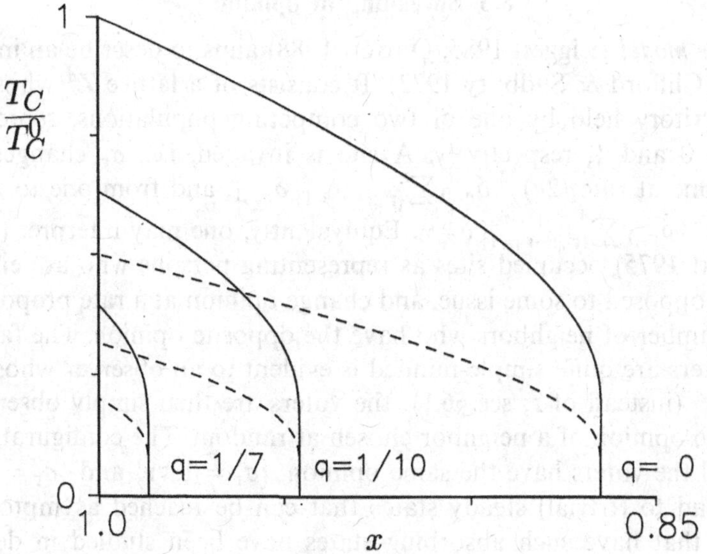

Fig. 8.11. Mean-field phase diagram for a two-dimensional lattice that evolves according to the Metropolis rate and distribution (8.29) for several values of the dilution variable q, as indicated, as a function of x which represents p or $1-p$ according to whether one considers the ferromagnetic or antiferromagnetic cases, respectively. For the asymmetric distribution corresponding to $J_2 = \frac{1}{2}J_1$, the solid lines represent the ferromagnetic limit and the dashed lines represent the antiferromagnetic limit. The solid lines also accurately represent both cases for $J_1 = J_2$. Here, the critical temperature is in units of the Bethe–Peierls equilibrium result.

paramagnetic and ferromagnetic regions corresponds to the line

$$16\phi(2) + 16\phi(1) - 4\phi(-1) - \phi(-2) = 0, \tag{8.32}$$

and antiferromagnetic states are only expected below the line

$$\phi(2) + 4\phi(1) - 16\phi(-1) - 16\phi(-2) = 0. \tag{8.33}$$

The phase diagram is illustrated in figure 8.11 for some representative cases. Requiring $T_C(p) = 0$ for any p in (8.32) implies a lack of ferromagnetic order for $q > q_0(p) = (37p - 32)/(37p - 5)$. That is, $q_0(p = 1) = \frac{5}{32}$ locates the onset of *percolation*. Further study of these interesting systems, including the case in which an external field is present, may be found in Alonso & Marro (1992a).

8.3 Spreading of opinion

The *voter model* (Liggett 1985, Durrett 1988) aims to describe an invasion process (Clifford & Sudbury 1973). It consists of a lattice $\mathbf{Z}^d$ whose sites $\mathbf{r}$ are territory held by one of two competing populations, represented by $\sigma_\mathbf{r} = 0$ and 1, respectively. A site is invaded, i.e., $\sigma_\mathbf{r}$ changes from zero to one at rate $(2d)^{-1}\,\delta_{\sigma_\mathbf{r},0}\sum_{\{\mathbf{r}',|\mathbf{r}-\mathbf{r}'|=1\}}\delta_{\sigma_{\mathbf{r}'},1}$, and from one to zero at rate $(2d)^{-1}\,\delta_{\sigma_\mathbf{r},1}\sum_{\{\mathbf{r}',|\mathbf{r}-\mathbf{r}'|=1\}}\delta_{\sigma_{\mathbf{r}'},0}$. Equivalently, one may interpret (Holley & Liggett 1975) occupied sites as representing persons who are either in favor or opposed to some issue, and change opinion at a rate proportional to the number of neighbors who have the opposite opinion. The fact that these voters are quite simple-minded is evident to an observer whose time scale is e^t (instead of t; see §6.1): the voters are then simply observed to adopt the opinion of a neighbor chosen at random. The configurations in which all the voters have the same opinion, $\{\sigma_\mathbf{r} = 1, \forall \mathbf{r}\}$ and $\{\sigma_\mathbf{r} = 0, \forall \mathbf{r}\}$, correspond to (trivial) steady states that can be reached asymptotically. Systems that have such absorbing states have been studied in detail in chapters 5 and 6. No further solutions may occur for $d \leqslant 2$, while the system exhibits a phase transition for $d \geqslant 3$ corresponding to the possible persistence of disagreements.

In the *majority vote process* (Liggett 1985, Gray 1985), each person looks at his or her neighbors, and adopts the majority opinion with probability p and the minority opinion with probability $1-p$. That is, the dynamical rule (which is often adopted in real life with p near or equal to 1) is even simpler than the above. An interesting geographical modification is the *Toom model*[6] in which the voter opinion changes to the majority of itself and its northern and eastern neighbors with probability $1-q-p$, to one with probability p, and to zero with probability q. Thus, a bias is introduced if the two noise parameters p and q take different values.

The zero-T limit of the nonequilibrium Ising glass corresponds to a majority vote process. That is, the competition in (8.24) reduces for $T \to 0$ for any d to

$$c\left(\mathbf{s}^\mathbf{r} \mid \mathbf{s}\right) = \begin{cases} p & \text{if } \theta(\mathbf{s}) > 0 \\ 1-p & \text{if } \theta(\mathbf{s}) < 0 \\ 1 & \text{if } \theta(\mathbf{s}) = 0, \end{cases} \tag{8.34}$$

where $\theta(\mathbf{s})$ is defined in (8.25). (The probability that $s_\mathbf{r}$ flips when the neighboring configuration has an equal number of plus and minus signs is 1 here instead of $1-p$ or $\frac{1}{2}$ in more familiar cases; see table 8.2.) Likewise, one has

$$c\left(\mathbf{s}^\mathbf{r} \mid \mathbf{s}\right) = \begin{cases} p & \text{if } \theta(\mathbf{s}) > 0 \\ 1 & \text{if } \theta(\mathbf{s}) \leqslant 0 \end{cases} \tag{8.35}$$

[6] Toom (1980); see also van Enter, Fernández, & Sokal (1993).

Table 8.2. Rates of the form (8.37) that characterize several stochastic lattice systems, namely, the zero-temperature limit of the indicated two familiar cases, and three majority vote processes defined in: (*a*) Liggett (1985), (*b*) equation (8.34), and (*c*) Garrido & Marro (1989), and de Oliveira (1992).

Rate for indicated value of $\theta(\mathbf{s})$			Case
$\theta(\mathbf{s}) > 0$	$\theta(\mathbf{s}) = 0$	$\theta(\mathbf{s}) < 0$	
0	1	1	Metropolis for $T = 0$
0	1	2	Kawasaki for $T = 0$
$p(1-p)^{-1}$	1	1	(*a*)
p	1	$1-p$	(*b*)
$2p$	1	$2(1-p)$	(*c*)

for the dilute nonequilibrium Ising glass under the same conditions. This observation brings additional interest to the study of these rules.

The case (8.34), which does not satisfy detailed balance for $d > 1$, has been studied numerically by Marro *et al.* (1994) for lattices of 128^2 and 64^3 sites. A sharp phase transition has been reported at $p_0 \simeq 0.928$ and 0.835 for $d = 2$ and $d = 3$, respectively (§8.2). The transition is of second order, and inspection of the neighborhood of p_0 reveals that the critical exponents for both square and sc lattices have the same values (within statistical errors) as the corresponding (thermal) ones for the pure Ising model. Figure 8.12 depicts some of the evidence supporting this statement. Figure 8.13 suggests a singularity in the mean number of particle–particle NN pairs at p_0, and a very rapid increase for $p > p_0$; p_0 marks the onset of the ordered phase. Very near p_0 the curves agree well with the equilibrium thermal (for example, Onsager) ones after proper scaling of the temperature variable (see figure caption 8.13), but the agreement worsens systematically as one goes away from $p_0(d)$.

This phenomenon near $p_0(d)$ differs from the phase transition reported above for the nonequilibrium Ising glass. The latter is governed by a conflict between two values of J and the thermal bath, whilst the former is rather a geometrical effect ensuing directly from (8.34) (which corresponds to the zero-T limit of the nonequilibrium Ising glass). The peculiar behavior described in the previous section for the nonequilibrium Ising glass at low temperature for $p < p_0$ (and the reported apparent departure of critical behavior from the Ising case in this region) may reflect a competition between these two coexisting tendencies to ordering at low T.

On the other hand, it is remarkable that the critical exponents that are implied by (8.34) seem to be, in general, of the Ising variety for any

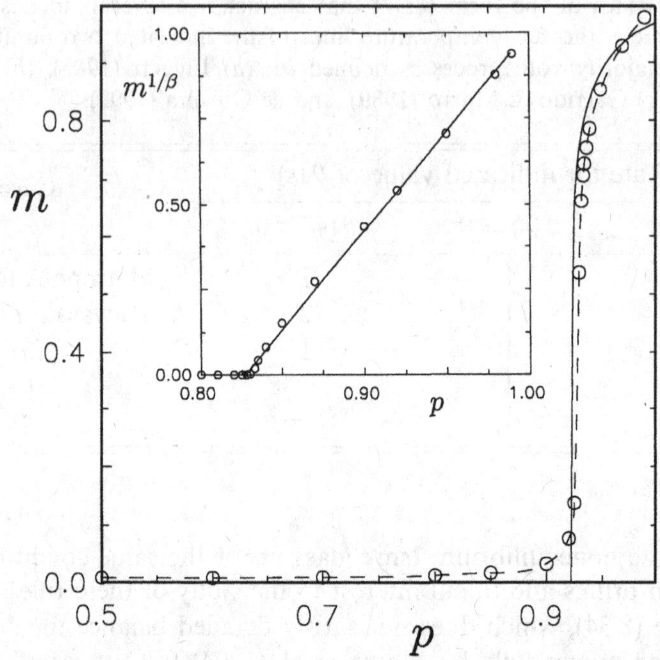

Fig. 8.12. The order parameter $m = L^{-2} \sum_{\mathbf{r}} s_{\mathbf{r}}$ for the steady state of the $L \times L$ lattice evolving according to the *majority vote* rule (8.34), as a function of p. The solid line in the main graph represents the Onsager solution of the Ising model for $N \to \infty$ shifted (arbitrarily) by $p = aT + b$ ($a = 0.133, b = 1.23$). The inset is a plot of $m^{1/\beta}$, with $\beta = \frac{5}{16}$ for $d = 3$.

$1 \leqslant d \leqslant 3$. This is in spite of the fact that the underlying mechanism differs essentially from the one in the pure Ising model. As described in the introduction to this chapter, other nonequilibrium systems that exhibit complex properties also have Ising critical behavior. In addition to the two-temperature systems (see this chapter introduction), we mention models studied in Tomé & Dickman (1993), Stauffer & Sashimi (1993), Heringa *et al.* (1994), and Santos & Teixeira (1995), for instance. Perhaps these cases are examples of the finding (Grinstein *et al.* 1985) that the ordinary Ising fixed point is locally stable in nonconservative systems with the up-down symmetry. However, this result ensues from a perturbative treatment which does not exclude the existence of other stable fixed points with their own domain of attraction. In fact, some examples of this have been described throughout this book. It is worth mentioning here, for example, the case in which one adds a random spin flip process to a majority rule (Lebowitz & Saleur 1986) that has been reported (Wang & Lebowitz 1988) to induce classical critical behavior for $d = 3$, and a

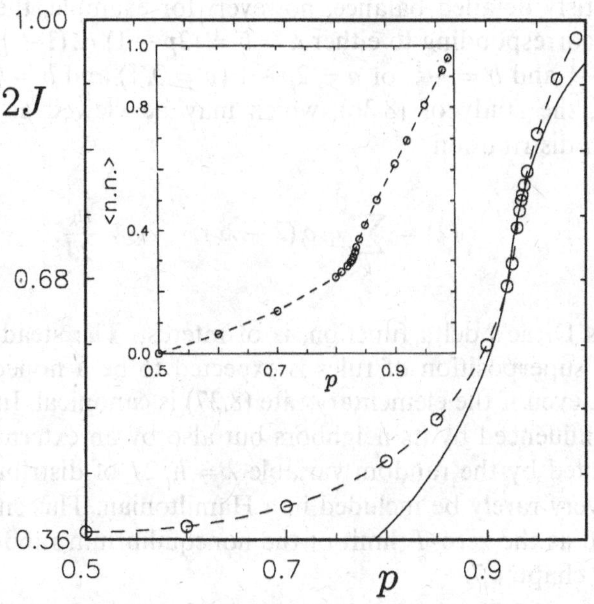

Fig. 8.13. The mean number of particle–particle NN pairs as a function of p for $d = 2$ (main graph) and 3 (inset). The solid line in the main graph is the Onsager result (shifted as for figure 8.12).

reaction–diffusion system (chapter 4) that seems to yield a different critical behavior (Menyhard 1994) (see Bassler & Schmittmann 1994). Therefore, the question of nonequilibrium *universality classes* and the defining of the range of relevance of the Ising critical point deserve further investigation.

This has motivated the study in Alonso, López-Lacomba, & Marro (1996) of a rule that may induce several different dynamic symmetries by varying its parameters. The case investigated is

$$c(\mathbf{s}^{\mathbf{r}} \mid \mathbf{s}) = 1 + a\sum_k q_k \mathrm{sgn}\left[s_{\mathbf{r}}(\varepsilon_{\mathbf{r}} + \lambda_k)\right] + b\sum_k q_k\, \mathrm{sgn}^2\left[s_{\mathbf{r}}(\varepsilon_{\mathbf{r}} + \lambda_k)\right], \quad (8.36)$$

where $\varepsilon_{\mathbf{r}} \equiv \frac{1}{2}\sum_{\{\mathbf{r}';|\mathbf{r}-\mathbf{r}'|=1\}} s_{\mathbf{r}'}$, i.e., $\theta(\mathbf{s}) = 2s_{\mathbf{r}}\varepsilon_{\mathbf{r}}$. Here, $\mathrm{sgn}(X)$ is the sign of X with $\mathrm{sgn}(0) = 0$, and a, b, J, h_k, and q_k, with $k = 1, 2, \ldots$, are parameters. One may interpret (8.36) as a superposition of rules

$$c(\mathbf{s}^{\mathbf{r}} \mid \mathbf{s}) = 1 + a\,\mathrm{sgn}(s_{\mathbf{r}}\varepsilon_{\mathbf{r}}) + b\,\mathrm{sgn}^2(s_{\mathbf{r}}\varepsilon_{\mathbf{r}}). \quad (8.37)$$

This is an interesting example of a hard rate. That is, (8.37) reduces to the zero-T limit of the Metropolis rate for $a = b = -\frac{1}{2}$ and of the Kawasaki rate for $a = -1$ and $b = 0$. These are canonical cases that imply a tendency as $t \to \infty$ towards the equilibrium state for $T \to 0$ and energy $H(\mathbf{s}) = -J\sum_{\{\mathbf{r},\mathbf{r}';|\mathbf{r}-\mathbf{r}'|=1\}} s_{\mathbf{r}}s_{\mathbf{r}'}$. Other familiar cases of (8.37) do not satisfy detailed balance, however: for example, the majority vote processes corresponding to either $a = b = (2p-1)/2(1-p)$ with $p \neq 0, 1$, or $a = p - \frac{1}{2}$ and $b = -\frac{1}{2}$, or $a = 2p - 1$ ($p \neq 0, 1$) and $b = 0$; see table 8.2. Therefore, the study of (8.36), which may be viewed as an average of (8.37) with distribution

$$g(\lambda) = \sum_k q_k \delta(\lambda - \lambda_k), \quad \lambda_k = \frac{h_k}{2J}, \tag{8.38}$$

where δ is Dirac's delta function, is of interest. The steady state implied by such a superposition of rules is expected to be a nonequilibrium one, in general, even if the elementary rate (8.37) is canonical. In fact, $s_{\mathbf{r}}$ is then not only influenced by its neighbors but also by an external agent, which is represented by the random variable $\lambda = h/2J$ of distribution $g(\lambda)$, that can only very rarely be included in a Hamiltonian. This situation may be interpreted as the zero-T limit of the nonequilibrium random-field model studied in chapter 7.

Consider (8.36) with $g(\lambda) = (q/2)\,\delta(\lambda - \lambda_0) + (q/2)\,\delta(\lambda + \lambda_0) + (1-q)\,\delta(\lambda)$. That is, (using the magnetic language) the applied field takes on the values $\pm\lambda_0$ or 0 at random with the indicated probabilities. This turns out to reduce the rate for $d = 1$ to $c(\mathbf{s}^{\mathbf{r}} \leftarrow \mathbf{s}) = \alpha_0 + \alpha_1 s_{\mathbf{r}} \varepsilon_{\mathbf{r}} + \alpha_2 \varepsilon_{\mathbf{r}} + \alpha_3 \varepsilon_{\mathbf{r}}^2$, and one finds by the method of §7.5 that an effective Hamiltonian exists with zero effective field if and only if $\alpha_2 = 0$. For $b \neq 0$, this implies that the system may be described by an effective temperature only for $\lambda_0 \neq 1$. Consequently, the most involved one-dimensional behavior might be expected to correspond to $b \neq 0$ with $\lambda_0 = 1$. The analytical study by Alonso *et al.* (1996) for $d > 1$ indicates that no effective Hamiltonian then exists except for a few cases of doubtful physical significance. It also suggests it would be of special interest to study the case $\lambda_0 = d > 1$.

The choice $\lambda_0 = d = 2$ has been studied in MC simulations for $a = b = -\frac{1}{2}$ (denoted hereafter as *rule M*) and $a = -1$ and $b = 0$ (denoted hereafter as *rule K*). The main conclusion is that the system exhibits a second-order phase transition with a critical point of the Ising variety at $q = q_C$, where $q_C = 0.146 \pm 0.002$ and 0.215 ± 0.005 for *rule M* and *rule K*, respectively. One also finds that the asymptotic behavior $m \sim \epsilon^{1/8}$, with m the magnetization and $\epsilon \equiv 1 - q q_C^{-1}$, holds for $\epsilon \lesssim 0.8$ for *rule M* and $\epsilon \lesssim 0.4$ for *rule K* (while the same holds only for $\epsilon \lesssim 0.1$ in the Onsager case with $\epsilon \equiv 1 - T T_C^{-1}$); see figure 8.14. This is perhaps an indication that the nature of correlations varies significantly from one case to the other, modifying the range of validity of homogeneity of the relevant functions.

The fact that observing the Ising regime (which is supposed to be an equilibrium feature) requires one to approach the critical point in

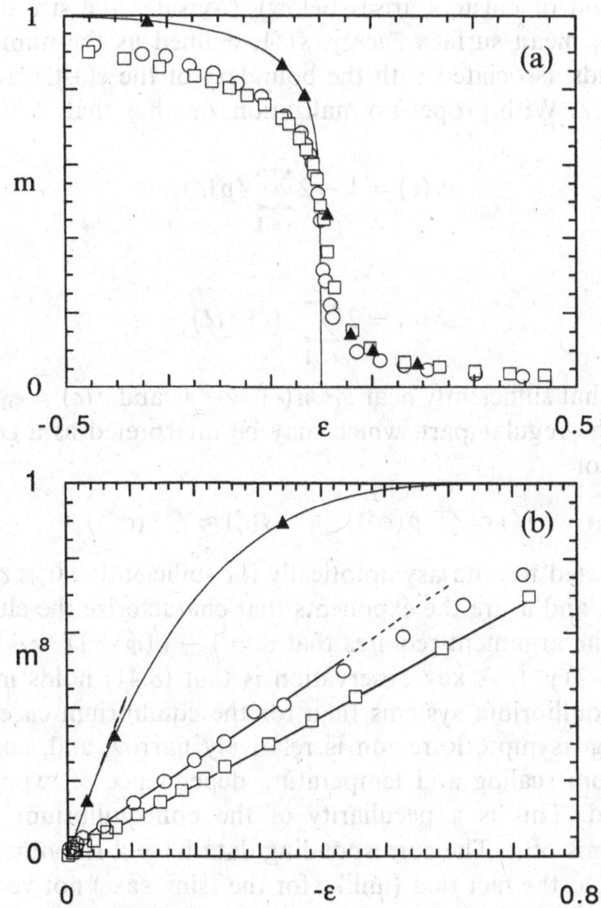

Fig. 8.14. (a) The magnetization m versus $\epsilon = (q - q_C)\, q_C^{-1}$ for the M (□) and K (o) rules (as defined in the main text) for the 128×128 lattice, and the exact Onsager solution (solid line), and corresponding MC data for the 64×64 lattice (▲) with $\epsilon = (T - T_C)\, T_C^{-1}$. (b) The same data showing consistency of all cases with two-dimensional Ising behavior, $\beta = \frac{1}{8}$; linear fits illustrate the extension of the Ising critical region.

terms of a natural definition for ϵ,[7] more closely in equilibrium than in nonequilibrium is interesting. One may expect to obtain some information about crossover phenomena from the study of cluster properties just below criticality, $q < q_C$. Let us define a cluster as any set of ℓ spins aligned in a given direction that have at least another NN aligned spin (no need for any

[7] As shown explicitly in §7.6 — see also below — certain very simple, quasi canonical systems allow a definition of ϵ depending on the dynamic rule that would obscure such a property.

other definition of clusters arises below). Consider the size distribution, $p(\ell)$, and the mean surface *energy*, $s(\ell)$, defined as the number of NN up-down bonds associated with the boundary of the cluster averaged for each value of ℓ. With proper normalization, one has that

$$m(\epsilon) = 1 - 2\sum_{\ell=1}^{\infty} \ell p(\ell), \tag{8.39}$$

and

$$e(\epsilon) = 2\sum_{\ell=1}^{\infty} s(\ell) p(\ell). \tag{8.40}$$

One expects that sufficiently near q_C $m(\epsilon) \approx \epsilon^{\beta}\tilde{m}$, and $e(\epsilon) \approx e_0(\epsilon) + \epsilon^{1-\alpha}\tilde{e}$, where e_0 is the regular part, which may be interpreted as a consequence of the behavior

$$p(\ell) \approx \ell^{-\tau}\tilde{p}(\epsilon\ell^{y}), \qquad s(\ell) \approx \ell^{\sigma}\tilde{s}(\epsilon\ell^{y}), \tag{8.41}$$

which is expected to hold asymptotically for sufficiently large ℓ and small ϵ.[8] Here, τ, y, and σ are the exponents that characterize the cluster distribution, and the argument requires that $\sigma = 1 - y(\phi - 1)$, $\alpha + \beta + \phi = 2$, and $\beta = (\tau - 2)\, y^{-1}$. A key observation is that (8.41) holds more generally for nonequilibrium systems than for the equilibrium case where the corresponding asymptotic region is relatively narrow and, consequently, deviations from scaling and temperature dependence of exponents have been reported. This is a peculiarity of the nonequilibrium system described in terms of q. The corresponding data have a smoother statistical behavior due to the fact that (unlike for the Ising case) not very ramified, say, percolating clusters occur that produce deviations from scaling behavior slightly below q_C. Figure 8.15 illustrates the situation. The scaling properties are better checked by considering the moment $\sum_{\ell} \ell^{\tau} p(\ell)$. For example, one may try to scale data according to the function

$$F(\epsilon n^{y}) \equiv \epsilon^{1/y} \sum_{\ell=1}^{n} \ell^{\tau} p(\ell;\epsilon) \tag{8.42}$$

with proper values for the exponents y and τ. Figure 8.16 illustrates the result. The best values obtained from this analysis are $\tau = 2.054 \pm 0.005$, $y = 0.44 \pm 0.01$, and $\sigma = 0.68 \pm 0.04$.

[8] We refer the reader for further details to Fisher (1967), Domb, Schneider, & Stoll (1975), Binder (1976), and Cambier & Nauenberg (1986). The last of these describes a controversy about the properties of (equilibrium) *clusters* and their definition. See also Stella & Vanderzande (1989), Marko (1992), Prakash *et al.* (1992), and Vanderzande (1992), for instance.

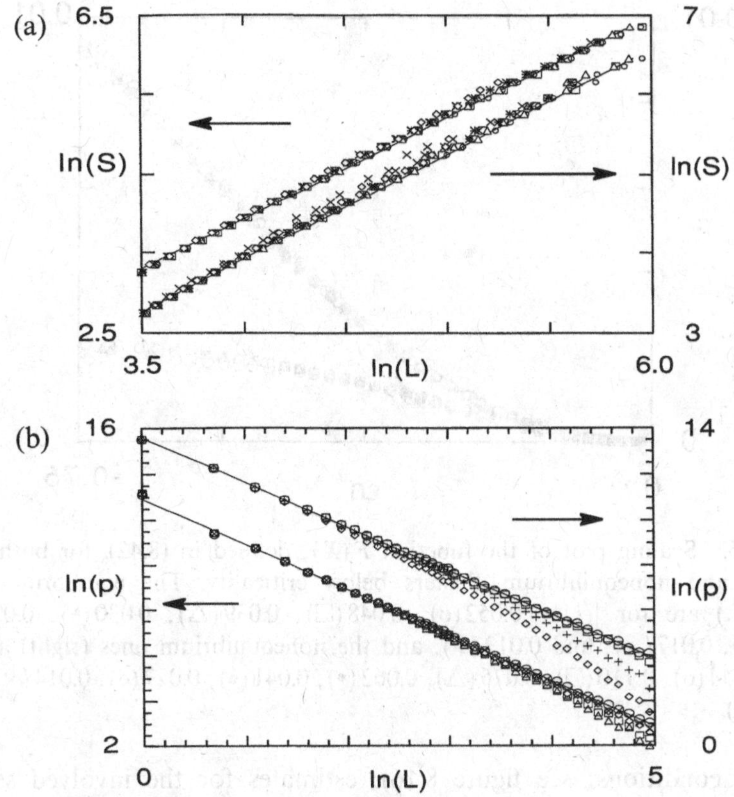

Fig. 8.15. Log–log plots to illustrate the size dependence of the properties of clusters of aligned spins as obtained for the *M rule* (different symbols correspond to $-\epsilon = 0.013, 0.027, 0.041, 0.062, 0.075, 0.144$) and for the equilibrium case (different symbols are for $-\epsilon = 0.004, 0.009, 0.022, 0.030, 0.039, 0.052$). (a) The total *surface* $S(\ell)$ of all clusters of size ℓ for $20 < \ell < 300$: the upper and lower sets correspond, respectively, to nonequilibrium and to equilibrium. The best fit here gives $\sigma = 0.68$. (b) The total number of clusters of size ℓ, $P(\ell)$: lower and upper sets correspond here to the *M rule* and equilibrium, respectively. The best fit gives $\tau = 2.05$. (From Alonso *et al.* (1996).)

Summing up, two different (nonequilibrium) dynamical rules seem to induce the Ising critical behavior that characterizes the equilibrium case (for $d = 2$). However, the associated steady states differ from each other, for example, the nonequilibrium phase transitions occur for $q = q_C = 0.146 \pm 0.002$ and 0.215 ± 0.005, respectively, and the *region of influence* of the Ising critical point varies from $\epsilon \lesssim 0.4$ for *rule K* to $\epsilon \lesssim 0.8$ for *rule M*, perhaps reflecting some differences in the nature of correlations. The scaling proposal (8.41)–(8.42) holds for nonequilibrium clusters (though the scaling function F varies essentially from equilibrium to nonequi-

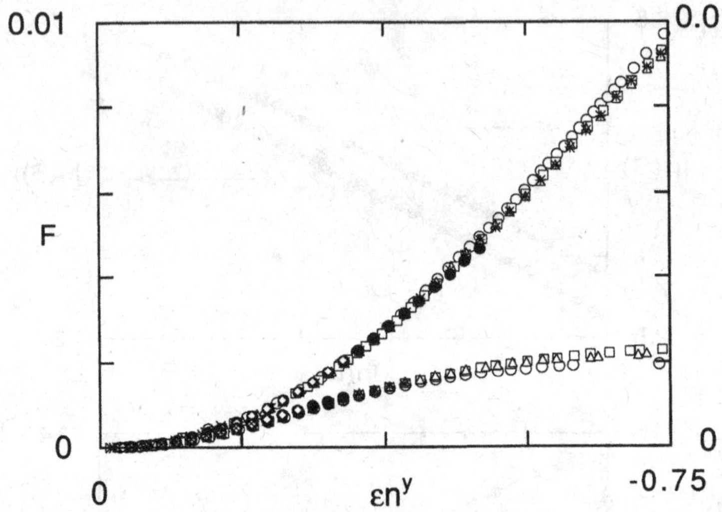

Fig. 8.16. Scaling plot of the function $F(X)$, defined in (8.42), for both equilibrium and nonequilibrium clusters below criticality. The equilibrium clusters (left) are for $|\epsilon| = 0.052\,(\text{o})$, $0.048\,(\square)$, $0.039\,(\triangle)$, $0.030\,(*)$, $0.026\,(\bullet)$, $0.021\,(\diamond)$, $0.017\,(\times)$, and $0.012\,(+)$, and the nonequilibrium ones (right) are for $|\epsilon| = 0.144\,(\text{o})$, $0.110\,(\square)$, $0.075\,(\triangle)$, $0.062\,(*)$, $0.041\,(\bullet)$, $0.027\,(\diamond)$, $0.014\,(\times)$, and $0.007\,(+)$.

librium conditions; see figure 8.16); estimates for the involved scaling parameters (that turn out to be model-independent) are consistent with $\beta = (\tau - 2)\, y^{-1} \simeq \frac{1}{8}$. The peculiar nature of the nonequilibrium system is confirmed by the fact that large and ramified clusters (of aligned spins) have not been observed just below criticality, while percolating clusters have sometimes hampered confirmation of scaling in equilibrium.

8.4 Kinetic ANNNI models

This section is devoted to a case of conflicting dynamics which reduces under two given limits to the ordinary Ising model and to the ANNNI model, respectively.[9] The system exhibits several spatially modulated phases and impure critical points. Furthermore, it is shown that, for any $d \geq 1$, the steady-state probability distribution for a class of transition rates has the quasi canonical structure defined in chapter 7.

The one-dimensional ANNNI model (§7.2) consists of a linear Ising chain with $s_z = \pm 1$ at each site $z = 1, 2, \ldots$. Typically, interactions are ferromagnetic, $J > 0$, between any pair of NN spins, s_z and s_{z+1}, and anti-

[9] We follow here López-Lacomba & Marro (1994b).

ferromagnetic, $J_{NNN} < 0$, between any pair of *next* NN (denoted NNN) spins, s_z and s_{z+2}. The d-dimensional (sc) case is defined as consisting of (one-dimensional) ANNNI chains that connect to each other by NN interactions of strength J_0 along any of the $2(d-1)$ transverse directions — Elliot (1961), see Selke (1992) for a review. This is supposed to be convenient to describe, for example, spatially modulated phenomena in certain magnetic substances such as erbium.

A MKD version of this quenched case corresponds to the rule

$$c(\mathbf{s}^{\mathbf{r}} \mid \mathbf{s}) = \int_{-\infty}^{+\infty} \mathrm{d}J_{NNN}\, p(J_{NNN})\, \varphi\left(\beta \delta H_{J_{NNN}}\right). \tag{8.43}$$

Here, J_{NNN} is a random variable of distribution $p(J_{NNN})$ normalized to unity, $\varphi(X)$ is defined in (7.47), and $\delta H_{J_{NNN}} = H(\mathbf{s}^{\mathbf{r}}; J; J_{NNN}) - H(\mathbf{s}; J; J_{NNN})$ with the *energy function* given, for example, by

$$H(\mathbf{s}; J; J_{NNN}) = -J \sum_{|\mathbf{r}-\mathbf{r}'|=1} s_{\mathbf{r}} s_{\mathbf{r}'} - J_{NNN} \sum_{|\mathbf{r}-\mathbf{r}'_z|=2} s_{\mathbf{r}} s_{\mathbf{r}'_z}, \tag{8.44}$$

where the last sum is over the NNN of $\mathbf{r} = (x, y, z)$ along direction z, denoted $\mathbf{r}'_z$. Except that NN interactions are assumed to be generally independent of direction here (i.e., $J = J_0$; see, however, page 271 for a different case), this is *formally* identical to the Hamiltonian that characterizes the ordinary ANNNI model. However, only the parameter J in (8.44) is assumed to be constant: according to (8.43), J_{NNN} is a random variable that one may interpret as varying (stochastically) during the evolution.

The function $\varphi(X)$ is arbitrary. Assuming that $\varphi(X)$ satisfies detailed balance, the same holds in some cases for the superposition (8.43). This occurs, for instance, for the 'trivial' case $J_{NNN} \equiv 0$, which transforms (8.44) into the Ising Hamiltonian, and for $p(J_{NNN}) = \delta(J_{NNN} - J_0)$, which reduces the nonequilibrium system to the simplest kinetic version of the ANNNI model one may think of. That is, the latter leads asymptotically to the equilibrium state for temperature T and energy (8.44) with J_{NNN} =const. The situation is less simple otherwise, namely, when J_{NNN} is sampled from a more general distribution $p(J_{NNN})$ at each step of the evolution.

Quasi canonical cases for arbitrary d

Consider the rate function in (7.48), namely, $\varphi(X) \propto \exp\left(-\frac{1}{2}X\right)$, which is a soft case as defined in §7.8. The steady state implied asymptotically is $P_{st}(\mathbf{s}) = Z^{-1} \exp[-E(\mathbf{s})]$ with $Z = \sum_{\mathbf{s}} \exp[-E(\mathbf{s})]$. Here, $E(\mathbf{s})$ has the structure (8.44) but involves different coupling constants, in general; one may express this by writing $E(\mathbf{s}) = H(\mathbf{s}; K_1; K_2)$, and one obtains

$$K_1 = \beta J, \quad \tanh(2K_2) = \frac{[[\sinh(2\beta J_{NNN})]]}{[[\cosh(2\beta J_{NNN})]]}, \tag{8.45}$$

where $[[\cdots]]$ denotes the average with $p(J_{NNN})$ in (8.43). This steady state is unique given that (global) detailed balance holds when (7.48) is used in (8.43). The proof is a matter of algebra (a simple way to verify this is by writing $E(\mathbf{s}) = H(\mathbf{s}; K_1; K_2)$, and looking for the result (8.45) after requiring that $c(\mathbf{s^r} \mid \mathbf{s}) \exp[-E(\mathbf{s})] = c(\mathbf{s} \mid \mathbf{s^r}) \exp[-E(\mathbf{s^r})]$). This quasi canonical property holds for any d for the choice (7.48). Almost any other function $\varphi(X)$ would imply full nonequilibrium behavior in general, however. This occurs, for instance, if $\varphi(X)$ in (8.43) corresponds to the Metropolis algorithm (7.51).

The above result allows a detailed analytical study of this d-dimensional case (which cannot be made for other rates). Consider first $d = 1$, the solution of which does not require any simplifying assumption; furthermore, the ANNNI model is then expected to exhibit most relevant features given its axial nature for any d. A specific question is the phase diagram implied by (8.45), and its comparison with the one for the ordinary, quenched case with J_{NNN} =const. A way to address this is by noticing the formal similarity between the two problems, and using consequently the known results for the quenched case by Stephenson (1970), Hornreich *et al.* (1979), and Tanaka, Morita & Hiroike (1987). That is, one may write $\mathbf{Z} = \mathrm{Tr}\left(\mathbf{M}^N\right)$ for a chain of N spins with periodic boundary conditions, where $\mathbf{M}$ is the transfer matrix of eigenvalues

$$\left.\begin{aligned} \lambda_{1,2} &= e^{K_2}\left[\cosh K_1 \pm \sqrt{\sinh^2 K_1 + e^{-4K_2}}\right], \\ \lambda_{3,4} &= e^{K_2}\left[\sinh K_1 \pm \sqrt{\cosh^2 K_1 - e^{-4K_2}}\right]. \end{aligned}\right\} \tag{8.46}$$

The correlation function is

$$G(n) = \langle s_z s_{z+n} \rangle = \frac{1}{2\lambda_1^n}\left[(1+\varepsilon)\lambda_3^n + (1-\varepsilon)\lambda_4^n\right], \tag{8.47}$$

where $\varepsilon \equiv \frac{1}{2}\sinh(2K_1)\left[\left(\sinh^2 K_1 + e^{-4K_2}\right)\left(\cosh^2 K_1 - e^{-4K_2}\right)\right]^{-1/2}$. Therefore, one may distinguish two cases:

(i) For $\cosh K_1 > e^{-2K_2}$, the four eigenvalues of $\mathbf{M}$ are real. One obtains $G(n) \sim \frac{1}{2}(1+\varepsilon)\left(\lambda_3/\lambda_1\right)^n$ as $n \to \infty$, and the inverse correlation length is $\xi^{-1} \sim \ln\left(\lambda_1/\lambda_3\right)$.

(ii) For $\cosh K_1 < e^{-2K_2}$, two eigenvalues of $\mathbf{M}$ are real, and the other two are complex. Then $G(n) \sim |1+\varepsilon|\,|\lambda_3/\lambda_1|^n \cos(nk+\theta)$, where

$k \equiv \tanh^{-1}(\mathrm{Im}\lambda_3/\mathrm{Re}\lambda_3)$, $\theta \equiv \tanh^{-1}[\mathrm{Im}(1+\varepsilon)/\mathrm{Re}(1+\varepsilon)]$, and $\xi^{-1} \sim \ln|\lambda_1/\lambda_3|$.

The quenched ANNNI model simply corresponds to $K_1 = \beta J > 0$ and $K_2 = \beta J_{NNN}$, with J_{NNN} a given constant. This implies that $G(n)$ has a pure exponential decay for (i), while it exhibits oscillations of wave vector k whose amplitude decays exponentially for (ii). The boundary between the two regions, known as the *disorder line,* occurs for β satisfying $\cosh(\beta J) = \exp(-2\beta J_{NNN})$; consequently, it is required that $J_{NNN} < 0$, and the boundary temperature depends on the value of $\alpha \equiv -J_{NNN}J^{-1} > 0$. It follows that

$$\xi \sim \begin{cases} \frac{1}{2}\epsilon^{-\nu} & \text{with } \nu = 1 \quad \text{for} \quad \alpha < \frac{1}{2} \\ 2\epsilon^{-\nu} & \text{with } \nu = \frac{1}{2} \quad \text{for} \quad \alpha > \frac{1}{2}, \end{cases} \tag{8.48}$$

where $\epsilon \equiv \exp[-2\beta|J + 2J_{NNN}|]$. For $\alpha = \frac{1}{2}$ and $T = 0$ (the so-called *frustration point*), the correlation length remains finite, which corresponds to the disappearance of the pure critical point due to competition between NN and NNN interactions.

The essentials of this picture persist for (8.45) but the details are more varied due to complex dependence of K_2 on temperature.[10] The situation now depends strongly on $p(J_{NNN})$. In order to parallel the results for the ordinary ANNNI case, let us illustrate the solution for

$$p(J_{NNN}) = p_+\delta(J_{NNN} - J_+) + p_-\delta(J_{NNN} + J_-) + (1-p)\,\delta(J_{NNN}), \tag{8.49}$$

where $J_+, J_- > 0$, and $p = p_+ + p_-$. Condition $\beta \to \infty$ leads to $\exp(-4K_2) \sim (p_-/p_+)\exp[2\beta(J_- - J_+)]$, and $\beta \to 0$ leads to $K_2 \sim \beta(p_+J_+ - p_-J_-)$. That is, this version of the kinetic ANNNI system behaves at high temperature as the quenched case with $J_{NNN} = p_+J_+ - p_-J_-$, while the low-temperature behavior is equivalent to having $J_{NNN} = \frac{1}{2}(J_+ - J_-)$ instead. It is also noteworthy that the effective NNN interaction $K_2(T)$ may change sign. That is, a changeover from ferromagnetic to antiferromagnetic effective coupling occurs for both low and high enough T ; at low T, ferromagnetism is exhibited for $\eta \equiv -\frac{1}{2}(J_+ - J_-)J^{-1} < 0$, while $\eta > 0$ leads to antiferromagnetic behavior; at high T, one has ferromagnetism or antiferromagnetism according to whether $\gamma \equiv J_+/J_- > p_-/p_+$ or $\gamma < p_-/p_+$, respectively. This and, more generally, the complex dependence of $\beta^{-1}K_2$ on T are the reasons for the qualitative differences observed below between the quenched and nonequilibrium models.

---

[10] Furthermore, a more varied situation ensues if the choice for $\varphi(X)$ in (8.43) is other than (7.48), i.e., if GDB and (8.45) do not hold.

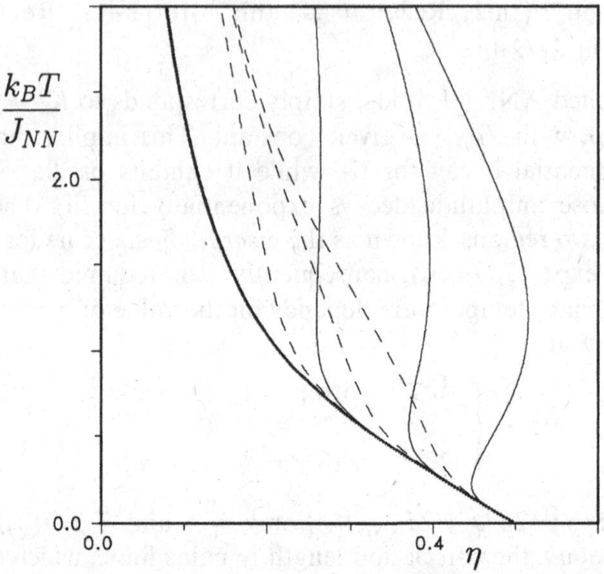

Fig. 8.17. The *disorder line,* which separates a region to the left in which correlations exhibit pure exponentially decay with distance (i.e., $k \equiv 0$) from a region to the right in which spatial correlations decay oscillating with wave vector k, for the one-dimensional kinetic ANNNI system. The bold full line is for the ordinary ANNNI model; the broken lines are for $p_+ = p_- = 0.5$ and, from left to right, for $\gamma \equiv J_+/J_- = 0.75$, 0.5, and 0.25; the solid lines are for the same values of γ but for $p_+ = p_- = 0.1$. Here, $\eta \equiv -\frac{1}{2}(J_+ - J_-)J^{-1}$. (From López-Lacomba & Marro (1994a).)

The only close similarity between these two cases occurs for $\eta > 0$ and $p_+ = p_- = \frac{1}{2}$. The ordinary ANNNI model is then recovered with $J_{NNN} = \frac{1}{2}(J_+ - J_-)$ for any temperature, i.e., $\beta^{-1}K_2$ becomes independent of T for this case. Otherwise, the differences are important even for the simple choices (7.48) and (8.49). Figures 8.17 and 8.18 illustrate the phase diagram as one varies parameters γ, p_+ and p_-. The situation in figure 8.17 is relatively conventional. It is surprising, however, that the disorder line for $p_+ = p_- = 0.1$ (but not for $p_+ = p_- = \frac{1}{4}$) indicates the existence, for certain values of η, of two (three if $\gamma = \frac{1}{4}$) transitions between normal and modulated phases. More intriguing is the situation in figure 8.18, where different curves correspond to different relations between γ and p_-/p_+; K_2 is antiferromagnetic for $\gamma < 0.5$ but ferromagnetic for $\gamma > 0.5$ at high enough temperature (whereas it always remains antiferromagnetic at low T). This is similar to the case in figure 8.17 for $\eta \simeq 0.5$, while the system can only exhibit modulations at low enough T and becomes *normal* for high T if $\eta \gg 0.5$ and γ is large enough. It would be interesting to know

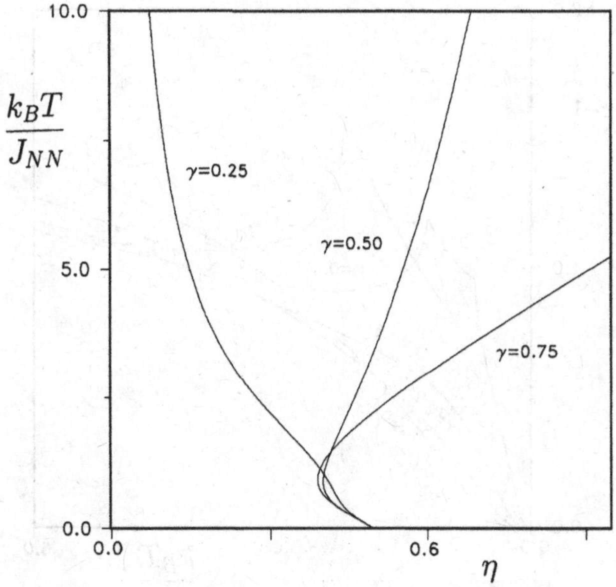

Fig. 8.18. The same as figure 8.17 but for $p_+ = 0.5$ and $p_- = 0.25$ for indicated values of γ.

whether this behavior, which does not have a counterpart in the quenched model, occurs in nature. Figure 8.19 illustrates the correlation length for $\gamma = 0.75$. It is notable that correlations for $\eta = 0.3$ remain monotonic for any T, while for $\eta = 0.4$ a transition occurs to modulated, and then back to monotonic behavior, as T is increased, and both the second transition and the critical point do not occur for $\eta = 0.3$.

The most interesting impure behavior occurs for $\beta \to \infty$, as suggested already by figure 8.19, near the origin. More explicitly, one obtains $K_2 \approx \frac{1}{4} \ln (p_+/p_-) + \frac{1}{2}\beta (J_+ - J_-)$ as $\beta \to \infty$. Therefore, the correlation length diverges as

$$\xi \sim \begin{cases} \dfrac{p_+}{2p_-}\epsilon^{-1} & \text{for} \quad \eta > 0.5 \\ \\ \dfrac{2p_-}{p_+}\epsilon^{-1/2} & \text{for} \quad \eta < 0.5, \end{cases} \tag{8.50}$$

where $\epsilon \equiv \exp\left[-2\beta \left|J_+ - J_- + J\right|\right]$. A correspondence between these cases and the canonical ones in (8.48) can be established if one introduces a *distance* ϵ to the critical point, which is system-dependent. There is also an interesting fact concerning the *frustration point* for $\eta = 0.5$ (see figure 8.19): The usual (thermal) critical point does not occur, but one

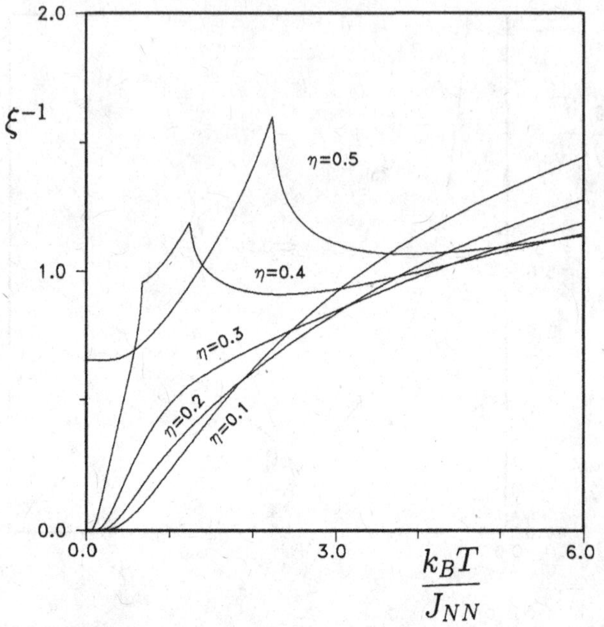

Fig. 8.19. The inverse correlation length as a function of $(\beta J)^{-1}$for the one-dimensional kinetic ANNNI model for $p_+ = 0.5$, $p_- = 0.25$, and $\gamma = 0.75$, for different values of η. The maxima locate the disorder line where the decay of correlations changes from monotonic to oscillatory. The *frustration point* corresponds to $\eta = 0.5$, where no thermal critical point occurs (see the main text, however).

gets

$$\xi^{-1} \sim \begin{cases} \ln \dfrac{1+\sqrt{1+\varkappa}}{1+\sqrt{1-\varkappa}} & \text{for } \varkappa < 1 \\ \ln\left(1+\sqrt{2}\right) & \text{for } \varkappa = 1 \\ \ln\left\{\sqrt{\varkappa^{-1}}+\sqrt{1+\varkappa^{-1}}\right\} & \text{for } \varkappa > 1, \end{cases} \tag{8.51}$$

where $\varkappa \equiv 4p_-/p_+$. That is, *percolative* critical behavior occurs at zero temperature for either $p_+ \to 0$ or $p_- \to 0$ (only) if $p_+ \neq 4p_-$. Other thermodynamic quantities may easily be obtained from (8.45) and (8.46) by standard methods.

Some information follows from (8.45) for $d > 1$, for example, after using known results for the ordinary case. It is known from both numerical and approximate analytical methods that the two-dimensional quenched ANNNI model may exhibit four different phases:

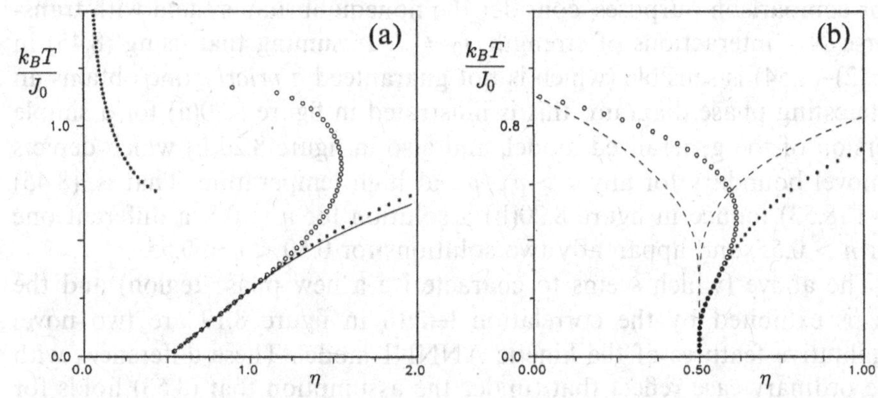

Fig. 8.20. The phase diagram for the two-dimensional kinetic ANNNI system obtained for $J = J_0/10$ after using the exact result (8.45) in equations (8.52) (empty circles), (8.53) (full line), and (8.54) (asterisks). (a) The case $\gamma = 0.75$, and $p_+ = p_- = 0.25$; the broken line corresponds to the ordinary ANNNI model. (b) The same but $p_+ = 0.5$ and $p_- = 0.25$; see the main text for details. (From López-Lacomba & Marro (1994a).)

(1) a paramagnetic phase, for any $\alpha \equiv -J_{NNN}J^{-1}$ at high enough T;

(2) a ferromagnetic phase, for $\alpha < \frac{1}{2}$ at low T;

(3) the [2, 2] antiphase in which two spins up alternate with two spins down in the ground state, for $\alpha > 0.5$ at low T (see figure 8.20(a) where the boundaries between these three phases are represented by dashed lines); and

(4) an incommensurate modulated phase with continuously varying wavevector, for $0.5 < \alpha < 0.75$ at moderate T between (1) and (3).

The boundaries of (2) and (3) have been estimated (Hornreich *et al.* 1979, Kroemer & Pesch 1982) for $d = 2$ as

$$\sinh(2K_1 + 4K_2)\sin(2K_0) = 1, \tag{8.52}$$

and

$$\exp(2K_0) = \frac{1 - \exp(4K_2)}{[1 - \exp(-K_1 + 2K_2)]\,[1 - \exp(K_1 + 2K_2)]}, \tag{8.53}$$

respectively, if transverse NN interactions are of strength $J_0 \neq J$. Here, $K_1 = \beta J$, $K_2 = \beta J_{NNN}$, and $K_0 = \beta J_0$. Villain & Back (1981) confirmed the above, and concluded that the boundary between (4) and (3) at low enough temperature is characterized by

$$K_1 + 2K_2 = -2\exp(-2K_0). \tag{8.54}$$

For comparison purposes, consider the nonequilibrium system with transverse NN interactions of strength $J_0 \neq J$. Assuming that using (8.45) in (8.52)–(8.54) is sensible (which is not guaranteed *a priori*), one obtains an interesting phase diagram; this is illustrated in figure 8.20(a) for a simple version of the generalized model, and also in figure 8.20(b) which depicts a novel boundary for any $\gamma > p_+/p_-$ at high temperature. That is, (8.45) and (8.53) induce in figure 8.20(b) a solution for $\eta < 0.5$, a different one for $\eta > 0.55$, and apparently two solutions for $0.50 < \eta < 0.55$.

The above (which seems to characterize a new phase region) and the cusps exhibited by the correlation length in figure 8.19 are two novel qualitative features of the kinetic ANNNI model. These differences with the ordinary case reflect that (under the assumption that (8.53) holds for the kinetic model) the effective NNN interaction, K_2, depends on T, η, and γ, and may even change sign. That is, the resulting situation is rather involved even though it corresponds to the choice $\varphi(X) \propto \exp\left(-\frac{1}{2}X\right)$, which produces factorization of interactions between the given spin and its surroundings (§7.8). As a matter of fact, two different temperatures may be defined here, namely, the canonical temperature T, which is associated with the NN term of $E(\mathbf{s})$, and K_2/β associated with the NNN term. Consequently, even this apparently simple case of the kinetic model cannot be characterized by a canonical steady distribution function. (This is in contrast with the situation for the ordinary one-dimensional system for which one may interpret K_2/β as an effective temperature and J_{NN} as an external field after making the substitution $s_z s_{z+1} \rightarrow \sigma_z$.)

8.5 Proton glasses and neural noise

The conflicting dynamics considered above assumes that one of the parameters characterizing the potential energy is a random variable, which aims to represent ideally diffusion of impurities or the action of an external agent. It is natural to generalize this by assuming that the competition involves two or more parameters. Consider, for example, the Ising Hamiltonian (7.5) and the possibility that both the interaction strength, J, and the applied field, h, may vary at random during the evolution or, more precisely, (7.2)–(7.4) with $p(\vec{\zeta}) = f(J)\,g(h)$, where $f(J)$ and $g(h)$ are normalized distributions. The symmetry (7.4) does not necessarily imply detailed balance for the superposition (7.3) and, consequently, a nonequilibrium steady state is reached asymptotically. In particular, the nonequilibrium Ising glass in §8.1 corresponds to $p(\vec{\zeta}) = f(J)\,\delta(h)$ with a realization for $f(J)$ such as the one in (8.10), for instance, and the NRFM in §7.1 ensues for $p(\vec{\zeta}) = \delta(J)\,g(h)$ with some appropriate realization for $g(h)$. Therefore, one may imagine two simple realizations of the general case of interest here: (i) a spin-glass material exposed to a random (or very rapidly fluctuating) magnetic field, or (ii) a disordered system with fast diffusion of impurities consisting of both exchanges (as in the nonequilibrium Ising glass) and local fields (as in the NRFM). One expects that such situations may bear some relevance for randomly diluted magnetic or electric alloys, in which random local fields are reported to be always present along with random interactions, and for neural networks. For example, a close correlation may be drawn between the situation in (ii) and the one in *proton glasses* (Cowley, Ryan, & Courtens 1986, Hutton *et al.* 1991, Oresic & Pirc 1994); a similar situation occurs in models with neuronal activity noise (Garrido & Marro 1991b, Ma *et al.* 1992, Torres *et al.* 1997) whose study requires more realistic synaptic couplings, which are allowed by the present formalism.

The emergent behavior, including critical properties, which in some cases reminds one of the situation for the NRFM in chapter 7, has been described in López-Lacomba & Marro (1994a) for various dynamics. An interesting case corresponds to the one-dimensional lattice evolving *via* the elementary rate

$$c\left(\mathbf{s};\vec{\zeta};r\right) = 1 - c_1 s_r + c_2\left(s_{r-1} + s_{r+1}\right) - c_3 s_r\left(s_{r-1} + s_{r+1}\right) + c_4 s_{r-1} s_r s_{r+1}, \tag{8.55}$$

$r = 1, 2, \ldots$, with the Glauber coefficients (see equations (7.42) and (7.52) and table 7.1 in §7.4):

$$\left.\begin{aligned} c_1 = \tanh(\beta h), \quad c_2 = \tfrac{1}{2}\tanh(\beta h)\tanh(2\beta J), \\ c_3 = \tfrac{1}{2}\tanh(2\beta J), \quad c_4 = 0. \end{aligned}\right\} \tag{8.56}$$

Assuming distributions

$$g(h) = \frac{p_0}{2}\delta(h - (\eta + \kappa)) + \frac{p_0}{2}\delta(h - (\eta - \kappa)) + (1 - p_0)\delta(h - \eta) \tag{8.57}$$

and

$$f(J) = \frac{q_0}{2}\delta(J - (j + \imath)) + \frac{q_0}{2}\delta(J - (j - \imath)) + (1 - q_0)\delta(J - j), \tag{8.58}$$

with $j > \imath > 0$ (which implies that $f(J) = 0$ for all $J \leqslant 0$), the method in §§7.5 and 7.6 indicates that, in general, a nontrivial function $E(\mathbf{s})$ exists, defined as in (7.60) and (7.68). For $\eta \geq \kappa$ no critical point exists, while interesting critical behavior may be exhibited otherwise. That is, it follows for $\kappa > \eta > 0$ that

$$\xi \sim \exp\left(2\beta \min\{j - \imath, \kappa - \eta\}\right) \tag{8.59}$$

as $\beta \to \infty$, and ξ diverges also in some cases as $q_0 \to 0$.

The d-dimensional case (§7.8)

$$c\left(\mathbf{s};\vec{\zeta};\mathbf{r}\right) \propto \left(e^{h}\prod_{\hat{\mathbf{u}}} e^{J_{\hat{\mathbf{u}}}s_{\mathbf{r}+\hat{\mathbf{u}}}}\right)^{-\beta s_{\mathbf{r}}}, \tag{8.60}$$

where a product over the NN sites of $\mathbf{r}$ (denoted $\mathbf{r}+\hat{\mathbf{u}}$ with $\hat{\mathbf{u}} = \pm\hat{\imath}, \pm\hat{\jmath}, \pm\hat{\mathbf{k}}$) is indicated, may also be worked out in detail. That is, under assumption (7.60) one obtains for this rate that

$$E(\mathbf{s}) = -\sum_{\mathbf{r}} s_{\mathbf{r}} \left[\sum_{\hat{\mathbf{u}}} K_{\hat{\mathbf{u}}} s_{\mathbf{r}+\hat{\mathbf{u}}} + h_e\right], \tag{8.61}$$

with

$$\tanh(h_e) = \frac{\langle \sinh(\beta J)\rangle}{\langle \cosh(\beta J)\rangle} \tag{8.62}$$

and

$$\tanh(K_{\hat{\mathbf{u}}}) = \frac{\langle \sinh(\beta J_{\hat{\mathbf{u}}})\rangle}{\langle \cosh(\beta J_{\hat{\mathbf{u}}})\rangle}. \tag{8.63}$$

This unique result is a consequence of the fact that the effective rate, (7.3), which is implied by (8.60) is simply proportional to

$$\left[\left\langle e^{h}\right\rangle \prod_{\hat{\mathbf{u}}} \exp(J_{\hat{\mathbf{u}}} s_{\mathbf{r}+\hat{\mathbf{u}}})\right]^{-\beta s_{\mathbf{r}}}. \tag{8.64}$$

Therefore, a lack of correlations between bonds with different neighbors is required, as stated by $p(\vec{\zeta}) = \tilde{p}(h, J_{\hat{\mathbf{u}}}) = g(h)\prod_{\hat{\mathbf{u}}} f_{\hat{\mathbf{u}}}(J_{\hat{\mathbf{u}}})$. In other words, only products of NN pairs of spins are involved in the rate, even for $d > 1$, in such a way that a simple factorization is allowed which in a sense amounts to disregarding, for some distributions $p(\vec{\zeta})$, all but two-body correlations within the dynamical process. In spite of this simple, *quasi canonical* structure, no straightforward conclusion is allowed in this case concerning, for example, phase transitions: getting more information would require the use of approximate methods. Further situations for which the nonequilibrium steady-state distribution can be obtained exactly for $d > 1$ have been studied in Garrido & Muñoz (1995, Muñoz & Garrido 1994), see also §§7.8 and 8.4)

8.6 Interpreting thermal baths

The concept of a thermal bath is insensitive to microscopic details, and may, in most instances, simply be characterized by the temperature parameter β. The thermal bath that enters the definition of many microscopic systems in this book may be imagined as consisting of a set of variables, $\vec{\zeta}$, or *phonon subsystem*. This is in some way coupled to the *spin subsystem* of energy $H_{spin}(\mathbf{s})$, for example, the Ising Hamiltonian (7.1), thus inducing stochastic changes such as the ones described by the master equation, (7.2). If a tendency to evolve with time towards the Gibbs state $\mu_{st}(\mathbf{s})$ needs to be guaranteed, it is sufficient to require detailed balance, as in (7.40). Further details are irrelevant and, therefore, one may imagine that the bath microscopic state changes stochastically on a microscopic time scale as, for example, stated in (7.11). One has no criteria to specify the associated transition rate $\varpi(\vec{\zeta} \to \vec{\zeta}' \mid \mathbf{s})$. However, one case of the generalized model introduced in §7.2 may be regarded as the (microscopic) realization of a thermal bath (Garrido & Marro 1994).

Consider (7.9)–(7.11) with rates $c(\mathbf{s};\vec{\zeta};\mathbf{r})$ and $\varpi(\vec{\zeta} \to \vec{\zeta}' \mid \mathbf{s})$ whose specific form is unknown *a priori*. As discussed in §7.3, these imply within the limit $\Gamma \to \infty$ that $\vec{\zeta}$ evolves on a microscopic time scale towards its stationary state, while $\mathbf{s}$ is driven on a macroscopic time scale by an effective stochastic process, as in (7.26), which is characterized by the rate $c_{eff}(\mathbf{s} \to \mathbf{s}^{\mathbf{r}})$. Let us assume also that this rate satisfies detailed balance, and $\mu_{st}(\mathbf{s}) \propto \exp\left[-\beta H_{spin}(\mathbf{s})\right]$ is reached asymptotically in time for $\Gamma \to \infty$. Furthermore, one may assume for simplicity that: (a) $\vec{\zeta}$ evolves independently of $\mathbf{s}$, as stated in (7.12), and (b) only transitions $\mathbf{s} \to \mathbf{s}^{\mathbf{r}}$ decreasing the total energy for the given $\vec{\zeta}$ are allowed.

The total energy is $H(\mathbf{s},\vec{\zeta}) = H_{spin}(\mathbf{s}) + H_{bath}(\vec{\zeta}) + H_{int}(\mathbf{s},\vec{\zeta})$, where $H_{bath}(\vec{\zeta})$ is the phonon subsystem energy and $H_{int}(\mathbf{s},\vec{\zeta})$ is the energy associated with the interaction between the phonon and spin subsystems. One may write for the sake of simplicity that

$$H_{int}\left(\mathbf{s},\vec{\zeta}\right) = -\frac{1}{2}\lambda^{-1}\zeta\sum_{\mathbf{r}} s_{\mathbf{r}}, \tag{8.65}$$

i.e., the phonon subsystem is characterized by a single variable, ζ, and λ measures the spin–phonon coupling. Consistent with (b), one may require that

$$c\left(\mathbf{s};\vec{\zeta};\mathbf{r}\right) = f_{\mathbf{r}}\left(\mathbf{s};\vec{\zeta}\right)\left\{1 - \operatorname{sgn}\left[H\left(\mathbf{s}^{\mathbf{r}};\vec{\zeta}\right) - H\left(\mathbf{s};\vec{\zeta}\right)\right]\right\}, \tag{8.66}$$

where $\operatorname{sgn}(x) = 1$ for $x \geq 0$ and $\operatorname{sgn}(x) = -1$ for $x < 0$. On the other hand, one may assume that, if isolated, i.e., $H_{int}(\mathbf{s},\vec{\zeta}) = 0$, the steady-state distribution of the bath is

$$P_{st}(\vec{\zeta}) \propto \exp\left[-H_{bath}(\vec{\zeta})\right], \tag{8.67}$$

and the associated dynamics may be obtained from the detailed balance condition, namely,

$$\varpi\left(\vec{\zeta} \to \vec{\zeta}'\right) P_{st}\left(\vec{\zeta}\right) = \varpi\left(\vec{\zeta}' \to \vec{\zeta}\right) P_{st}\left(\vec{\zeta}'\right). \tag{8.68}$$

Under the above conditions, the unknown rates may be specified. That is, (a) and (b), as well as (8.65)–(8.68), are convenient to determine a specific family of transition rates out of infinitely many leading asymptotically to the steady state $\mu_{st}(\mathbf{s}) \propto \exp\left[-\beta H_{spin}(\mathbf{s})\right]$. However, one still needs to determine $f_{\mathbf{r}}(\mathbf{s};\vec{\zeta})$ and $H_{bath}(\vec{\zeta})$ which characterize completely processes (8.66) and (8.68), respectively. With this aim, one realizes that the fact that $c_{eff}(\mathbf{s} \to \mathbf{s}^{\mathbf{r}})$ needs to satisfy detailed balance for $\Gamma \to \infty$ implies a relation between $f_{\mathbf{r}}(\mathbf{s};\vec{\zeta})$ and $H_{bath}(\vec{\zeta})$, namely,

$$f_{\mathbf{r}}(\mathbf{s};\vec{\zeta}) = f(\mathbf{s};\mathbf{r}) \left[P_{st}(\vec{\zeta})\,\mathrm{e}^{\zeta}\left(1+\mathrm{e}^{-\zeta}\right)^{2}\right]^{-1}, \tag{8.69}$$

where

$$f(\mathbf{s};\mathbf{r}) = \int_{-\infty}^{+\infty} \mathrm{d}\vec{\zeta}\, P_{st}(\vec{\zeta}) f_{\mathbf{r}}\left(\mathbf{s};\vec{\zeta}\right). \tag{8.70}$$

Therefore, the type of thermal bath, namely, $H_{bath}(\vec{\zeta})$, determines the microscopic mechanisms. For example, the choice $P_{st}(\vec{\zeta}) = \left[2\cosh\left(\zeta/2\right)\right]^{-2}$ immediately leads to

$$c\left(\mathbf{s};\vec{\zeta};\mathbf{r}\right) = f(\mathbf{s};\mathbf{r})\left(1 - \operatorname{sgn}\left\{\lambda\left[H_{spin}(\mathbf{s}^{\mathbf{r}}) - H_{spin}(\mathbf{s})\right] + \zeta s_{\mathbf{r}}\right\}\right) \tag{8.71}$$

and

$$\varpi\left(\vec{\zeta} \to \vec{\zeta}'\right) = \varphi\left(2\ln\frac{\cosh\zeta/2}{\cosh\zeta'/2}\right), \tag{8.72}$$

where $\varphi(x) = \varphi(-x)\,\mathrm{e}^{-x}$. That is, (7.9)–(7.11) with (8.71)–(8.72) is a realization of the familiar concept of a thermal bath for $\Gamma \to \infty$. It would be interesting to study the generalization of this for finite values of Γ, rendering the time scales for the evolution of $\mathbf{s}$ and $\vec{\zeta}$ comparable to one another.

9
Particle reaction models

When we extend the basic CP, a host of intriguing questions arise: How do multi-particle rules and diffusion affect the phase diagram? When should we expect a first-order transition? Can multiple absorbing configurations or conservation laws change the critical behavior? In this chapter we examine a diverse collection of models whose behavior yields some insight into these issues.

9.1 Multiparticle rules and diffusion

In the CP described in chapter 6, the elementary events (creation and annihilation) involve single particles. What happens if one of the elementary events involves a cluster of two or more particles? Consider pairwise annihilation: in place of $\bullet \to \circ$ (as in the CP), $\bullet\,\bullet \to \circ\,\circ$. (In other words, a pair of particles at neighboring sites can annihilate one another, but there is no annihilation of isolated particles.) In fact, we have already seen one instance of pairwise annihilation: in the ZGB model (chapter 5), adsorption of O_2 destroys a pair of vacancies. The transition to the absorbing O-saturated state corresponds to (and belongs to the same class as) the CP. Another example is a version of the CP (the 'A2' model), in which annihilation is pairwise (Dickman 1989a,b). Here again the transition to the absorbing state is continuous, and the critical exponents are the same as for the CP.

Evidently pairwise annihilation does not alter the nature of the phase transition in the CP, any more than diffusion does (see §6.7). However, consider the effect of diffusion in a model with pairwise annihilation. If the density is low, and diffusion is so rapid that the density of pairs is $\mathcal{O}(\rho^2)$, then particles rarely annihilate, and might be expected to survive even for very small λ. The pair annihilation model (PAM) provides a test of this notion. In this model, hopping, creation, and annihilation occur with probabilities D, $(1-D)\lambda/(1+\lambda)$, and $(1-D)/(1+\lambda)$, respectively. (Creation and hopping events are the same as in the DCP.) In an annihilation event, a NN pair of sites is selected at random, and if both are occupied, the two particles are removed.

To begin we work out the mean-field theory (at the pair level — see §6.7), for this model on a d-dimensional sc lattice. Consider the transition (• ○) → (○ •), and its mirror image. With n_1 and n_2 representing the number of occupied neighbors in the original and final positions, respectively, the rate for such transitions (per bond) is

$$R = 2D\binom{2d-1}{n_1}\binom{2d-1}{n_2}\frac{u^{n_1}(\rho-u)^{2d-n_1+n_2}(1-2\rho+u)^{2d-n_2-1}}{[\rho(1-\rho)]^{2d-1}}, \tag{9.1}$$

where ρ and u represent the particle and doubly-occupied pair densities, respectively. The change in the number of doubly-occupied NN pairs is $\Delta N_p = n_2 - n_1$. Summing over n_1 and n_2, and recalling that $u = N_p/Nd$, one finds the rate of change of u due to diffusion. Adding the contributions due to creation and annihilation, we obtain the mean-field equations:

$$\frac{d\rho}{dt} = \tilde{\lambda}(\rho - u) - 2(1-\tilde{\lambda})u, \tag{9.2}$$

and

$$\frac{du}{dt} = \frac{\tilde{\lambda}}{d}\frac{(\rho-u)}{1-\rho}[2d(\rho-u)+1-2\rho+u] - \frac{(1-\tilde{\lambda})u}{d}\left[1+\frac{2(2d-1)u}{\rho}\right] + \frac{2\tilde{D}(2d-1)}{d}\frac{(\rho-u)(\rho^2-u)}{\rho(1-\rho)}, \tag{9.3}$$

where $\tilde{\lambda} \equiv \lambda/(1+\lambda)$, $\tilde{D} \equiv D/(1-D)$, and a factor of $1-D$ has been absorbed into a rescaled time variable. The active steady-state solution is

$$\overline{\rho} = \frac{\lambda}{\lambda+2}\frac{F-2}{F}, \tag{9.4}$$

where $F(\lambda,\tilde{D}) \equiv (4d-3)\lambda + 4\tilde{D}(2d-1)(1+\lambda)$. The stationary NN pair density is

$$\overline{u} = \frac{\lambda}{\lambda+2}\overline{\rho}. \tag{9.5}$$

This solution exists for $\lambda > \lambda_c$, which is a decreasing function of the diffusion rate D, falling to zero for $D > D_c(d)$. In one dimension, for example,

$$\lambda_c = \begin{cases} 2(1-3D)/(1+3D), & D < 1/3 \\ 0, & D > 1/3. \end{cases} \tag{9.6}$$

This result suggests that competition between diffusion and another elementary process can change the phase diagram qualitatively, at a *finite* diffusion rate, as found for reaction–diffusion models in chapter 4. Katori & Konno (1992) showed that in one dimension, the mean-field picture is *incorrect*: λ_c is strictly greater than zero for any finite D. While the validity of the simple mean-field approach is certainly questionable, the results motivate our studying other multi-particle processes.

More dramatic changes are observed in the diffusive *triplet annihilation* model (TAM) (Dickman 1989b, 1990b). The only difference from the PAM rules is that the annihilation process is: • • • → ∘ ∘ ∘. Pair mean-field theory predicts a reentrant phase diagram (see §7.8 for a discussion of a reentrant phase diagram in a spin model). In one dimension, the mean-field equations are

$$\frac{d\rho}{dt} = \tilde{\lambda}(\rho - u) - \frac{3(1-\tilde{\lambda})u^2}{\rho}, \tag{9.7}$$

and

$$\frac{du}{dt} = \frac{\tilde{\lambda}(\rho - u)(1-u)}{1-\rho} - \frac{2(1-\tilde{\lambda})u^2(\rho+u)}{\rho^2} + \tilde{D}\frac{(\rho-u)(\rho^2-u)}{\rho(1-\rho)}. \tag{9.8}$$

For small D an active state exists for λ *outside* $[\lambda_-, \lambda_+]$, where

$$\lambda_\pm = \frac{3}{10}\left[1 - 6\tilde{D} \pm \sqrt{1 - 12\tilde{D} - 44\tilde{D}^2}\right]. \tag{9.9}$$

For $D = D_c \simeq 0.0672$ the solutions coalesce, and for $D > D_c$ an active solution is possible for *any* positive λ; the population survives for arbitrarily small creation rates. Below D_c, survival is possible provided λ is either sufficiently large or sufficiently *small.* The former case is analogous to the phase transition in the CP. However, the small-λ active phase is something new. Here particles evade extinction by reproducing at a low rate, allowing sufficient time for clusters to disperse between creation events. These rather surprising predictions are borne out by simulations, as illustrated in figure 9.1.

One might expect that changing the annihilation step — making the rate proportional to the third power of density, in the mean-field approximation — would change the critical exponents. However, the entire phase boundary is a line of DP-like critical points (Dickman 1990b). (Precisely what happens as one approaches the origin of the λ–D plane has yet to be explored.) For $D > 0$ the transition at $\lambda = 0$ is mean-field-like, with $\overline{\rho} \propto \lambda^2$, as expected, since the system is thoroughly mixed between successive creation events. (In the site approximation, $d\rho/dt \approx \tilde{\lambda}\rho - 3\rho^3$ for small ρ.) We can draw several lessons from this example. First, the annihilation rate (away from $\lambda = 0$) is proportional to ρ not ρ^3, because the

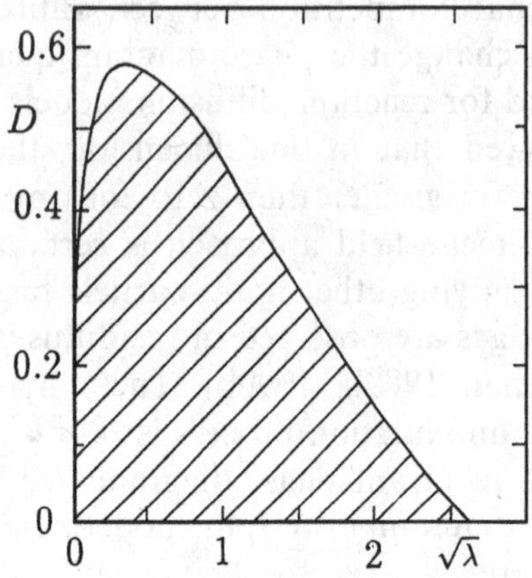

Fig. 9.1. Phase diagram of the one-dimensional TAM from simulations. There is no active steady state within the shaded region.

creation process generates strong correlations among particle positions. The second point concerns the continuum description of the TAM, along the lines of equation (6.38). As we saw, the transition from the vacuum to the active state occurs when the coefficient, $a(\lambda, D)$, of the linear term exceeds a critical value. For $D < D_c$, this coefficient is not monotonic, i.e., it must be a *decreasing* function of the creation rate as $\lambda \to \lambda_-$. Finally, it is interesting that in one dimension, diffusive breakup of clusters (at a finite hopping rate), can drive λ_c to zero when annihilation involves trimers, in contrast with pairwise annihilation, for which Katori & Konno (1992) established that λ_c is always greater than zero.

9.2 First-order transitions

All of the single-component models considered so far show continuous transitions. A number of two-component systems — the ZGB model, and the monomer–monomer or AB model for example — have discontinuous transitions. Can a single-component model have a first-order transition into an absorbing state? At first glance this appears straightforward. If we arrange for low-density states to be intrinsically unstable whilst high-density states are viable, then we should observe a first-order transition into the vacuum. A plausible way to destabilize low-density states is to require *pairs* of particles for creation, so that the rate of particle production is $\propto \rho^2$ in mean-field approximation. This is the rationale for Schlögl's 'second model,' in which two particles are required to create a third (Schlögl 1972).

Schlögl's model was originally formulated at the level of rate equations.

A simple lattice model along these lines involves creation by a pair of adjacent particles (• • ∘ → • • •), at rate λ, and annihilation as in the basic CP. (Another pair-creation scheme, • ∘ • → • • •, yields rather different results: in one dimension there is no active steady state, since gaps of two or more sites can never be reoccupied.) The one-site mean-field equation for this model,

$$\frac{d\rho}{dt} = \lambda\rho^2(1-\rho) - \rho, \tag{9.10}$$

admits an active steady state only for $\lambda \geq 4$:

$$\overline{\rho} = \frac{1}{2}\left(1 + \sqrt{1 - 4/\lambda}\right). \tag{9.11}$$

At $\lambda = 4$ the stationary density jumps from 0 to 1/2. Thus the transition is first order if the system is so well stirred that correlations can be ignored. In the *pair* approximation this transition — still at $\lambda = 4$, in one dimension — is *continuous*, with $\overline{\rho} \approx \sqrt{\lambda - 4}$. However, if we allow diffusion at *any* nonzero rate, pair mean-field theory predicts a jump in density. These simple approximations are not very conclusive, and so we turn to simulations. The first to perform such a study was Grassberger (1982), who found that in one and two dimensions the transition is *continuous*, and in the DP class. (He did observe a first-order transition in a four-dimensional version of the model.) Continuity of the transition in one dimension, even at quite high diffusion rates ($D = 0.95$) has been confirmed in time-dependent simulations (Dickman & Tomé 1991).

The first single-component lattice model with a discontinuous transition to an absorbing state was devised by Bidaux, Boccara, & Chaté (BBC) (1989). Their model is a probabilistic cellular automaton in which the state (vacant or occupied) of site **x** at time $t+1$ depends on $S_{\mathbf{x}}$, the number of occupied sites in the set comprising **x** and its eight closest neighbors, at time t.[1] Consider first the deterministic cellular automaton rule

$$\sigma_{\mathbf{x}}(t+1) = \begin{cases} 1, & \text{if } 3 \leqslant S_{\mathbf{x}}(t) \leqslant 6 \\ 0, & \text{otherwise.} \end{cases} \tag{9.12}$$

This rule incorporates a threshold for creation, as well as saturation, averting a filled-lattice absorbing state. In mean-field approximation, the threshold causes low-density populations to die off. In the probabilistic cellular automaton, the rule given above is 'diluted' so that occupancy (for $S_{\mathbf{x}}$ within the stated range), occurs only with probability p, independently at each site and time. (Sites which were to become vacant according to equation (9.12) do so, regardless of dilution.) For small values of the

[1] In one dimension, this is the set $\{\sigma_{x-4}, ..., \sigma_{x+4}\}$ while in two dimensions it consists of the NNs and second neighbors of **x**.

creation probability p we expect the vacuum to be the unique stationary state. Mean-field theory predicts a discontinuous transition to an active state as p is increased. Simulations confirm the existence of a transition, but show that in one dimension the BBC model has a *continuous* transition (at $p_c \simeq 0.7216$), which is in the DP class (Jensen 1991). Bidaux *et al.* showed that the transition is first order for $d \geq 2$. Their method is worth commenting on.

Due to finite size, simulations of models with continuous transitions also show a jump in density: below a certain creation rate, depending on the system size, fluctuations into the vacuum become likely. This makes it hard to distinguish a weak first-order transition from a continuous one. However, discontinuous transitions are also attended by metastability and hysteresis; if two phases can coexist, there is a region of parameter space where one phase is metastable, and may persist until the globally stable phase is nucleated by a rare fluctuation. Thus hysteresis is a useful criterion for distinguishing a first-order transition in simulations. Transitions to an absorbing state present a special difficulty, however: the absorbing state admits no fluctuations! The solution is to introduce a small ($\mathcal{O}(10^{-2})$) probability, ϵ, of *spontaneous* creation. As we saw in chapter 6, a source removes critical singularities, much as an external field does in the Ising model. However, if the vacuum and active states are separated by a first-order transition, introducing a weak source leads to distinct high and low density phases. Bidaux *et al.* (1989) found hysteresis between high- and low-density phases as they varied p (at fixed ϵ).

For both the pair-creation and the BBC models there is a spatial dimension d' (two for BBC, four for pair creation), below which the transition is continuous, in qualitative disagreement with mean-field predictions. Long-range interactions and rapid diffusion also tend to drive a system toward mean-field-like behavior. Restricting our attention to short-range interactions and finite diffusion rates, we can ask whether any one-dimensional model exhibits a first-order transition to an absorbing state. It turns out that if creation depends on *three-particle* clusters, the transition to the vacuum can be first order.[2]

In the *triplet creation model* (Dickman & Tomé 1991), creation ($\bullet\bullet\bullet\circ \to \bullet\bullet\bullet\bullet$), annihilation ($\bullet \to \circ$), and hopping occur with probabilities $(1-D)\lambda/(1+\lambda)$, $(1-D)/(1+\lambda)$, and D, respectively. Simulations yield the phase diagram shown in figure 9.2: for $D < D_t \approx 0.87$, there is a line of DP-like critical points separating the active state and the vacuum. At

[2] Note the parallel with the TAM discussed above: in one dimension, the phase diagram does not change qualitatively when diffusion competes with pairwise creation or annihilation, but it does change dramatically when three-particle clusters are involved. This is presumably because reunions of three random walkers are much less frequent than reunions of pairs.

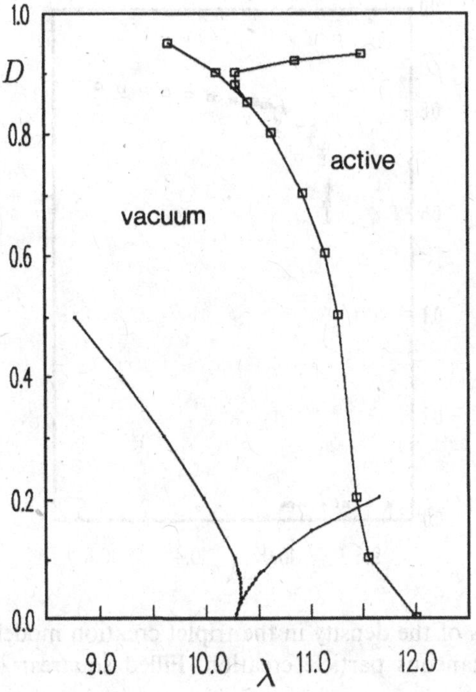

Fig. 9.2. Phase diagram of the triplet creation model in one dimension. Squares: λ_+, and λ_- from simulations. Solid lines (without squares): four-site mean-field theory prediction.

D_t, the critical line terminates at a tricritical point, the phase boundary branching into a pair of spinodals, $\lambda_+(D)$ and $\lambda_-(D)$, which represent stability limits for the two phases. For $\lambda < \lambda_-$, there is no active steady state, whilst only for $\lambda > \lambda_+$ is there any possibility of surviving, starting from an isolated particle. For λ between the spinodal values, the active steady state is attainable provided the initial density is sufficiently large. As in the BBC model, hysteresis between high- and low-density states is found (for $D > D_t$) when spontaneous creation is permitted at a small rate ϵ (see figure 9.3). Increasing ϵ should eventually lead to a critical point belonging to the Ising universality class (Tomé & Dickman 1993).

Does the transition in the one-dimensional BBC model become first order for sufficiently rapid diffusion? This question was studied by Ódor, Boccara, & Szabó (1993). They found that for NN hopping, the change to a first-order transition happens when the diffusion rate is about 40 times the reaction rate. For long-range hopping (particles may jump to any vacant site), this change occurs at a much smaller diffusion rate, about 0.015 times the reaction rate. The work on the diffusive BBC and

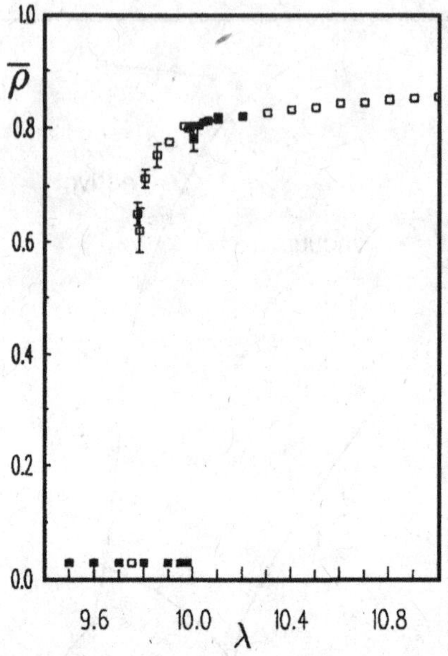

Fig. 9.3. Hysteresis of the density in the triplet creation model with a small rate (ϵ = 0.01) of spontaneous particle creation. Filled squares: λ increasing; open squares: λ decreasing.

triplet creation models suggests that to observe a first-order transition in a single-component, one-dimensional model, it is necessary to have a creation rule involving three (or more) particles, and a substantial rate of diffusion.

9.3 Multiple absorbing configurations: statics

Although we have so far considered systems with a single absorbing configuration, it is easy to devise models having a multitude of them; examples arise naturally in catalysis. Some of the questions motivating their study are: What are typical absorbing configurations like? Are they (statistically) well defined and reproducible? Do models with multiple absorbing configurations belong to a new universality class, distinct from that of DP? How are critical dynamics (for example, spreading from a seed) affected by the multiplicity of absorbing configurations?

We begin with a simple example, the *pair* CP (Jensen 1993a). The rules for the one-dimensional version are: choose a site x at random; do nothing unless both x and $x+1$ are occupied. If they are, then with probability p remove both particles; with probability $1-p$, choose $x-1$ or $x+2$ (with equal probability), and if this site is vacant, place a new particle there.

The pair CP shows the familiar continuous transition between an active and an absorbing state. However, notice that *any* configuration devoid of NN pairs is absorbing. (This would not be so if we allowed diffusion!) For a chain of N sites, the number of absorbing configurations exceeds $2^{N/2}$. On the other hand, the absorbing *state* is unique, for the density and correlation functions are reproducible functions of p. The active state (characterized by a nonzero density of pairs) is of course also unique. The exponents β, $\nu_{||}$, $\nu_{\perp}$ take DP values (Jensen 1993a). A field-theoretic model predicting DP-like static behavior in models such as the pair CP was proposed by Muñoz *et al.* (1996).

Next we consider several surface catalysis models with multiple absorbing states. The dimer–dimer model (Albano 1992) describes a surface reaction in which both species, A_2 and B_2, require NN pairs of vacancies to adsorb onto the surface. The basic steps are:

(i) $A_2 + 2v \rightarrow 2A_{ads}$;

(ii) $B_2 + 2v \rightarrow 2B_{ads}$;

(iii) $A_{ads} + B_{ads} \rightarrow AB_{ads} + v$;

(iv) $AB_{ads} + B_{ads} \rightarrow B_2A + 2v$.

With probability p_B, the next molecule arriving at the surface is B_2. Reactions between adsorbed species occur as soon as an A–B or AB–B pair is formed.[3] Configurations without NN vacancy pairs are absorbing. In fact, an active state is possible only for $p_B = 2/3$: there is a first-order transition between a mixed A/AB/B absorbing state and a B-rich absorbing state at this value. (Typical absorbing configurations have about 10% of the sites vacant.) Albano (1992) also studied two variants of the dimer–dimer model: one in which diffusion of B is permitted, and a second including both B diffusion and recombination/desorption of B. The latter modification prohibits BB pairs; when formed, either by diffusion or by nonreactive adsorption, such pairs immediately leave the surface. Allowing B diffusion ensures that the absorbing state for $p_B > 2/3$ contains B only, whilst for $p_B < 2/3$ the system reaches an absorbing configuration free of AB pairs. With diffusion and recombination/desorption of B, there is no longer any B-saturated phase, and one finds instead a *continuous* transition between an A/AB absorbing state and an active one at $p_B = p_c \simeq 0.7014$. An obvious choice for the order parameter is the fraction of NN vacancy pairs; another possibility is the difference $\Delta\theta$ between the sum of the A and AB concentrations and the value at saturation (0.907). Albano found $\Delta\theta \propto (p_B - p_c)^{\beta}$, with $\beta = 0.50(3)$, signaling a possible departure from the DP value, $\beta \simeq 0.58$.

[3] With A representing oxygen and B hydrogen, the above scheme provides a rough description of the surface-catalyzed reaction: $H_2 + (1/2)O_2 \rightarrow H_2O$.

A closely related model, again presenting a multitude of absorbing configurations, is the dimer–trimer model (Köhler & ben-Avraham 1991, ben-Avraham & Köhler 1992a). The rules are analogous to those of the ZGB model, if we replace CO by the dimer A_2, and O_2 by B_3. Here are the basic reaction steps in the dimer–trimer model:

(i) $A_2 + 2v \to 2A_{ads}$;
(ii) $B_3 + 3v \to 3B_{ads}$;
(iii) $A_{ads} + B_{ads} \to AB + 2v$.

Processes (i) and (ii) are attempted with probabilities p and $1-p$, respectively. Process (iii) occurs immediately an AB pair is formed. Neither diffusion nor nonreactive desorption are permitted. As in the dimer–dimer model, any configuration free of vacancy pairs is absorbing.

Köhler & ben-Avraham studied the model on a triangular lattice, with the trimer requiring a vacancy cluster in the form of a unit equilateral triangle for adsorption. For small p, a poisoned phase (mixed Bs and isolated vacancies) followed by a continuous transition (at $p = p_1 = 0.3403(2)$) to an active phase, then a discontinuous transition (at $p = 0.4610(8)$) to an A-poisoned phase (again with a certain fraction of isolated vacancies). Of prime interest is the critical point at p_1: does it belong to the DP class? Köhler & ben-Avraham's study yielded exponents somewhat different from the DP values, but more detailed simulations provide convincing evidence of DP behavior (Jensen 1994b).

A final example is a model for the reaction $CO + NO \to CO_2 + \frac{1}{2} N_2$ (Yaldram & Khan 1991, 1992; Yaldram *et al.* 1993). CO, as usual, adsorbs at a single vacant site, whilst NO, which dissociates upon adsorption, requires a pair of vacancies. The reaction steps are:

(i) $CO + v \to CO_{ads}$;
(ii) $NO + 2v \to N_{ads} + O_{ads}$;
(iii) $CO_{ads} + O_{ads} \to CO_2 + 2v$;
(iv) $2N_{ads} \to N_2 + 2v$.

CO adsorption is attempted with probability p, NO adsorption with probability $1-p$. Steps (iii) and (iv) occur the moment CO–O or N–N pairs are formed. On the triangular lattice, the possible steady states are an O–N absorbing phase (no vacancies, CO, or N–N pairs), an active phase, and a CO–N absorbing phase.[4] The transition from O–N absorbing to active, at $p = 0.185(5)$, is continuous, that from active to

[4] For a detailed analysis of the phase diagram on the *square lattice*, including the effects of nonreactive desorption, see Meng, Weinberg, & Evans (1994).

CO–N absorbing, at $p = 0.354(1)$, discontinuous (Brosilow & Ziff 1992). Detailed simulations showed that the critical exponents agree with those of DP in 2+1 dimensions (Jensen 1994a).

Given these results, it is natural to expect that continuous transitions into an absorbing state fall generically in the DP class, even for models exhibiting multiple absorbing configurations. This hypothesis receives further support from studies of several one-dimensional models with multiple absorbing states. We have already mentioned the pair CP as an example of a model with many absorbing configurations. Another model devised in order to study the effect of multiple absorbing states is the *dimer reaction*, an elaboration of a simpler model introduced by Browne & Kleban (1989). In the Browne & Kleban model particles rain down on the lattice at a constant rate, adsorbing (provisionally) at vacant sites. Suppose a particle has just landed at site $\mathbf{x}$. If all neighbors of $\mathbf{x}$ are vacant, the particle stays there. However, any of the neighbors are occupied, the particle remains only with probability p; with probability $1-p$ there is a reaction between the new arrival and one of the neighboring particles (chosen at random, if there is a choice), which clears both from the lattice. For small p the process attains an active steady state, but as $p \to p_c$ there is a continuous transition to the (unique) absorbing state: all sites filled (in one dimension, $p_c = 0.2765(5)$). As one would expect, the transition belongs to the DP class (Aukrust, Browne, & Webman 1989, 1990).

The one-dimensional *dimer reaction* is similar to the Browne & Kleban model, but with extended hard cores — particles may not occupy neighboring sites — and an extended range for reactions. Particles may only adsorb at vacant sites with both neighbors vacant; such sites are said to be *open*. (If we think of the sites as corresponding to *bonds* in the *dual* lattice, the particles correspond to dimers.) Suppose a particle has just arrived at site x. If all of the second and third neighbors (i.e., $x-3$, $x-2$, $x+2$, and $x+3$) are vacant, the new particle must remain. However, if any of these four sites are occupied, the new particle remains only with probability p; with probability $1-p$ it reacts with one of the other particles, and both are removed. Second neighbors take priority in the reaction: the new particle can react with a third neighbor only if both second neighbor sites are empty. The reaction rules are illustrated in figure 9.4. Although reactions with third neighbors are an unwelcome complication, they are essential: without them there is no active steady state even for $p = 0$. (Second-neighbor priority does *not* appear to be necessary for existence of an active steady state.)

In the dimer reaction any configuration devoid of open sites is absorbing. Of these absorbing states the one with maximal particle density consists of alternating vacant and occupied sites; in the one with minimal density, occupied sites alternate with pairs of vacant sites. The obvious

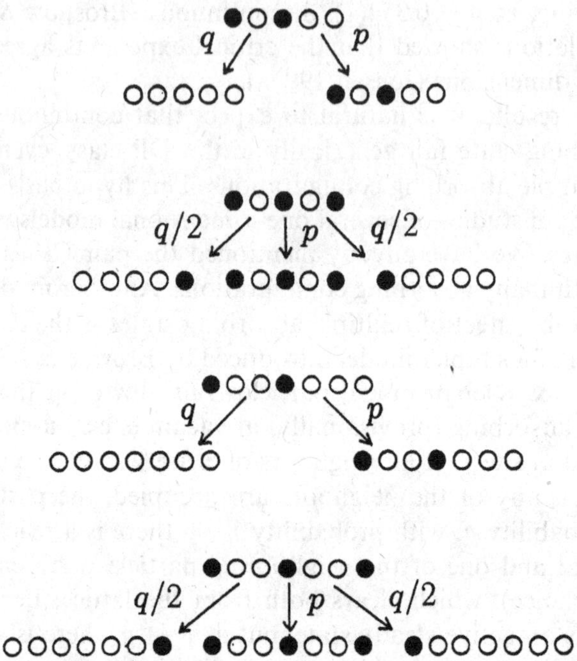

Fig. 9.4. Dimer reaction rules. In each case the new particle is at the central site. $q \equiv 1 - p$.

choice for the order parameter is the density of open sites, ρ_o. Simulations show ρ_o decreasing continuously from 0.6 when $p = 0$, to zero as p approaches $p_c = 0.2640$. At the same time ρ_p, the particle density, increases from zero to about 0.418(1) at the transition. Studies of the stationary density, and of the half-life τ_H (§6.5), show that the dimer reaction has a continuous transition, with the exponents characterizing static critical behavior ($\beta, \nu_{\|}, \nu_{\perp}$) taking the expected DP values (Jensen & Dickman 1993c).

One further example is the *threshold transfer process* (Mendes *et al.* 1994). In the (one-dimensional) threshold transfer process sites may be vacant, or singly or doubly occupied, corresponding to $\sigma_x = 0, 1$, or 2. In each cycle of the evolution, a site x is chosen at random, and if $\sigma_x = 0$, then $\sigma_x \to 1$ with probability r; if $\sigma_x = 1$, then $\sigma_x \to 0$ with probability $1-r$. (0 and 1 sites are left unchanged with probabilities $1-r$ and r, respectively.) Doubly-occupied sites are created and destroyed *via* particle transfers: if $\sigma_x = 2$, then if $\sigma_{x-1} < 2$, one particle moves to that site; independently, a particle moves from x to $x+1$ if $\sigma_{x+1} < 2$. σ_x is diminished accordingly in this deterministic, particle-conserving transfer. Survival of the doubly-occupied sites hinges on the process $(1, 2, 1) \to (2, 0, 2)$ — the only event in which the number of doubly-occupied sites increases. The likelihood of this process depends in turn upon r, which controls the particle density. In the absence of doubly-occupied sites, the threshold transfer process falls into an absorbing subspace of the dynamics, in which the density approaches r. For $r > r_c = 0.6894(3)$, a reactive state occurs characterized by a nonzero density, ρ_2, of doubly-occupied sites. Simulations indicate that the transition is continuous, and yield an order parameter exponent $\beta = 0.279(5)$, showing that this model too belongs to the DP class.

In the examples cited in this section, the number of absorbing configurations grows exponentially with system size. Yet all of these models have the same static critical exponents as DP. They share an important feature of having many absorbing configurations, but only a single absorbing *state*, with well-defined statistical properties. Fluctuations in intensive properties (for example, particle density) are quite small in the absorbing state — they fall off with system size as $L^{-1/2}$ — indicating that correlations are of short range. The dynamics of each model (at given parameter values) leads to a specific probability distribution on the set of possible absorbing configurations. This distribution is not known *a priori*, except for the threshold transfer process, where the absorbing subspace obeys a trivial equilibrium dynamics. Absorbing configurations generated at other, noncritical, parameter values also have well-defined statistical properties. For example, when $p = 1$ (no reactions) the dimer reaction becomes dimer *random sequential adsorption*, for which $\rho_p = (1 - e^{-2})/2 = 0.43233...$ (Flory 1939, Widom 1966, Evans 1993b). In any case, the average properties of absorbing configurations are well-defined functions of the control parameter, and so the absorbing state is unique.

9.4 Multiple absorbing configurations: spreading dynamics

Suppose one wants to study the evolution of a model with many absorbing configurations (the spread of pairs in the pair CP, for example), starting from a single seed. This, as we have seen, is a very useful way to locate the critical point, and yields the exponents δ, η, and z. But how should we select the initial condition? In time-dependent simulations of the CP, we begin with one particle in the vacuum. For the pair CP we likewise begin with a single NN pair; but what about the rest of the lattice? We could leave it empty (as in the CP), place an alternating sequence of particles and vacancies, or use a jammed random sequential adsorption configuration, among other possibilities. A special, 'natural' class of absorbing configurations are those generated by the critical dynamics. To be more precise, the set of natural initial configurations is defined as follows. Starting in a uniform, far-from-absorbing state (all sites occupied, in the pair CP), run the model at the critical point, until it becomes trapped in an absorbing configuration. Then place one order parameter seed (a particle pair, in the pair CP) anywhere in the system. Repeating this process many times generates the ensemble of natural initial configurations.

To study spreading we must first characterize the absorbing configurations. Consider, for example, the dimer reaction. Absorbing configurations have the form ...OGOGOGO..., where 'O' is an occupied site and 'G' is a 1- or 2-site gap. Thus each absorbing configuration is specified by a sequence $g_1, ..., g_N$ of binary random variables, $g_i = 1, 2$, being the length of the gap between the ith and $(i+1)$st particles. To describe the statistics of natural absorbing configurations, we require the probability $p(1) = \mathrm{Prob}[g=1]$ and the pair correlation $C(n) =< g_i g_{i+n} > - < g_i >^2$. Simulations show that at the critical point, $p(1) = 0.625(3)$, corresponding to a particle density of $\phi_{nat} = 0.418$, and that gaps are only weakly correlated: $C(1) = 0.048(2)$, and $C(i) \approx 0$ for $i \geq 2$. So natural absorbing configurations are well characterized by the 1-site probabilities and the NN correlation, or equivalently by the transition probabilities $p(i|j), (i, j = 1, 2)$ between successive gaps. Since correlations are very weak, we can mimic such configurations by a simple random process, using the transition probabilities $p(i|j)$ to generate the sequence of gaps. (This is much faster than 'growing' an absorbing configuration, and also permits us to vary correlations between successive gaps whilst maintaining a constant particle density.)

Starting from randomly generated configurations with gap frequencies and correlations matching those of natural absorbing configurations, the spreading exponents δ, η, and z assume DP values. This is not unexpected, since the static critical behavior in these models is the same as in DP. In fact, the exponents are insensitive to changes in the NN gap–gap correlation, so long as the initial particle density ϕ_0 is held constant. (The influence of *long-range* correlations has yet to be examined, however.) When we investigate other kinds of initial configurations, there is a surprise: the exponents δ and η *depend continuously* on ϕ_0! (Note that varying the initial density causes no detectable shift in the critical point — one always observes power-law spreading at the same value of p.) For the maximally occupied initial state $(\cdots \circ \bullet \circ \bullet \circ \circ \circ \bullet \circ \bullet \circ \cdots)$, with $\phi_0 = \frac{1}{2}$, one finds $\delta \approx 0.20$, $\eta \approx 0.26$, and $z \approx 1.246$. (The high particle density renders spreading more difficult, hence the increase in δ and the decrease in η.) Starting in the minimally occupied absorbing configuration ($\phi_0 = \frac{1}{3}$), the exponents are skewed in the opposite direction, as shown in table 9.1 and figure 9.5.

From studies of the pair CP, the dimer reaction and the threshold transfer process (Jensen & Dickman 1993c, Mendes *et al.* 1994) — all one-dimensional models — we conclude that exponents δ and η vary linearly, and over a wide range, as ϕ_0 is varied. (The change in z is much smaller.) Thus we have a family of critical spreading processes with tunable

Table 9.1. Critical exponents for the dimer reaction.

ϕ_0	δ	η	z	β'
0.333	0.107(2)	0.362(3)	1.267(6)	0.182(10)
0.380	0.133(2)	0.327(3)	1.258(10)	0.241(6)
0.418	0.160(2)	0.304(4)	1.252(6)	0.275(2)
0.500	0.205(5)	0.250(5)	1.240(6)	0.357(10)

exponents. Only the process starting from the natural initial density has a stationary counterpart, however. A glance at table 9.1 reveals that the hyperscaling relation,

$$4\delta + 2\eta = dz \tag{9.13}$$

is violated for $\phi_0 \neq \phi_{nat}$. Does this signal a breakdown of the scaling hypothesis, or is some sort of generalization feasible?

It turns out that the scaling argument presented in §6.4 for the CP and related models (all having unique absorbing configurations) can be generalized to describe exponents dependent on initial density (Mendes *et al.* 1994). We assume that the order parameter density and survival probability obey the same scaling forms as were postulated in §6.4 for the CP:

$$\rho(x,t) \simeq t^{\eta - dz/2} F(x^2/t^z, \Delta t^{1/\nu_{||}}), \tag{9.14}$$

and

$$P(t) \simeq t^{-\delta} \Phi(\Delta t^{1/\nu_{||}}). \tag{9.15}$$

Since the stationary distribution is unique, the static scaling of the order parameter density should obey

$$\rho(x,t) \to P_\infty \Delta^\beta, \tag{9.16}$$

as $t \to \infty$, with β the usual DP exponent, independent of ϕ_0. However, there is now no reason to suppose that $P_\infty \propto \overline{\rho}$, when $\phi_0 \neq \phi_{nat}$. In fact if this were so, we would have $\delta \nu_{||} = \beta$, implying that the exponents satisfy equation (9.13), for arbitrary ϕ_0. Since they do not, the exponent governing the ultimate survival probability must also depend upon ϕ_0, i.e.,

$$P_\infty \propto \Delta^{\beta'}, \tag{9.17}$$

with

$$\beta' = \delta \nu_{||}. \tag{9.18}$$

The arguments applied to the CP in chapter 6 again imply that

$$z = 2\nu_\perp / \nu_{||}. \tag{9.19}$$

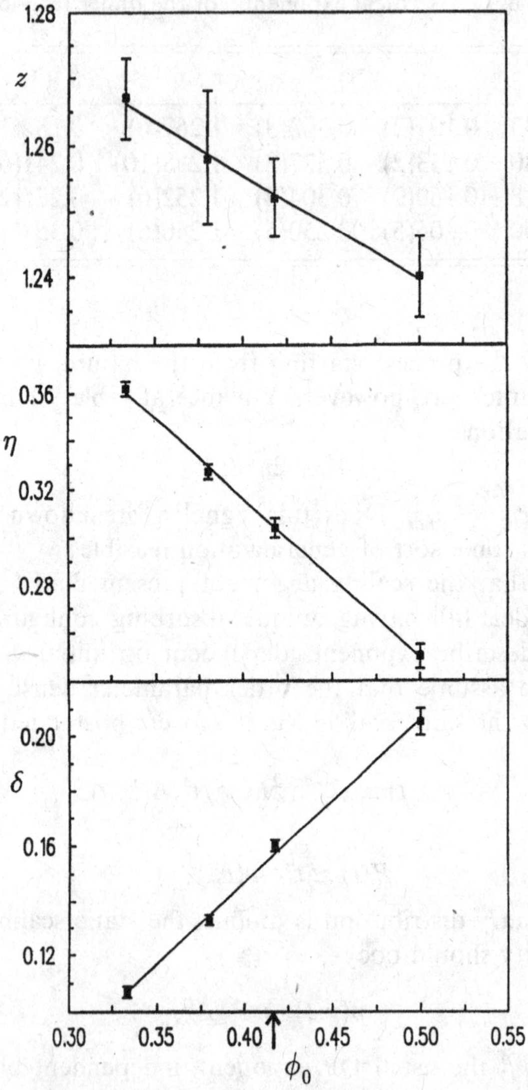

Fig. 9.5. Critical exponents δ, η, and z in the dimer reaction versus initial particle density ϕ_0. The arrow marks the natural initial density.

The asymptotic behavior of the order parameter density is now $\rho(x,t) \to \Delta^{\beta+\beta'}$, hence $F(0,y) \propto y^{\beta+\beta'}$ for large y, which entails the generalized hyperscaling relation

$$2\left(1+\frac{\beta}{\beta'}\right)\delta + 2\eta = dz. \tag{9.20}$$

This exponent relation is confirmed by simulations of the dimer reaction and the threshold transfer process (see table 9.1).

Further examination of the data indicates that $\delta + \eta$ is also very nearly constant. Now $\delta + \eta$ is the exponent governing the population growth in *surviving* critical trials. Its independence of ϕ_0 suggests that asymptotic properties of surviving trials are not affected by the initial density, a conclusion supported by the absence of any detectable shift in the critical point as ϕ_0 is varied. One may also note that as $t \to \infty$, only a negligible fraction of the population is actually in contact with the external density ϕ_0. Deep inside the active region, the particle density must approach the natural value. This also implies that z, which describes surviving trials exclusively, should be constant. (Note also that the data are consistent with $\nu_{||} = \beta'/\delta$ equal to its DP value of about 1.74.)

In equilibrium critical phenomena, the appearance of continuously variable exponents is usually associated with a *marginal parameter*, one that does not change under a renormalization group transformation (Fisher 1984). Can we find such a marginal parameter here? In the pair CP, dimer reaction, and threshold transfer process a second variable, ϕ, is coupled to the order parameter. One may view the initial density ϕ_0 as playing a role analogous to that of a marginal parameter, for it is never forgotten (i.e., it retains its value as the evolution is viewed on larger length and time scales) since to survive, the process must repeatedly invade new territory.

A two-dimensional version of the dimer reaction has also been investigated (Dickman 1996). This *second neighbor reaction* has the same rules as the dimer reaction (again, particles may not occupy neighboring sites), except that reactions occur only between particles occupying second-neighbor sites on the square lattice (i.e., between sites $\mathbf{x}=(i,j)$ and $\mathbf{y}=(i \pm 1, j \pm 1)$). As in the dimer reaction there are many absorbing configurations, this time with particle densities ranging from $1/2$ to $1/5$. Simulations show that the stationary critical behavior is DP-like.

Spreading in the second neighbor reaction holds several surprises. As expected, studies at the natural particle density ($\phi_{nat} = 0.33443$) yield the same exponents as for the two-dimensional CP. However, for $\phi_0 \neq \phi_{nat}$ the critical point, p_c (determined as the value for which power-law spreading obtains), differs from the bulk critical value. The shift in p_c implies that critical processes starting from non-natural absorbing configurations do not evolve to the bulk critical state, which exists uniquely at $p_{c,bulk}$. In absorbing configurations with $\phi_0 < \phi_{nat}$, critical spreading occurs at $p = p_c > p_{c,bulk}$, where an active steady state is not possible. As the population of open sites spreads, it leaves in its wake a region devoid of open sites, with particle density ϕ_f, slightly greater than ϕ_{nat}. After some time, this region cannot be reactivated. (For $p > p_{c,nat}$, the survival probability decays exponentially even when $\phi_0 = \phi_{nat}$.) Thus for $p_{c,nat} < p < p_c(\phi_0)$, the active region is confined to an expanding ring; it is a kind of 'chemical wave' which converts one type of absorbing configuration to another as it passes. The spreading exponents are much changed from their DP values: for $\phi_0 = 1/5$, $\delta \simeq 0.08$, $\eta \simeq 0.64$, $z \simeq 1.84$, and $\beta' \simeq 0.13$.

Critical spreading with $\phi_0 > \phi_{nat}$ presents a different picture. Here the critical point lies *below* $p_{c,bulk}$, which means that surviving trials consist of an expanding region with a small but finite density of open sites. The corresponding exponents are $\delta \simeq 1.30$, $\eta \simeq -0.1$, $z \simeq 0.9$, and $\beta' \simeq 2$. It is evident that a nonnatural environment has a much more dramatic effect on spreading in two dimensions than in one. Another surprising feature of the second neighbor reaction is that environments with the same particle density but different symmetries give rise to different sets of spreading exponents. Clearly much remains to be learned regarding critical spreading in models with multiple absorbing states.

9.5 Branching annihilating random walks

Conservation laws have profound consequences on dynamics. One knows, for example, that systems with a conserved order parameter density fall into different dynamic universality classes (Hohenberg & Halperin 1977) and exhibit more extended correlations (chapter 2) than their nonconserving counterparts. In this section we examine several models with a continuous transition to the vacuum, models affected in a fundamental way by a conservation law. In the *branching annihilating random walk* (BAW) model (Bramson & Gray 1985), particles hop about on a lattice, annihilate upon contact, and produce 'offspring' (further particles) at neighboring sites. At each step of the process a particle (located, say, at site x), is selected at random; it either hops (with probability p) to a randomly chosen neighboring site, or (with probability $1 - p$) creates n new particles which are placed (one each) at the site(s) nearest x. (Placement of offspring is random, if there is a choice. For example, if $n = 1$ the new particle appears at a randomly chosen NN; in one dimension, when $n = 2(4)$, the new particles appear at the two (four) sites nearest the parent.) Whenever, due either to hopping or to branching, a particle arrives at an occupied site, pairwise annihilation ensues. Clearly the vacuum is absorbing in BAWs, just as in the CP. In general, there is an active stationary state for small p, the density vanishing continuously as $p \to p_c^-$. (The exception is $n = 2$ in one-dimension, which has no active state for $p > 0$ (Sudbury 1990, Takayasu & Inui 1992).) There is numerical evidence for mean-field critical behavior (i.e., $\beta = 1$) in $d \geq 2$ (Takayasu & Tretyakov 1992).

Note a crucial difference between even-n and odd-n BAWs: in the former, particle number is conserved modulo 2 ('parity conservation'). For even-n BAWs, the vacuum is accessible only if the particle number is even. Odd-n BAWs aren't restricted in this manner, and it comes as no surprise that they belong to the DP universality class (Jensen 1993b). (The PAM discussed in §9.1 is closely related to the $n = 1$ BAW.)

Even-n BAWs, however, exhibit *non-DP* critical behavior, evidently as a consequence of parity conservation. A revealing observation is that minute violations of parity conservation (in the form of a very small rate for single-particle annihilation) throw the even-n BAW back into the DP class (Jensen 1993b). On the other hand, a model with global, but not local parity conservation exhibits a DP-like transition (Inui, Tretyakov & Takayasu, 1995). Jensen (1994c) derived precise critical exponents for an even-n BAW ($n = 4$, in one dimension): $\beta = 0.93(5)$; $\delta = 0.286(1)$; $\eta = 0.000(1)$; $z = 1.143(3)$; $\nu_{||} = 3.25(10)$; $1/\delta_h = 0.222(2)$. (The critical value of p is 0.7215(5).) These values are quite far from their DP counterparts (see table 6.1), and, remarkably, are consistent with simple rational values: $\delta \simeq 2/7$; $z \simeq 8/7$; $\nu_{||} \simeq 13/4$; $\beta \simeq 13/14$; $\delta_h \simeq 9/2$; and $\eta = 0$. They also satisfy the hyperscaling relation $4\delta + 2\eta = dz$.

The dynamic exponents quoted above obtain for even particle-number initial conditions. If the particle number is odd, all trials survive indefinitely ($P(t) = P_\infty = 1$), and so the only sensible value for δ and for β' (9.17) is zero. In this case Jensen's simulations yield $\eta = 0.282(5)$ (consistent with 2/7), and z essentially the same as quoted above. The generalized hyperscaling relation (9.20) holds in the form

$$\delta + \beta/\nu_{||} + \eta = dz/2, \tag{9.21}$$

where we used (9.18) to replace δ/β'.

The need to study $n = 4$ BAWs in order to have a parity-conserving model with an active state is an unwelcome complication. Zhong & ben-Avraham (1995) showed that an $n = 2$ BAW with partial reaction probabilities exhibits an active state, and confirmed the exponent values found by Jensen.

If the difference between the critical exponents of parity-conserving BAWs and those of the CP is obvious, the reason for the difference is not. (Why, for example, is η zero for even-numbered populations?) The correlation-time exponent, $\nu_{||}$, is nearly twice the DP value. This presumably reflects the conservation law: similar trends are found in the Ising model, where, for example, the dynamic critical exponent $z \approx 2$ for nonconserving dynamics in two dimensions, whilst $z \approx 4$ if magnetization is conserved (Hohenberg & Halperin 1977).

Several other models appear to show the same critical behavior as parity-conserving BAWs. In fact, the first examples of this class were devised some time ago by Grassberger, Krause, & von der Twer (1984),

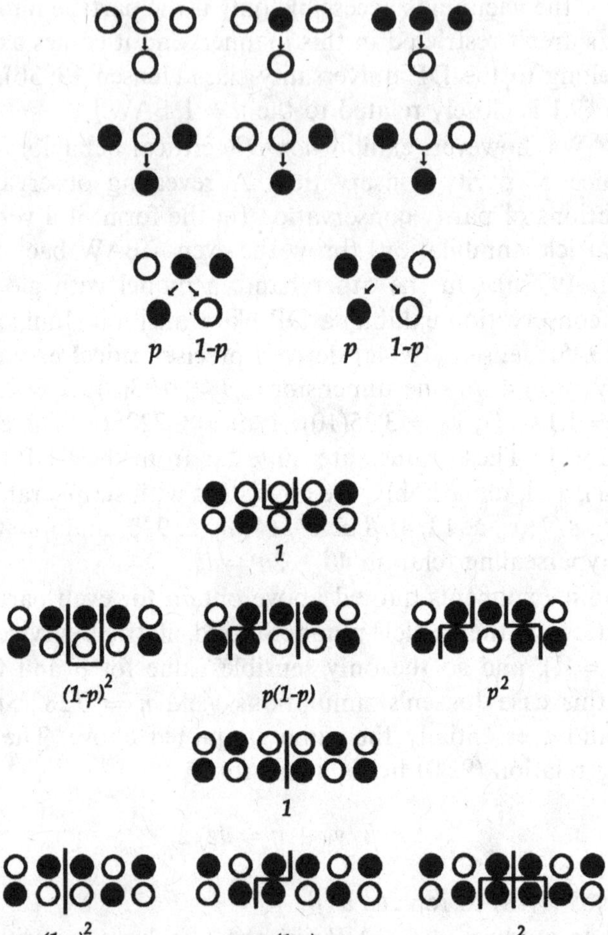

Fig. 9.6. (a) Kink cellular automaton rules. Transitions in the first two rows are deterministic. In the bottom row the branching probabilities are indicated. (b) Kink transitions in the kink cellular automaton: vertical lines represent kinks, and the expression below each picture denotes the transition probability.

and studied further by Grassberger (1989b). Here we focus on one of these models, which we refer to as the *kink cellular automaton* . This is a one-dimensional stochastic cellular automaton in which the state (occupied or not) of site x at time t depends on the states of $x-1$, x, and $x+1$ at time $t-1$. The rules are illustrated in figure 9.6. Inspection shows that there are two absorbing 2-cycles: checkerboard space-time patterns which differ by a unit shift. (In contrast with the multiple absorbing configurations considered in §§9.3 and 9.4 which possess no special symmetry, in the even-

n BAW we have a pair of absorbing states which differ by a symmetry operation of the underlying lattice.) If we regard a string of sites with alternating occupancy ($\cdots \bullet \circ \bullet \circ \bullet \cdots$) as a domain, then any pair of like neighbors is a domain wall or kink. These play the same role as particles in the BAW: when all kinks have vanished the system consists of a single domain, and is trapped in an absorbing 2-cycle. Figure 9.6(b) shows the rules in terms of kinks, allowing us to identify the elementary events: hopping, pairwise annihilation, and branching with $n = 2$. With increasing p, not only does branching occur more frequently, but an effective repulsion, $\propto p(1-p)$, between kinks at neighboring sites comes into play. In this way the kink cellular automaton can evade the vacuum, unlike the $n = 2$ BAW. The dynamics of the kink cellular automaton is essentially the same as for parity-conserving BAW (domain walls can only be created or destroyed in pairs!) so their critical behavior should be the same.

Simulations of the kink cellular automaton by Grassberger and coworkers place the critical point at $p_c \simeq 0.540$, and yield $\beta = 0.94(6)$ and $\nu_{||} = 3.3(2)$. Time-dependent studies starting with a single kink (the initial configuration $\cdots \bullet \circ \bullet \circ \circ \bullet \circ \bullet \circ \cdots$), yield $\eta \simeq 0.27$. These exponents agree quite nicely with those found in parity-conserving BAWs, supporting the expectation of a common critical behavior in the two models. It seems reasonable to conclude that the kink cellular automaton and parity-conserving BAWs belong to the same universality class, distinct from DP, which might be called 'critical domain dynamics.'

Two further examples of this new class have been studied. The *interacting monomer–dimer model* (Kim & Park 1994) is a one-dimensional version of the ZGB model discussed in chapter 5, except that a pair of monomers cannot reside at adjacent sites, and a dimer (BB) cannot adsorb adjacent to a B on the lattice. (In other words, the sequences AA and BBB are disallowed. However, A is allowed to adsorb between an A and a B: the immediate reaction eliminates the possibility of an AA pair.) There are two absorbing configurations — perfect domains of alternating A-filled and vacant sites, in a striped space-time pattern — that again differ by a unit shift. The transition to the absorbing state is continuous, occurring *via* pairwise annihilation of domain walls. The second example (Menyhárd 1994) belongs to the class of reaction–diffusion models studied in chapter 4. This is a one-dimensional nonequilibrium kinetic Ising model (NIM) with spin flips (Glauber dynamics) at zero temperature in competition with NN exchange (Kawasaki dynamics) operating at infinite temperature. The interactions are ferromagnetic and domain walls or kinks now correspond to up-down (↑↓) pairs. At zero temperature the allowed spin flips are ↑↑↓→↑↓↓ (corresponding to diffusion of a kink) and ↑↓↑→↑↑↑, (pairwise annihilation). Branching with

$n = 2$ is achieved via exchange events of the form ↑↑↓↓→↑↓↑↓. Denoting the rates of diffusion, annihilation, and exchange by p_D, p_A, and p_{ex}, respectively, the state of the system is determined by two parameters, for example p_{ex} and $r \equiv p_D/p_A$. A critical line in the p_{ex} - r plane separates the active (multi-domain) state from the absorbing single-domain state. Simulations of the interacting monomer–dimer model and NIM yield exponents consistent with those of parity-conserving BAWs and the kink cellular automaton.

At least four realizations of the critical domain dynamics universality class have been identified. The common feature of these models is that the change in the number of particles is always even. (Here 'particle' means the entity representing the order parameter: a random walker in BAWs, a kink or domain wall in the kink cellular automaton and the interacting monomer–dimer model.) The approach to the absorbing state occurs through the expansion of a single domain to fill the system. Close to the critical point, where the particle density is small, coalescence of domains depends on diffusive encounters between particles which were initially well separated. We expect critical domain dynamics to be the generic critical behavior for models obeying local parity conservation, just as DP is generic for models free of this constraint. Of course, a wider range of models needs to be explored before one can have confidence in this conjecture. There are also some fundamental open questions regarding models with parity conservation. First, all of the models studied so far are one-dimensional. What happens in higher dimensions? As mentioned, it appears likely that parity-conserving BAWs exhibit mean-field behavior in two or more dimensions. The interacting monomer–dimer model and NIM are readily extended to higher dimensions, where the approach to the absorbing state again depends on the expansion of a single domain. However, for $d \geq 2$ this looks nothing like a BAW: the evolution takes place on $(d-1)$-dimensional domain walls, and the 1:1 correspondence between kinks and random walkers is lost. One has, therefore, no reason to expect models like the interacting monomer–dimer model to be in the same class as BAWs in two or more dimensions.

Only very recently has a field-theoretic analysis of BAWs been devised (Cardy & Täuber 1996). This theory, which is cast in terms of a 'second-quantized' formalism rather than as a stochastic partial differential equation, confirms that odd-n BAWs belong to the DP universality class for $d \geqslant 2$. (In higher dimensions the model remains in the active phase for arbitrarily small branching rates.) A proper theory of the even-n case demands that the renormalization group equations preserve parity conservation, and predicts a non-DP transition below a critical dimension of $\approx 4/3$. Thus far, however, the theory does not yield accurate values for most of the critical exponents.

9.6 Cyclic models

Models with cyclic dynamics have been proposed for chemical and ecological systems, and elections (Tainaka 1988, 1993, 1994; Itoh & Tainaka 1994). The basic process is quite simple. Each site of a lattice is in one of three states, denoted X_1, X_2, and X_3. At each step, a NN pair of sites is chosen at random, and if the two sites are in different states, they change according to the rules:

(i) $X_1 + X_2 \to 2X_1$;

(ii) $X_2 + X_3 \to 2X_2$;

(iii) $X_3 + X_1 \to 2X_3$.

(These rules are similar to the paper–scissors–stone game, in which each object dominates its successor in a cyclic fashion.) Denoting site probabilities by p_i, we have, in the mean-field (random mixing) approximation

$$\frac{dp_1}{dt} = p_1(p_2 - p_3), \tag{9.22}$$

the other equations being given by cyclic permutation of indices. There are three absorbing states ($p_i = 1$), and a neutrally stable stationary solution ($p_i = 1/3$). For general (normalized) initial conditions, the solution tends to a limit cycle about the stationary solution. In the pair mean field approximation, solutions for nonsymmetric initial conditions develop wild oscillations (with each p_i in turn assuming a value ≈ 1) of ever-increasing period.

Simulations yield a rather different picture. In one dimension (starting, for example, from a random configuration), domains (all sites in the same state) grow until an (absorbing) single-domain configuration is attained. In two dimensions domains do not grow indefinitely; their typical size is roughly ten lattice spacings. The stationary state is characterized by continuous growth and shrinkage of domains. The likelihood of evolving to an absorbing state is very small (presumably it decreases exponentially with system size). Site or pair mean-field theories yield reasonable predictions for the phase diagram of the CP and catalysis models; here they fail utterly. It seems likely that a theory of cyclic models should be cast in terms of topological defects such as vortices and strings (Tainaka 1994).

Further surprises appear when the rules are perturbed (Tainaka 1993). Suppose we add to the original model the process

(iv) $X_3 \to X_1$

at rate b. (That is, a fraction of steps $\propto b$ consist in picking a site, and changing it to X_1 if it is found in state X_3.) This breaks the cyclic symmetry, and appears to favor X_1 at the expense of X_3. Simulations reveal, however, that the stationary value of p_3 *grows* with b! The stationary concentrations of X_1 and X_2 decrease with increasing b, and at a critical value $b_c \simeq 0.41$, p_2 falls to zero. The nature of this (continuous) absorbing state transition has yet to be explored in detail.

We have surveyed, without delving into great detail, several factors which influence the phase diagrams and critical properties of models with absorbing states. We began with reasonably well-understood issues, and moved on to some that are currently under intensive study. The problems of nonuniversal spreading, of parity-conserving models, and of cyclic dynamics present many avenues for further investigation, at the level of specific models as well as more general, continuum descriptions.

References

Abarbanel H. D. I., Bronzan J. B., Sugar R. L., and White A. R. 1975: *Phys. Rep.* **21**, 119.

Achahbar A., and Marro J. 1995: *J. Stat. Phys.* **78**, 1493.

Achahbar A., Alonso J. J., and Muñoz M. A. 1996: *Phys. Rev. E* **54**, 4838.

Achahbar A., Garrido P. L., and Marro J. 1996: *Molec. Phys.* **88**, 1157.

Acharyya M. 1997: preprint.

Acharyya M., and Chakraborty B. K. 1995: *Phys. Rev. B* **52**, 6550.

Adler J., and Duarte J. A. M. S. 1987: *Phys. Rev. B* **35**, 7046.

Adler J., Berger J., Duarte J. A. M. S., and Meir Y. 1988: *Phys. Rev. B* **37**, 7529.

Aertsens M., and Naudts J. 1991: *J. Stat. Phys.* **62**, 609.

Aharony A. 1978: *Phys. Rev. B* **18**, 3318

Albano E. V. 1990: *J. Phys. A* **23**, L545.

Albano E. V. 1992: *J. Phys. A* **25**, 2557.

Albano E. V. 1994a: *J. Phys. A* **27**, 431.

Albano E. V. 1994b: *Surf. Sci.* **306**, 240.

Alexander F. J., Edrei I., Garrido P. L., and Lebowitz J. L. 1992: *J. Stat. Phys.* **68**, 497.

Alexander F. J., Laberge C. A., Lebowitz J. L., and R. H. P. Zia 1996: *J. Stat. Phys.* **82**, 1133.

Alonso J. J., and Marro J. 1992a: *Phys. Rev. B* **45**, 10408.

Alonso J. J., and Marro J. 1992b: *J. Phys.: Condens. Matt.* **4**, 9309.

Alonso J. J., Garrido P. L., Marro J., and Achahbar A. 1995: *J. Phys. A* **28**, 4669.

Alonso J. J., López-Lacomba A. I., and Marro J. 1996: *Phys. Rev. E* **52**, 6006.

Alonso J. J., Marro J., and González-Miranda J. M. 1993: *Phys. Rev. E* **47**, 885.

Al'tshuler L., and Lee P. A. 1988: *Physics Today* **41** (December), 36.

Amar J. G. 1989: *Molec. Phys.* **67**, 739.

Amit D. J. 1984: *Field Theory, the Renormalization Group, and Critical Phenomena*, World Scientific, Singapore.

Arnold L., and Theodosopulu M. 1980: *Adv. Appl. Prob.* **12**, 367.

Aronson D. G., and Weinberger H. F. 1975: in *Partial Differential Equations and Related Topics*, J.A. Goldstein ed., Springer-Verlag, Berlin.

Aukrust T., Browne D. A., and Webman I. 1989: *Europhys. Lett.* **10**, 249.

Aukrust T., Browne D. A., and Webman I. 1990: *Phys. Rev. A* **41**, 5294.

Bagnoli F., Bulajich R., Livi R., and Maritan A. 1992: *J. Phys. A* **25**, L1071.

Bagnoli F., Droz M., and Frachebourg L. 1991: *Physica A* **179**, 269.

Bak P., Tang C., and Weisenfeld K. 1988: *Phys. Rev. A* **38**, 364.

Bakaev A. V., and Kabanovich V. I. 1995: *Int. J. Mod. Phys. B* **9**, 3181.

Baker G. A. 1990: *Quantitative Theory of Critical Phenomena*, Academic Press, New York.

Bär M., Gottschalk N., Eiswirth M., and Ertl G. 1994: *J. Chem. Phys.* **100**, 1202.

Baram A., and Kutasov D. 1989: *J. Phys. A* **22**, L251.

Barber M. N., and Ninham B.W. 1970: *Random and Restricted Walks; Theory and Applications*, Gordon and Breach, New York.

Bassler K. E., and Schmittmann B. 1994: *Phys. Rev. Lett.* **73**, 3343.

Bassler K. E, and Zia R. K. P. 1994: *Phys. Rev. E* **49**, 5871; *J. Stat. Phys.* **40**, 499 (1995).

Bates J. B., Wang J. C., and Dudney N. J. 1982: *Physics Today* **35** (July), 46.

Baxter R. J. 1982: *Exactly Solved Models in Statistical Mechanics*, Academic Press, London.

Bellon P., and Martin G. 1989: *Phys. Rev. B* **39**, 2403.

Ben Jacob E., Brand H., Dee G., Kramer L., and Langer J. S. 1985: *Physica D* **14**, 348.

ben-Avraham D. 1993: *J. Phys. A* **26**, 3725.

ben-Avraham D., and Köhler J. 1992a: *Phys. Rev. A* **45**, 8358.

ben-Avraham D., and Köhler J. 1992b: *J. Stat. Phys.* **65**, 839.

ben-Avraham D., Considine D., Meakin P., Redner S., and Takayasu H. 1990a: *J. Phys. A* **23**, 4297.

ben-Avraham D., Redner S., Considine D. B., and Meakin P. 1990b: *J. Phys. A* **23**, L613.

Bergersen B., and Rácz Z. 1991: *Phys. Rev. Lett.* **67**, 3047.

Berker A. N. 1993: *Physica A* **194**, 72.

Beyeler H. U. 1981: in *Physics in One Dimension*, J. Bernasconi and T. Schneider eds., Springer-Verlag, Berlin.

Beysens D., and Gbadamassi M. 1980: *Phys. Rev. A* **22**, 2250.

Bezuidenhout C., and Grimmett G. 1990: *Ann. Probab.* **18**, 1462.

Bhattacharjee S. M., and Nagle J. F. 1985: *Phys. Rev. A* **31**, 3199.

Bidaux R., Boccara N., and Chaté H. 1989: *Phys. Rev. A* **39**, 3094.

Binder K. 1976: *Ann. Phys. (N.Y.)* **98**, 390.

Binder K. (ed.) 1979: *Monte Carlo Methods in Statistical Physics*, Springer-Verlag, Berlin.

Binder K. (ed.) 1987: *Applications of Monte Carlo Methods in Statistical Physics*, Springer-Verlag, Berlin.

Binder K., and Wang J. S. 1989: *J. Stat. Phys.* **55**, 87.

Binder K., and Young A. P. 1986: *Rev. Mod. Phys.* **58**, 801.

Binney J. J., Dowrick N. J., Fisher A. J., and Newman M. E. J. 1992: *The Theory of Critical Phenomena: An Introduction to the Renormaliation Group*, Oxford Univ. Press, Oxford.

Birgeneau R. J., Cowley R. A., Shirane G., and Yoshizawa H. 1982: *J. Stat. Phys.* **48**, 1050.

Blease J. 1977: *J. Phys. C* **10**, 917, 923, 3461.

Bonnier B., Hontebeyrie M., and Meyers C. 1993: *Physica A* **198**, 1.

Boyce J. B., and Huberman B. A. 1979: *Phys. Rep.* **51**, 189.

Bramson M., and Gray L. 1985: *Z. Warsch. verw. Gebiete* **68**, 447.

Bramson M., Calderoni P., de Masi A., Ferrari P. A., Lebowitz J. L., and Schonmann R. H. 1986: *J. Stat. Phys.* **45**, 905.

Bramson M., Durrett R., and Schonmann R. H. 1991: *Ann. Prob.* **19**, 960.

Braun P., and Fähnle M. 1988: *J. Stat. Phys.* **52**, 775.

Bricmont J., and Kupiainen A. 1987: *Phys. Rev. Lett.* **59**, 1829.

Bricmont J., and Kupiainen A. 1988: *Commun. Math. Phys.* **116**, 870.

Broadbent S. R., and Hammersley J. M. 1957: *Proc. Camb. Phil. Soc.* **53**, 629.

Brosilow B. J., and Ziff R. M. 1992a: *Phys. Rev. A.* **46**, 4534.

Brosilow B. J., and Ziff R. M. 1992b: *J. Catal.* **136**, 275.

Brosilow B. J., Gulari E., and Ziff R. M. 1993: *J. Chem. Phys.* **98**, 674.

Brower R. C., Furman M. A., and Moshe M. 1978: *Phys. Lett. B* **76**, 213.

Browne D. A., and Kleban P. 1989: *Phys. Rev. A* **40**, 1615.

Bruinsma R. 1984: *Phys. Rev. B* **30**, 289.

Bruinsma R., and Aeppli G. 1983: *Phys. Rev. Lett.* **50**, 1494.

Buldyrev S. V., Barabási A.-L., Caserta F., Havlin S., Stanley H. E., and Vicsek T. 1992: *Phys. Rev. A* **45**, R8313.

Burger J. M. 1974: *The Nonlinear Diffusion Equation*, Reidel, Dordrecht.

Cambier J. L., and Nauenberg M. 1986: *Phys. Rev. B* **34**, 8071.

Campos I., Tarancón A., Clérot F., and Fernández L. A. 1995: *Phys. Rev. E* **52**, 5946.

Cannella V., and Mydosh J. A. 1972: *Phys. Rev. B* **6**, 4220.

Cardy J. L. 1984: *Phys. Rev. B* **29**, 505.

Cardy J. L. (ed.) 1988: *Finite-Size Scaling*, North-Holland, Amsterdam.

Cardy J. L., and Sugar R. L. 1980: *J. Phys. A* **13**, L423.

Cardy J. L., and Täuber U. C. 1996: *Phys. Rev. Lett.* **77**, 4780.

Carlow G. R., and Frindt R. F. 1994: *Phys. Rev. B* **50**, 11107.

Casties A., Mai J., and von Niessen, W. 1993: *J. Chem. Phys.* **99**, 3082.

Chan C. K., and Lin L. 1990: *Europhys. Lett.* **11**, 13.

Chaté H., and Manville P. 1986: *Phys. Rev. Lett.* **58**, 112.

Chaté H., and Manville P. 1990: in *New Trends in Nonlinear Dynamics and Pattern-Forming Phenomena*, P. Coullet and P. Huerre eds., Plenum, New York.

Cheng Z., Garrido P. L., Lebowitz J. L., and Vallés J. L. 1991: *Europhys. Lett.* **14**, 507.

Chopard B., and Droz M. 1988: *J. Phys. A* **21**, 205.

Chopard B., and Droz M. 1991: *Europhys. Lett.* **15**, 459.

Chopard B., Droz M., and Kolb M. 1989: *J. Phys. A* **22**, 1609.

Chowdhury D., and Stauffer D. 1986: *J. Stat. Phys.* **44**, 211.

Christmann K., and Ertl G. 1973: *Z. Naturforsch. Teil. A* **28**, 1144.

Clifford P., and Sudbury A. 1973: *Biometrika* **60**, 581.

Considine D., Takayasu H., and Redner S. 1990: *J. Phys. A* **23**, L1181.

Cornell S., Droz M., Dickman R., and Marques M. C. 1991: *J. Phys. A* **24**, 5605.

Cowley R. A., Ryan T. W., and Courtens E. 1986: *Z. Phys. B* **65**, 181.

Cox M. P., Ertl G., and Imbihl R. 1985: *Phys. Rev. Lett.* **54**, 1725.

Creutz M. 1986: *Ann. Phys. (N.Y.)* **167**, 62.

Cross M. C., and Hohenberg P. C. 1993: *Rev. Mod. Phys.* **65**, 851.

Dab D., Boon J-P., and Li Y-X. 1991: *Phys. Rev. Lett.* **66**, 2535.

Dagotto E. 1996: *Physics World* **9** (April), 22.

Dahmen K., and Sethna J. P. 1996: *Phys. Rev. B* **53**, 14872.

de Dominicis C., and Peliti L. 1978: *Phys. Rev. B* 18, 353.

de Gennes P. G., 1984: *J. Phys. Chem.* **88**, 6469.

de Groot S. R., and Mazur P. 1984: *Non-equilibrium Thermodynamics*, Dover, New York.

de Masi A., Ferrari P. A., and Lebowitz J. L. 1986a: *J. Stat. Phys.* **44**, 589.

de Masi A., Presutti E., and Scacciatelli E. 1986b: *Annales del I.H.P.* **25**, 1.

de Masi A., Presutti E., and Vares M. E. 1986c: *J. Stat. Phys.* **44**, 645.

de Oliveira M. J. 1992: *J. Stat. Phys.* **66**, 273.

de Oliveira M. J. 1994: *Physica A* **203**, 13.

de Oliveira M. J., Tomé T., and Dickman, R. 1992: *Phys. Rev. A* **46**, 6294.

De'Bell K., and Essam J. W. 1983: *J. Phys. A* **16**, 385.

Debye P., and Hückel E. 1923: *Phys. Z.* **24**, 185.

Derrida B., and Weisbuch G. 1987: *Europhys. Lett.* **4**, 657.

Derrida B., Domany E., and Mukamel D. 1992: *J. Stat. Phys.* **69**, 667.

Derrida B., Vannimenus J., and Pomeau Y. 1978: *J. Phys. C* **11**, 4749.

Deutscher G., Zallen R., and Adler J. (eds.) 1983: *Percolation Structures and Processes*, Hilger, Bristol.

Dickman R. 1986: *Phys. Rev. A* **34**, 4246.

Dickman R. 1987: *Phys. Lett. A* **122**, 463.

Dickman R. 1988: *Phys. Rev. A* **38**, 2588.

Dickman R. 1989a: *J. Stat. Phys.* **55**, 997.

Dickman R. 1989b: *Phys. Rev. B* **40**, 7005.

Dickman R. 1990a: *Phys. Rev. A* **41**, 2192.

Dickman R. 1990b: *Phys. Rev. A* **42**, 6985.

Dickman R. 1994: *Phys. Rev. E* **50**, 4404.

Dickman R. 1996: *Phys. Rev. A* **53**, 2223.

Dickman R., and Burschka M. 1988: *Phys. Lett. A* **127**, 132.

Dickman R., and Jensen I. 1991: *Phys. Rev. Lett.* **67**, 2391.

Dickman R., and Tomé T. 1991: *Phys. Rev. A* **44**, 4833.

Dickman R., and Tretyakov A. Yu. 1995: *Phys. Rev. E* **52**, 3218.

Dickman R., Wang J.-S., and Jensen I. 1991: *J. Chem. Phys.* **94**, 8252.

Dieterich W., Fulde P., and Peschel I. 1980: *Adv. Phys.* **29**, 527.

Dixon M., and Gillan M. J. 1982: in *Computer Simulation of Solids*, C. R. A. Catlow and W. C. Mackrodt eds., Springer-Verlag, Berlin, p. 275.

Doi M. 1976: *J. Phys. A* **9**, 1465, 1479.

Domany E., and Kinzel W. 1984: *Phys. Rev. Lett.* **53**, 447.

Domb C., and Green M. S. (eds.) 1972–76: *Phase Transitions and Critical Phenomena*, vols. 1–6, Academic Press, New York.

Domb C., and Lebowitz J. L. (eds.) 1977–95: *Phase Transitions and Critical Phenomena*, vols. 7–17, Academic Press, New York.

Domb C., Schneider T., and Stoll E. 1975: *J. Phys. A* **8**, L90.

Dorfman J. R., Kirkpatrick T. R., and Sengers J. V. 1994: *Ann. Rev. Phys. Chem.* **45**, 213.

Dotsenko V. 1994: *J. Phys. A* **27**, 3397.

Droz M., and McKane A. 1994: *J. Phys. A* **27**, L467.

Droz M., Rácz Z., and Schmidt J. 1989: *Phys. Rev. A* **39**, 2141.

Droz M., Rácz Z., and Tartaglia P. 1990: *Phys. Rev. A* **41**, 6621.

Droz M., Rácz Z., and Tartaglia P. 1991: *Physica A* **177**, 401.

Dumont M., Dufour P., Sente B., and Dagonnier R. 1990: *J. Catal.* **122**, 95.

Durrett R. 1988: *Lecture Notes on Particle Systems and Percolation*, Wadsworth & Brooks/Cole, Pacific Grove, CA

Durrett R., and Griffeath D. 1983: *Ann. Probab.* **11**, 1.

Duke T. A. J. 1989: *Phys. Rev. Lett.* **62**, 2877.

Dyson F. J. 1968: *Commun. Math. Phys.* **12**, 91, 212.

Ebeling W. 1986: *J. Stat. Phys.* **45**, 891.

Edwards S. F., and Anderson P. W. 1975: *J. Phys. F* **5**, 965.

Edwards S. F., and Anderson P. W. 1976: *J. Phys. F* **6**, 1927.

Ehsasi M., Matloch M., Frank O., Block J. H., Christmann K., Rys F. S., and Hirschwald W. 1989: *J. Chem. Phys.* **91**, 4949.

Eichhorn K., and Binder K. 1996: *Z. Phys. B* **99**, 413.

Elderfield D., and Vvedensky D. D. 1985: *J. Phys. A* **18**, 2591.

Elliot R. J. 1961: *Phys. Rev.* **124**, 346.

Engel T., and Ertl G. 1979: *Advances in Catalysis* **28**, Academic Press, New York.

Ertl G., and Koch J. 1972: in *Proceedings of the 5th International Congress on Catalysis*, Palm Beach, Fla., p. 969.

Ertl G., Norton P.R., and Rüstig J. 1982: *Phys. Rev. Lett.* **49**, 177.

Essam J. W. 1989: *J. Phys. A* **22**, 4927.

Essam J. W., De'Bell K., Adler J., and Bhatti F. M. 1986: *Phys. Rev. B* **33**, 1982.

Essam J. W., Guttmann A. J., and De'Bell K., 1988: *J. Phys. A* **33**, 3815.

Evans J. W. 1991: *Langmuir* **7**, 2514.

Evans J. W. 1992: *J. Chem. Phys.* **97**, 572.

Evans J. W. 1993a: *J. Chem. Phys.* **98**, 2463.

Evans J. W. 1993b: *Rev. Mod. Phys.* **65**, 1281.

Evans J. W., and Miesch M. S. 1991: *Phys. Rev. Lett.* **66**, 833; see also *Surf. Sci.* **245**, 401 (1991).

Evans J. W., and Ray T. R. 1994: *Phys. Rev. E* **50**, 4302.

Evans M. R., and Derrida B. 1994: *Acta Physica Slovaca* **44**, 331.

Evans M. R., Foster D. P., Godréche C., and Mukamel D. 1995: *Phys. Rev. Lett.* **74**, 208.

Eyink G. L., Lebowitz J. L., and Spohn H. 1996: *J. Stat. Phys.* **83**, 385.

Fan C., and McCoy B. M. 1969: *Phys. Rev.* **182**, 614

Feller W. 1957: *An Introduction to Probability Theory and its Applications* , Vol. 1. Wiley Interscience, New York.

Ferreira A. L. C., and Mendiratta S. K. 1993: *J. Phys. A* **26**, L145.

Fichthorn K. A., and Ziff R. M. 1986: *Phys. Rev. B* **34**, 2038.

Field R. J., and Burger M. (eds.) 1985: *Oscillations and Travelling Waves in Chemical Systems*, Wiley-Interscience, New York.

Fife P. C. 1979:*Mathematical Aspects of Reacting and Diffusing Systems*, Springer-Verlag, New York.

Fischer P., and Titulaer U. M. 1989: *Surf. Sci.* **221**, 409.

Fisher D. S. 1986: *Phys. Rev. Lett.* **56**, 419.

Fisher D. S., Grinstein G. M., and Khurana A. 1988: *Physics Today* **41** (December), 56.

Fisher M. E. 1967: *Physics (N.Y.)* **3**, 255.

Fisher M. E. 1971: in *Proceedings of the International Summer School 'Enrico Fermi', CourseLI*, Academic Press, New York.

Fisher M. E. 1984: in *Critical Phenomena*, Springer-Verlag, Berlin.

Fisher M. E., and Barber M. N. 1972: *Phys. Rev. Lett.* **28**, 1516.

Fishman S., and Aharony A. 1979: *J. Phys. C* **12**, L729.

Flory P. J. 1939: *J. Chem. Phys.* **61**, 1518.

Flyvbjerg H., and Petersen H. G. 1989: *J. Chem. Phys.* **91**, 461.

Frachebourg L., Krapivsky P. L., and Redner, S. 1995: *Phys. Rev. Lett.* **75**, 2891.

Fritz J., and Maes C. 1988: *J. Stat. Phys.* **53**, 1179.

Garrido P. L., and Marro J. 1987: *Physica A* **144**, 585.

Garrido P. L., and Marro J. 1989: *Phys. Rev. Lett.* **62**, 1929.

Garrido P. L., and Marro J. 1991a: *Europhys. Lett.* **15**, 375.

Garrido P. L., and Marro J. 1991b: in *Artificial Neural Networks*, A. Prieto ed., Springer-Verlag, Berlin, pp. 25–33.

Garrido P. L., and Marro J. 1992: *J. Phys. A* **25**, 1453.

Garrido P. L., and Marro J. 1994: *J. Stat. Phys.* **74**, 663.

Garrido P. L., and Muñoz M. A. 1993: *Phys. Rev E* **48**, R4153.

Garrido P. L., and Muñoz M. A. 1995: *Phys. Rev. Lett.* **75**, 1875.

Garrido P. L., Labarta A., and Marro J. 1987: *J. Stat. Phys.* **49**, 551.

Garrido P. L., Lebowitz J. L., Maes C., and Spohn H. 1990b: *Phys. Rev. A* **42**, 1954.

Garrido P. L., Linares R. J., Marro J., and Muñoz M. A. 1996: in *The Non Linearity and the Disorder,* L. Vázquez and S. Jiménez eds., Editorial Complutense S.A., Madrid.

Garrido P. L., Marro J., and Dickman R. 1990a: *Ann. Phys. (N.Y.)* **199**, 366.

Garrido P. L., Marro J., and González-Miranda J. M. 1989: *Phys. Rev. A* **40**, 5802.

Garrido P. L., Muñoz M. A., and de los Santos F. 1997: preprint; see also Garrido P. L., de los Santos F., and Muñoz M. A., *Phys. Rev E* **57**, 752 (1998).

Gates D. J. 1988: *J. Stat. Phys.* **52**, 245.

Gawedzki K., and Kupiainen A. 1986: *Nucl. Phys. B* **269**, 45.

Gillan M. J. 1985: *Physica B* **131**, 157.

Glauber R. J. 1963: *J. Math. Phys.* **4**, 294.

Glotzer S. C., Stauffer D., and Jan N. 1994: *Phys. Rev. Lett.* **72**, 4109.

Goh M. C., Goldburg W. I., and Knobler C. M. 1987: *Phys. Rev. Lett.* **58**, 1008.

Golchet A., and White J. M. 1978: *J. Catal.* **53**, 266.

González-Miranda J. M., Garrido P. L., Marro J., and Lebowitz J. L. 1987: *Phys. Rev. Lett.* **59**, 1934.

González-Miranda J. M., Labarta A., Puma M., Fernández J.F., Garrido P. L., and Marro J. 1994: *Phys. Rev. E* **49**, 2041.

Graham R. 1981: in *Order and Flustuations in Equilibrium and Nonequilibrium Statistical Mechanics*, G. Nicolis, G. Dewel, and J. W. Turner eds., John Wiley & Sons, New York.

Graham R., and Haken H. 1971: *Z. Physik* **243**, 289.

Grassberger P. 1982: *Z. Physik B* **47**, 465.

Grassberger P. 1989a: *J. Phys. A* **22**, 3673.

Grassberger P. 1989b: *J. Phys. A* **22**, L1103.

Grassberger P. 1995: *J. Stat. Phys.* **79**, 13.

Grassberger P., and Scheunert M. 1980: *Fortschr. Phys.* **28**, 547.

Grassberger P., and de la Torre A. 1979: *Ann. Phys. (N.Y.)* **122**, 373.

Grassberger P., Krause F., and von der Twer J. 1984: *J. Phys. A* **17**, L105.

Grassberger P., and Zhang Y. 1996: *Physica A* **224**, 169.

Gray L. 1985: in *Particle Systems, Random Media and Large Deviations*, R. Durret ed., Am. Math. Soc., Providence.

Gray P., and Scott S. K. 1990: *Chemical Oscillations and Instabilities*, Oxford Univ. Press, New York.

Gribov V. N. 1968: *Sov. Phys. JETP* **26**, 414.

Gribov V. N., and Migdal A. A. 1969: *Sov. Phys. JETP* **28**, 784.

Griffeath D. 1979: *Additive and Cancellative Interacting Particle Systems*, Springer-Verlag, Berlin.

Griffiths R. B. 1969: *Phys. Rev. Lett.* **23**, 17.

Grinstein G. 1985: in *Fundamental Problems in Statistical Mechanics VI*, E.G.D. Cohen ed., North-Holland, Amsterdam.

Grinstein G. 1991: *J. Appl. Phys.* **69**, 5441.

Grinstein G., and Mukamel D. 1983: *Phys. Rev. B* **27**, 4503.

Grinstein G., Jayaprakash C., and He Y. 1985: *Phys. Rev. Lett.* **55**, 2527.

Grinstein G., Jayaprakash C., and Socolar J. E. S. 1993: *Phys. Rev. E* **48**, R643.

Grinstein G., Lai Z. W., and Browne D.A. 1989: *Phys. Rev. A* **40**, 4820.

Grinstein G., Lee D. H., and Sachdev S. 1990: *Phys. Rev. Lett.* **64**, 1927.

Grynberg M. D., Newman T. J., and Stinchcombe R. B. 1994: *Phys. Rev. E* **50**, 957.

Gurevich Yu. Ya., and Kharkats Yu. 1986: *Phys. Rep.* **139**, 201.

Guttmann A. J. 1989: in *Phase Transitions and Critical Phenomena*, vol.13, C. Domb and J.L. Lebowitz eds., Academic Press, New York.

Haken H. 1978: *Synergetics*, Springer-Verlag, New York.

Haken H. 1983: *Advanced Synergetics*, Springer-Verlag, New York.

Hansen P. L., Lemmich J., Ipsen L. H., and Mouritsen O. G. 1993: *J. Stat. Phys.* **73**, 723.

Harris A. B. 1974: *J. Phys. C* **7**, 1671.

Harris R, and Grant M. 1988: *Phys. Rev. B* **38**, 9323.

Harris T. E. 1974: *Ann. Probab.* **2**, 969.

Harrison L. G. 1993: *Kinetic Theory of Living Patterns*, Cambridge Univ. Press, New York.

Hayes W. 1978: *Contemp. Phys.* **19**, 469.

Heeger A. J., Garito A. F., and Iterrande L. V. 1975: in *Low Dimensional Cooperative Phenomena*, H.J. Keller ed., Plenum Press, New York.

Helbing D., and Molnár P. 1995: *Phys. Rev. E* **51**, 4282.

Heringa J. R., Blöte H. W. J., and Hoogland A. 1994: *Int. J. Mod. Phys. C* **5**, 589.

Hernández-Machado A., Guo H., Mozos J. L., and Jasnow D. 1989: *Phys. Rev. A* **39**, 4783.

Heuer H. O. 1993: *J. Phys. A* **26**, L333.

Hibma T. 1980: *Solid State Commun.* **33**, 445.

Hill C.C., Zia R. K. P., and Schmittmann B. 1996: *Phys. Rev. Lett.* **77**, 514.

Hill J. P., Thurston T. R., Erwin R. W., Ramsted M. J., and Birgeneau R. J. 1991: *Phys. Rev. Lett.* **66**, 3281.

Hinrichsen H., Weitz J. S., and Domany E. 1996: preprint; see also *J. Stat. Phys.* **88**, 617 (1997).

Hohenberg P. C., and Halperin B. I. 1977: *Rev. Mod. Phys.* **49**, 435.

Holey T., and Fähnle M. 1990: *Phys. Rev. B* **41**, 11709.

Holley R., and Ligget T. M. 1975: *Ann. Probab.* **3**, 643.

Hornreich R. M., Liebmann R., Schuster H. G., and Selke W. 1979: *Z. Phys. B* **35**, 91.

Horváth V. K., Family F., and Vicsek T. 1990: *Phys. Rev. Lett.* **65**, 1388.

Hovi J. P., Vaari J., Kaukonen H. P., and Nieminen R. M. 1993: *Comp. Mat. Sci.* **1**, 33.

Hu X., and Kawazoe Y. 1994: *J. Appl. Phys.* **75**, 6486.

Hutton S. L., Fehst I., Böhmer R., Braume M., Mertz B., Lunkenheimer P., and Loidl A. 1991: *Phys. Rev. Lett.* **66**, 1990.

Hwa T., and Kardar M. 1989: *Phys. Rev. Lett.* **62**, 1813.

Imbrie J. Z. 1986: in *Critical Phenomena, Random Systems, Gauge Theories*, K. Osterwalder and R. Stora eds., Elsevier Sci. Pub., Amsterdam.

Imry Y. 1984: *J. Stat. Phys.* **34**, 849.

Imry Y., and Ma S. K. 1975: *Phys. Rev. Lett* . **35**, 1399.

Inui N., Tretyakov A. Yu., and Takayasu H. 1995: *J. Phys. A* **28**, 145.

Itoh Y., and Tainaka K. 1994: *Phys. Lett. A* **189**, 37.

Janssen H. K. 1981: *Z. Phys. B* **42**, 151.

Janssen H. K. 1985: *Z. Phys. B* **58**, 311.

Janssen H. K. 1997: *Phys. Rev. E* **55**, 6253.

Janssen H. K., and Schmittmann B. 1986: *Z. Physik B* **63**, 517; **64**, 503.

Jensen I. 1991: *Phys. Rev. A* **43**, 3187.

Jensen I. 1992: *Phys. Rev. A* **45**, R563.

Jensen I. 1993a: *Phys. Rev. Lett.* **70**, 1465.

Jensen I. 1993b: *Phys. Rev. E* **47**, 1.

Jensen I. 1993c: *J. Phys. A* **26**, 3921.

Jensen I. 1994a: *J. Phys. A* **27**, L61.

Jensen I. 1994b: *Int. J. Mod. Phys. B* **8**, 3299.

Jensen I. 1994c: *Phys. Rev. E* **50**, 3263.

Jensen I. 1996: *J. Phys. A* **29**, 7013.

Jensen I., and Dickman R. 1993a: *J. Phys. A* **26**, L151.

Jensen I., and Dickman R. 1993b: *J. Stat. Phys.* **71**, 89.

Jensen I., and Dickman R. 1993c: *Phys. Rev. E* **48**, 1710.

Jensen I., and Dickman R. 1994: *Physica A* **203**, 175.

Jensen I., and Fogedby H. C. 1991: *Phys. Rev. A* **42**, 1969.

Jensen I., and Guttmann A. J. 1996: *Nucl. Phys. B* **S47**, 835.

Jensen I., Fogedby H. C., and Dickman R. 1990: *Phys. Rev. A* **41**, 3411.

Kang H. C., and Weinberg W. H. 1993: *Phys. Rev. E* **48**, 3464.

Kang H. C., and Weinberg W. H. 1994: *J. Chem. Phys.* **100**, 1630.

Kang K., and Redner S. 1985: *Phys. Rev. A* **32**, 455.

Kardar M., Parisi G., and Zhang Y. C. 1986: *Phys. Rev. Lett.* **56**, 889.

Katori M. 1994: preprint.

Katori M., and Konno N. 1992: *Physica A* **186**, 578.

Katori M., and Konno N. 1993a: *J. Phys. A* **26**, 6597.

Katori M., and Konno N. 1993b: in *Formation, Dynamics and Statistics of Patterns*, K. Kawasaki and M. Suzuki, eds. World Scientific, Singapore.

Katz S., Lebowitz J. L., and Spohn H. 1984: *J. Stat. Phys.* **34**, 497.

Kawasaki K. 1966: *Phys. Rev.* **145**, 224.

Kawasaki K. 1972: in *Phase Transitions and Critical Phenomena*, vol. 4, C. Domb and M.S. Green eds., Academic Press, London.

Kemeny J. G., and Snell J. L. 1962: *Mathematical Models in the Social Sciences*, Ginn, Boston.

Kennedy J. H. 1977: in *Solid Electrolytes*, S. Geller ed., Springer-Verlag, Berlin, p.105.

Kerstein A. R. 1986: *J. Stat. Phys.* **45**, 921.

Kértesz J., and Vicsek T. 1995: in *Fractals in Science*, A. Bunde and S. Havlin eds., Springer-Verlag, Berlin.

Kikuchi R. 1951: *Phys. Rev.* **81**, 988.

Kim M. H., and Park H. 1994: *Phys. Rev. Lett.* **73**, 2579.

Kinzel W. 1983: in *Percolation Structures and Processes*, G.Deutscher, R. Zallen, and J. Adler, eds., Hilger, Bristol.

Kinzel W. 1985: *Z. Phys. B* **58**, 229.

Kirkpatrick T. R., Cohen E. G. D., and Dorfman J. R. 1980: *Phys. Rev. A* **26**, 950, 966, 972.

Klein L., Adler J., Aharony A., Harris A. B., and Meir Y. 1991: *Phys. Rev. B* **43**, 11249.

Köhler J., and ben-Avraham D. 1991: *J. Phys. A* **24**, L621.

Köhler J., and ben-Avraham D. 1992: *J. Phys. A* **25**, L141.

Kohring G. A., and Shreckenberg M. 1992: *J. Phys. I* **2** 2033.

Kolmogorov A. 1934: *Math. Ann.* **112,** 155.

Konno N. 1994: *Phase Transitions of Interacting Particle Systems*, World Scientific, Singapore.

Konno N., and Katori M. 1990: *J. Phys. Soc. Jpn.* **59**, 1581.

Korniss G., Schmittmann B., and Zia R. K. P. 1997: *J. Stat. Phys.* **86**.

Kramers H. A., and Wannier G. H. 1941: *Phys. Rev.* **60**, 252, 263.

Kreuzer H. J. 1981: *Nonequilibrium Thermodynamics and its Statistical Foundations,* Clarendon Press, Oxford.

Kroemer J., and Pesch W. 1982: *J. Phys. A* **15**, L25.

Krug J., and Spohn H. 1992: in *Solids Far From Equilibrium*, C. Godrèche ed., Cambridge Univ. Press, Cambridge.

Krug J., and Ferrari P. A. 1996: *J. Phys. A* **29**, L465.

Krug J., Lebowitz J. L., Spohn H., and Zhang M. Q. 1986: *J. Stat. Phys.* **44**, 535.

Kukla V., Kornatowski J., Demuth D., Girnus I., Pfeifer H., Rees L. V. C., Schunk S., Unger K. K., and Kärger J. 1996: *Science* **272**, 702.

Künsch H. 1984: *Z. Wahr. verw. Geb.* **66**, 407.

Kurtz T. 1970: *J. Appl. Prob.* **7**, 49.

Kurtz T. 1971: *J. Appl. Prob.* **8**, 344.

Kurtz T. 1978: *Stoch. Processes and their Appl.* **6**, 223.

Labarta A., Marro J., Martínez B., and Tejada J. 1988: *Phys. Rev. B* **38**, 500.

Labarta A., Marro J., and Tejada J. 1986: *Physica B* **142**, 31.

Landau D. P. 1976: *Phys. Rev. B* **13,** 2997.

Landau D. P. 1990: in *Finite Size Scaling and Numerical Simulation of Statistical Systems,* V. Privman, Ed., World Scientific, Singapore, p. 223.

Lauritsen K. B., and Fogedby H. C. 1993: *Phys. Rev. E* **47**, 1563.

Law B. M., Segrè P. N., Gammon R. W., and Sengers J. V. 1990: *Phys. Rev. A* **41**, 816.

Lebowitz J. L. 1986: *Physica A* **140**, 232.

Lebowitz J. L., and Penrose O. 1966: *J. Math. Phys.* **7**, 98.

Lebowitz J. L., and Saleur H. 1986: *Physica A* **138**, 194.

Lebowitz J. L., Presutti E., and Spohn H. 1988: *J. Stat. Phys.* **51**, 841.

Lee J. C. 1993: *Phys. Rev. Lett.* **70**, 3599.

Leung K.-t. 1991: *Phys. Rev. Lett.* **66**, 453.

Leung K.-t. 1992: *Int. J. Mod. Phys. C* **3**, 367.

Leung K.-t., and Cardy J. L. 1986: *J. Stat. Phys.* **44**, 567.

Leung K.-t., Schmittmann B., and Zia R. K. P. 1989: *Phys. Rev. Lett.* **62**, 1772.

Leung K.-t., Mon K. K., Vallés J. L., and Zia R. K. P. 1988: *Phys. Rev. Lett.* **61**, 1744.

Lieb E.H. 1976: *Rev. Mod. Phys.* **48**, 553.

Liebman R. 1986: *Statistical Mechanics of Periodic Frustrated Ising Systems*, Springer-Verlag, Berlin.

Liggett T. M. 1985: *Interacting Particle Systems*, Springer-Verlag, New York.

Liggett T. M. 1991: in *Random Walks, Brownian Motion, and Interacting Particle Systems*, R. Durrett and H. Kesten eds., Birkäuser, Boston.

Lo W. S., and Pelcovits R. A. 1990: *Phys. Rev. A* **42**, 7471.

López-Lacomba A. I., and Marro J. 1992: *Phys. Rev. B* **46**, 8244.

López-Lacomba A. I., and Marro J. 1994a: *Europhys. Lett.* **25**, 169.

López-Lacomba A. I., and Marro J. 1994b: *J. Phys. A* **27**, 1111.

López-Lacomba A. I., Garrido P. L., and Marro J. 1990: *J. Phys. A* **23**, 3809.

López-Lacomba A. I., Marro J., and Garrido P. L. 1993: *Phase Transitions* **42**, 141.

Ma S. K. 1976: *Modern Theory of Critical Phenomena*, Benjamin, Reading, Mass.

Ma Y., Zhang Y., and Gong C. 1992: *Phys. Rev. B* **46**, 11591.

Mai J., Casties A., and von Niessen W. 1993: *Chem. Phys. Lett.* **211**, 197.

Mai J., and von Niessen W. 1991: *Phys. Rev. A* **44**, R6165.

Mai J., and von Niessen W. 1993: *J. Chem. Phys.* **98**, 2032.

Maltz A., and Albano E. V. 1992: *Surf. Sci.* **277**, 414.

Marko J. F. 1992: *Phys. Rev. B* **45**, 5023.

Markowich P. A., Ringhofer C. A., and Schmeisser C. 1990: *Semiconductor Equations*, Springer, Vienna.

Marqués M. C. 1989: *J. Phys. A* **22**, 4493.

Marqués M. C. 1990: *J. Phys. A* **23**, 3389.

Marqués M. C. 1993: *J. Phys. A* **26**, 1559.

Marqués M. C., and Ferreira A. L. 1994: *J. Phys. A* **27**, 3389.

Marro J. 1976: *On the Kinetics of Phase Transitions in Binary Alloys*, University Microfilms, Ann Arbor, Michigan.

Marro J. 1996: in *Monte Carlo and Molecular Dynamics of Condensed Matter Systems*, K. Binder and G. Ciccotti eds., Italian Physical Society, Conf. Proceed. vol. 49, Editrice Compositori, Bologna, pp.843–58.

Marro J., and Achahbar A. 1997: preprint; see also *J. Stat. Phys.* **90**, 817 (1998).

Marro J., and Vacas J. A. 1997: *Phys. Rev. B* **56**, 8863.

Marro J., and Vallés J. L. 1987: *J. Stat. Phys.* **49**, 121.

Marro J., Achahbar A., Garrido P. L., and Alonso J. J. 1996: *Phys. Rev. E.* **53**, 6038.

Marro J., Fernández J. F., Gozález-Miranda J. M., and Puma M. 1994: *Phys. Rev. E* **50**, 3237.

Marro J., Garrido P. L., Labarta A., and Toral R. 1989: *J. Phys.: Condens. Matt.* **1**, 8147.

Marro J., Garrido P. L., and Vallés J. L. 1991: *Phase Transitions* **29**, 129.

Marro J., Labarta A., and Tejada J. 1986: *Phys. Rev. B* **34**, 347.

Marro J., Lebowitz J. L., and Kalos M. H. 1979: *Phys. Rev. Lett.* **43**, 282.

Marro J., Lebowitz J. L., Spohn H., and Kalos M. H. 1985: *J. Stat. Phys.* **38**, 725.

Marro J., Vallés J. L., and González-Miranda J. M. 1987: *Phys. Rev. B* **35**, 3372.

Martin P. C., Siggia E. D., and Rose H. A. 1978: *Phys. Rev. A* **8**, 423.

Matsushima T., Hashimoto H., and Toyoshima I. 1979: *J. Catal.* **58**, 303.

Mattis D. C. 1976: *Phys. Lett. A* **56**, 421.

Meakin P., and Scalapino D. J. 1987: *J. Chem. Phys.* **87**, 731.

Meier W. M., and Olson D. H. 1992: *Atlas of Zeolite Structure Types, Butterworth-Heinemann*, London.

Mendes J. F. F., and Lage E. J. S. 1993: *Phys. Rev. E* **48**, 1738.

Mendes J. F. F., Dickman R., Henkel M., and Marqués M. C. 1994: *J. Phys. A* **27**, 3019.

Meng B., Weinberg W. H., and Evans J. W. 1994: *J. Chem. Phys.* **101**, 3234.

Menyhárd N. 1994: *J. Phys. A* **27**, 6139.

Metropolis N., Rosenbluth A. W. and M. M., Teller A. H. and E. 1953: *J. Chem. Phys.* **21**, 1087.

Mézard M., Parisi G., and Virasoro M. A. 1987: *Spin Glass Theory and Beyond*, World Scientific, Singapore.

Moreira A. G. 1996: Ph. D. Thesis, Universidade Federal de Belo Horizonte.

Moreira A. G., and Dickman R. 1996: *Phys. Rev. E* **54**, R3090.

Mon K. K. 1993: *Phys. Rev. B* **47**, 5497.

Moshe M. 1978: *Phys. Rep.* **37**, 255.

Mueller C., and Tribe R. 1994: *Prob. Th. Rel. Fields* **100**, 131.

Muñoz M. A. 1994: Ph. D. Thesis, University of Granada.

Muñoz M. A., and Garrido P. L. 1994: *Phys. Rev. E* **50**, 2458.

Muñoz M. A., Grinstein G., Dickman R., and Livi R. 1996: *Phys. Rev. Lett.* **76**, 451.

Nagel K., and Paczuski M. 1995: *Phys. Rev. E* **51**, 2909.

Nicolis G., and Prigogine I. 1977: *Self-organization in Nonequilibrium Systems*, Wiley Interscience, New York.

Nguyen T. N., Lee P. A., and Loye H.-C. zur 1996: *Science* **271**, 489.

Newman M. E. J., Roberts B.W., Barkema G. T., and Sethna J. P. 1993: *Phys. Rev. B* **48**, 16533.

Noest A. J. 1986: *Phys. Rev. Lett.* **57**, 90.

Noest A. J. 1988: *Phys. Rev. B* **38**, 2715.

Normand C., Pomeau Y., and Velarde M. G. 1977: *Rev. Mod. Phys.* **49**, 581.

Obukhov S. P. 1980: *Physica A* **101**, 145.

Obukhov S. P. 1987: *JETP Lett.* **45**, 172.

Ódor G., Boccara N., & Szabó G. 1993: *Phys. Rev. E* **48**, 3168.

Ohtsuki T., and Keyes T. 1987a: *J. Chem. Phys.* **87**, 6060.

Ohtsuki T., and Keyes T. 1987b: *Phys. Rev. A* **35**, 2697; *ibid* **36**, 4434.

O'Keeffe M, and Hyde B. G. 1976: *Phil. Mag.* **33**, 219.

Olson M. A., and Adelman S. A. 1985: *J. Chem. Phys.* **83**, 1865.

Onuki A., Yamazaki K., and Kawasaki K. 1981: *Ann. Phys. (N.Y.)* **131**, 217.

Oresic M., and Pirc R. 1994: *J. Non-Crystaline Solids* **172**, 506.

Paczuski M., Maslov M., and Bak P. 1994: *Europhys. Lett.* **27**, 97.

Panagiotopoulos A. Z. 1987: *Molec. Phys.* **61**, 812; **62**, 701.

Panagiotopoulos A. Z. 1992: *Molec. Simul.* **9**, 1.

Parisi G. 1980: *Phys. Rep.* **67**, 97.

Park H., Köhler J., Kim I-M, ben-Avraham D., and Redner S. 1993: *J. Phys. A* **26**, 2071.

Pathria R. K. 1972: *Statistical Mechanics*, Pergamon Press, Oxford.

Pauli W. 1928: Festschrift zum 60. Geburtstag A. Sommerfeld, Hirzel, Leipzig, p. 30.

Peliti L. 1985: *J. Phys. (Paris)* **46**, 1469.

Peliti L. 1994: *Physics World* **7** (March), 24.

Penrose O., and Lebowitz J. L. 1971: *J. Stat. Phys.* **3**, 211.

Poland D. 1993: *Physica A* **193**, 1.

Pomeau Y. 1986: *Physica D* **23**, 3.

Pontin L. F., Segundo J. A. B., and Pérez J. F. 1994: *J. Stat. Phys.* **75**, 51.

Praestgaard E., Larsen H., and Zia R. K. P. 1994: *Europhys. Lett.* **25**, 447.

Prakash S., Havlin S., Schwartz M., and Stanley H. E. 1992: *Phys. Rev. A* **46**, R1724.

Price P. J. 1965: in *Fluctuation Phenomena in Solids*, R. E. Burgess ed., Academic Press, New York, p. 355.

Privman V. (ed.) 1990: *Finite Size Scaling and Numerical Simulation of Statistical Systems*, World Scientific, Singapore.

Privman V. (ed.) 1997: *Nonequilibrium Statistical Mechanics in One Dimension*, Cambridge Univ. Press, Cambridge.

Puri S., Binder K., and Dattagupta S. 1992: *Phys. Rev. B* **46**, 98.

Puri S., Parekh N., and Dattagupta S. 1994: *J. Stat. Phys.* **75**, 839.

Pynn R., and Skjeltorp A. (ed.) 1987: *Time Dependent Effects in Disordered Systems*, Plenum Press, New York.

Rácz Z., and Zia R. K. P. 1994: *Phys. Rev. E* **49**, 139.

Ray J. R., and Vashishta P. 1989: *J. Chem. Phys.* **90**, 6580.

Redner S. 1982: *Phys. Rev. B* **25**, 3243.

Rieger H. 1995: in *Annual Reviews of Computational Physics II*, D. Stauffer ed., World Scientific, Singapore.

Rovere M., Nielaba P., and Binder K. 1993: *Z. Phys. B* **90**, 215.

Ruelle D. 1969: *Statistical Mechanics: Rigorous Results*, W.A. Benjamin, New York.

Rujān P. 1987: *J. Stat. Phys.* **49**, 139.

Salomon M. B. 1979, *Physics of Superionic Conductors*, Springer-Verlag, Berlin.

Santos M. A., and Teixeira S. 1995: *J. Stat. Phys.* **79**, 963.

Sato H., and Kikuchi R. 1971: *J. Chem. Phys.* **55**, 677, 702.

Satulovsky J., and Albano E. V. 1992: *J. Chem. Phys.* **97**, 9440.

Schlögl F. 1972: *Z. Phys.* **253**, 147.

Schmittmann B., and Bassler K. E. 1996: *Phys. Rev. Lett.* **77**, 3581.

Schmittmann B., and Zia R. K. P. 1991: *Phys. Rev. Lett.* **66**, 357.

Schmittmann B., and Zia R. K. P. 1995: in *Phase Transitions and Critical Phenomena*, vol. 17, C. Domb and J.L. Lebowitz eds., Academic Press, London.

Selke W. 1992: in *Phase Transitions and Critical Phenomena*, vol.15, C. Domb and J.L. Lebowitz eds., Academic Press, London

Selke W., Shchur L. N., and Talapov A. L. 1994: in *Annual Review of Computational Physics I*, D. Stauffer ed. World Scientfic, Singapore, p.17.

Shapira S., Klein L., Adler J., Aharony A., and Harris A. B. 1994: *Phys. Rev. B* **49**, 8830.

Sherrington D., and Kirkpatrick S. 1975: *Phys. Rev. Lett.* **35**, 1972.

Silverberg M., and Ben-Shaul A. 1987: *J. Chem. Phys.* **87**, 3178.

Silverberg M., Ben-Shaul A., and Rebentrost F. 1985: *J. Chem. Phys.* **83**, 6501.

Smit B., Smedt P., and Frenkel D. 1989: *Molec. Phys.* **68**, 931.

Smoller, J. 1983: *Shock Waves and Reaction–Diffusion Equations*, Springer-Verlag, Berlin.

Sompolinsky H. 1988: *Physics Today* **41** (December), 70.

Spohn H. 1982: in *Stochastic Processes in Quantum Theory and Statistical Physics*, Springer-Verlag, Berlin.

Spohn H. 1989: private communication.

Spohn H. 1991: *Large Scale Dynamics of Interacting Particles*, Springer-Verlag, Berlin.

Stanley H. E. 1971: *Introduction to Phase Transitions and Critical Phenomena*, Oxford Univ. Press, New York.

Stauffer D. 1979: *Phys. Reports* **54**, 1.

Stauffer D., and Aharoni A. 1992: *Introduction to Percolation Theory*, Taylor and Francis, London.

Stauffer D., and Sashimi M. 1993: in *Physics Computing '92*, R. A. de Groot and J. Nadrchal eds., World Scientific, Singapore, p.169.

Stella A. L., and Vanderzande C. 1989: *Phys. Rev. Lett.* **62**, 1067; *J. Phys. A* **22**, L445.

Stephenson J. 1970: *Phys. Rev. B* **1**, 4405.

Stinchcombe R. B. 1983: in *Phase Transitions and CriticalPhenomena*, vol.7, C. Domb and J.L. Lebowitz eds., Academic Press, London.

Sudbury A. 1990: *Ann. Probab.* **18**, 581.

Suzuki M. 1986: *J. Phys. Soc. Jpn.* **55**, 4205.

Swendsen R. H. 1984: *Phys. Rev. B* **30**, 3866.

Swift J., and Hohenberg P. C. 1977: *Phys. Rev. A* **15**, 319.

Swift M. R., Maritan A., Cieplak M., and Banavar J. R. 1994: *J. Phys. A* **27**, 1575.

Szabó G. 1994: *Phys. Rev. E* **49**, 3508.

Szabó G., and Szolnoki A. 1990: *Phys. Rev. E* **41**, 2235.

Szabó G., and Szolnoki A. 1996: *Phys. Rev. E* **53**, 2196.

Szabó G., Szolnoki A., and Antal T. 1994: *Phys. Rev. E* **49**, 229, 2764.

Szabó G., Szolnoki A., and Bodócs L. 1991: *Phys. Rev. A* **44**, 6375.

Tainaka K. 1988: *J. Phys. Soc. Jpn.* **57**, 2588.

Tainaka K. 1993: *Phys. Lett. A* **176**, 303.

Tainaka K. 1994: *Phys. Rev. E* **50**, 3401.

Takayasu H., and Inui N. 1992: *J. Phys. A* **25**, L585.

Takayasu H., and Tretyakov A. Yu. 1992: *Phys. Rev. Lett.* **68**, 3060.

Tang L.-H., and Leschhorn H. 1992: *Phys. Rev. A* **45**, R8309.

Thompson C. J. 1972: *Mathematical Statistical Mechanics*, Princeton Univ. Press, Princeton, New Jersey.

Thorpe M. F., and Beeman D. 1976: *Phys. Rev. B* **14**, 188.

Tomé T., and Dickman R. 1993: *Phys. Rev. E* **47**, 948.

Tomé T., de Oliveira M. J., and Santos M. A. 1991: *J. Phys. A* **24**, 3677.

Toom A. L. 1980: in *Multicomponent Random Systems*, R. L. Dobrushin and Ya. G. Sinai eds., Dekker, New York.

Torres J., Garrido P.L., and Marro J. 1997: *J. Phys. A* **30**, 7801; see also Garrido P.L., Marro J., and Torres J., *Physica A*, to appear.

Toulouse G. 1977: *Commun. Phys.* **2**, 115.

Tretyakov A. Yu., and Takayasu H. 1991: *Phys. Rev. A* **44**, 8388.

Turing A. M. 1952: *Philos. T. Roy. Soc. B* **237**, 37.

Vallés J. L. 1992: *J. Phys. I France* **2**, 1361.

Vallés J. L., and Marro J. 1986: *J. Stat. Phys.* **43**, 441.

Vallés J. L., and Marro J. 1987: *J. Stat. Phys.* **49**, 89.

Vallés J. L., Leung K.-t., and Zia R. K. P. 1989: *J. Stat. Phys.* **56**, 43.

van Beijeren H., and Nolden I. 1987: in *Structure Dynamics Surfaces II*, W. Schommers and P. von Blanckenhagen eds., Springer-Verlag, Berlin.

van Beijeren H., and Schulman L. S. 1984: *Phys. Rev. Lett.* **53**, 806.

van Enter A., Fernández R., and Sokal A. 1993: *J. Stat. Phys.* **72**, 879.

van Hove L. 1950: *Physica* **16**, 137.

van Kampen N. G. 1981: *Stochastic Processes in Physics and Chemistry*, North-Holland, Amsterdam.

Vanderzande C. 1992: *J. Phys. A* **25**, L75.

Vargas R.A., Salamon M.B., and Flynn C.P. 1978: *Phys. Rev. B* **17**, 269.

Vega L., de Miguel E., Rull L. F., Jackson G., and McLure A. 1992: *J. Chem. Phys.* **96**, 2296.

Velarde M. G. (ed.) 1983: *Nonequilibrium Cooperative Phenomena in Physics and Related Fields*, Plenum Press, New York.

Vichniac G. 1984: *Physica D* **10**, 117.

Vicsek 1992: *Fractal Growth Phenomena*, World Scientific, Singapore.

Vilfan I., Zia, R. K. P., and Schmittmann B. 1994: *Phys. Rev. Lett.* **73**, 2071.

Villain J., and Bak P. 1981: *J. Phys. (Paris)* **42**, 657.

Vincent E., Hammann J., and Ocio M. 1992: in *Recent Progress in Random Magnets*, D. H. Ryan ed., World Scientific, Singapore.

Voigt C. A., and Ziff R. M. 1997: *Phys. Rev. E* **56**, R6241.

Wang J.-S. 1996: *J. Stat. Phys.* **82**, 1409.

Wang J.-S., and Chowdhury D. 1989: *J. de Physique* **50**, 2905.

Wang J.-S., and Lebowitz J. L. 1988: *J. Stat. Phys.* **51**, 893.

Wang J.-S., Binder K., and Lebowitz J. L. 1989: *J. Stat. Phys.* **56**, 783.

Wannier G. H. 1951: *Phys. Rev.* **83**, 281; **87**, 795 (1952); *Bell Syst. Techn J.* 32, 170 (1953)

Weher J. 1989: Ph. D. Thesis, Rutgers University, New Jersey.

Widom B. 1965: *J. Chem. Phys.* **43**, 3892, 3898.

Widom B. 1966: *J. Chem. Phys.* **44**, 3888.

Widom B. 1973: *J. Chem. Phys.* **58**, 4043.

Wiltzius P., Dierher S. B., and Dennis B. S. 1989: *Phys. Rev. Lett.* **62**, 804.

Wio H. S. 1994: *An Introduction to Stochastic Processes and Nonequilibrium Statistical Physics*, World Scientific, Singapore.

Wolfram S. 1983: *Rev. Mod. Phys.* **55**, 601.

Wolfram S. 1986: *Theory and Applications of Cellular Automata*, World Scientific, Singapore.

Wong A., and Chan M. H. W. 1990: *Phys. Rev. Lett.* 65, 2567.

Xu H.-J., Bergersen B., and Rácz Z. 1993: *Phys. Rev. E* **47**, 1520.

Yaldram K., and Khan K. M. 1991: *J. Catal.* **131**, 369.

Yaldram K., and Khan K. M. 1992: *J. Catal.* **136**, 279.

Yaldram K., Khan K. M., Ahmed N., and Khan M. A. 1993: *J. Phys. A* **26**, L801.

Yang C. N., and Lee T. D. 1952: *Phys. Rev.* **87**, 404, 412.

Yeung C., Mozos J. L. , Hernández-Machado A., and Jasnow D. 1993: *J. Stat. Phys.* **70**, 1149.

Yeung C., Rogers T., Hernández-Machado A., and Jasnow D. 1992: *J. Stat. Phys.* **66**, 1071.

Yokota T. 1988: *Phys. Rev. B* **38**, 11669.

Yu B., Browne D. A., and Kleban, P. 1991: *Phys. Rev. A* **43**, 1770.

Zhabotinsky A. M., and Zaikin A. N. 1973: *J. Theor. Biol.* **40**, 45.

Zhang M. Q., Wang J.-S., Lebowitz J. L., and Vallés J. L. 1988: *J. Stat. Phys.* **52**, 1461.

Zhdanov V. P., and Kasemo B. 1994: *Surf. Sci. Rep.* **20**, 111.

Zhong D., and ben-Avraham D. 1995: *Phys. Lett. A* **209**, 333.

Zhou J., Redner S., and Park H. 1993: *J. Phys. A* **26**, 4197.

Ziff R. M., and Brosilow B. J. 1992: *Phys. Rev. A* **46**, 4630.

Ziff R. M., and Fichthorn K. 1986: *Phys. Rev. B* **34**, 2038.

Ziff R. M., Fichthorn K., and Gulari E. 1991: *J. Phys. A* **24**, 3727.

Ziff R. M., Gulari E., and Barshad Y. 1986: *Phys. Rev. Lett.* **56**, 2553.

Ziman J. M. 1979: *Models of Disorder*, Cambridge Univ. Press. Cambridge.

Zinn-Justin J. 1990: *Quantum Field Theory and Critical Phenomena*, Clarendon Press, Oxford.

Index